BROADBAND COMMUNICATIONS, II

IFIP Transactions C: Communication Systems

International Federation for Information Processing

Technical Committee 6

Communication Systems

IFIP Transactions Abstracted/Indexed in:
INSPEC Information Services
Index to Scientific & Technical Proceedings®
CompuMath Citation Index®, Research Alert™, and SciSearch®

C-24

BROADBAND COMMUNICATIONS, II

Proceedings of the IFIP TC6 Second International Conference on
Broadband Communications
Paris, France, 2-4 March, 1994

Edited by

SAMIR TOHMÉ
TELECOM Paris
Paris, France

AUGUSTO CASACA
IST/INESC
Lisbon, Portugal

N·H

1994

ELSEVIER
AMSTERDAM • LAUSANNE • NEW YORK • OXFORD • SHANNON • TOKYO

ELSEVIER SCIENCE B.V.
Sara Burgerhartstraat 25
P.O. Box 211, 1000 AE Amsterdam, The Netherlands

Keywords are chosen from the ACM Computing Reviews Classification System, ©1991, with permission.
Details of the full classification system are available from
ACM, 11 West 42nd St., New York, NY 10036, USA.

ISBN: 0 444 81834 0
ISSN: 0926-549X
Transferred to digital printing 2006

This book is printed on acid-free paper.

Preface

The second Conference on Broadband Communications organized by IFIP Technical Committee 6, aimed at presenting and discussing the state-of-the-art and the evolution of broadband communications, both in the public and private networks.

The papers selected for presentation in the Conference cover the issues of switching, resource management, network control, communication services, testing, quality of service, traffic characteristics, traffic control and protocols.

Invited talks on the fields of multimedia communication and traffic control, and panels on the introduction of ATM were also included in the program.
The program was the result of the work of the members of the scientific program committee, organizing committee and the referees. The conference was made possible by the sponsoring organizations IFIP, TELECOM Paris, the Louis Leprince-Ringuet Foundation and the CNET.

Samir Tohmé
TELECOM Paris
Conference Chairman

Augusto Casaca
IST/INESC
Conference Co-Chairman

Committees

General Chair

S. Tohmé (ENST, F)
A. Casaca (IST/INESC, P)

Scientific Program Committee

S. Alexander (BT Develop. and Procur., UK)
J. P. Coudreuse (France Telecom CNET, F)
A. Danthine (Univ. Liège, B)
A. Forcina (SIP, I)
Y. Inoue (NTT, J)
V. B. Iversen (Techn. Univ. of Denmark, DK)
M. Kawarasaki (NTT Net. Arch. Lab., J)
D. Khakhar (Lund Univ., S)
H. Korezlioglu (ENST, F)
P. Kühn (IND, Univ. Stuttgart, D)
L. Mason (INRS, CND)
M. De Prycker (Alcatel Bell, B)
G. Pujolle (Univ. Paris 6, F)
A. K. Reilly (Bellcore, USA)
J. Riera (ETSI Telecom., E)
H. Rudin (IBM Res. Lab., CH)
H. Seguin (France Telecom CNET, F)
O. Spaniol (Techn. Univ. Aachen, D)
P. Schicker (Consultant, CH)
P. K. Verma (AT&T Bell Labs, USA)

Organizing Committee

D. Seret (ENST, F) (chair)
K. Chen (ENST, F), R. Coelho (ENST, F)
M. Gagnaire (ENST, F), D. Kofman (ENST, F)
C. Rigault (ENST, F)

Table of Contents

Switching

Traffic

Resource Management and Network Control

List of reviewers

A-B

M. AJMONE MARSAN (Politecnico di Torino, Italy)
M. d'AMBROSIO (CSELT, Italy)
E. ALTMAN (INRIA, France)
A. ALVES (IST/INESC, Portugal)
H. ARAI (NTT, Japan)
Y. BAGUETTE (Univ. Liège, Belgium)
D. BEAUMONT (BT Labs, United Kingdom)
C. BISDIKIAN (IBM Yorktown heights, USA)
P. BOYER (CNET Lannion, France)

C-D-E

A. CASACA (IST/INESC, Portugal)
R. CASTANET (ENSERB, France)
K. CHEN (ENST, France)
R. COELHO (ENST, France)
P. CROCETTI (Italtel, Italy)
A. DANTHINE (Univ. Liège, Belgium)
R. DSSOULI (Montreal Univ., Canada)
B. EVANS (Surrey Univ., United Kingdom)

F-G-H-I-J

A. FASBENDER (AAchen Univ., Germany)
S. FDIDA (Paris V Univ., France)
M. FERRANDIZ (HP Labs, United Kingdom)
I. GALLAGHER (BT Labs, United Kingdom)
A. FORCINA (SIP, Italy)
J. FORNE (Barcelona Univ. Spain)
M. GAGNAIRE (ENST, France)
I. GALLAGHER (BT Labs, United Kingdom)
G. GALLASSI (Italtel, Italy)
P. GODLEWSKI (ENST, France)
A. GRAVEY (CNET Lannion, France)
R. GUSELLA (HP Labs, USA)
V. B. IVERSEN (Techn. Univ. of Denmark, Denmark)
K. JACOBS (Aachen Univ., Germany)
R. JAIN (DEC, USA)

K-L-M-O

D. KHAKHAR (Lund Univ., Sweden)
D. KOFMAN (ENST, France)
H. KOREZLIOGLU (ENST, France)
H. KRÖNER (Stuttgart Univ., Germany)
P. KÜHN (Stuttgart Univ., Germany)
J-Y. LE BOUDEC (IBM Zürick, Switzerland)
J. MACKENZIE (BT Labs, United Kingdom)
P. MARTINI (Paderborn Univ., Germany)
L. MASON (INRS Telecommunications, Canada)
J. OTTENSMEYER (Paderborn Univ., Germany)

P-R

P. PARMENTIER (Alcatel, France)
C. PARTRIDGE (BBN, USA)
A. PATTAVINA (Cefriel, Italy)
G-H. PETIT (Alcatel Bell, Belgium)
M. de PRYCKER (Alcatel Bell, Belgium)
G. PUJOLLE (Paris VI Univ., France)
J. RIERA (ETSI Telecom., Spain)
G. RIGOLIO (Italtel, Italy)
C. ROSENBERG (Polytechn. Montreal, Canada)
H. RUDIN (IBM Rüschlikon, Switzerland)
J. RUELA (IST/INESC, Portugal)
M. RÜMEKASTER (Paderborn Univ., Germany)
E. RÜTSCHE (EURECOM, France)

S-T

H. SEGUIN (France Telecom, France)
D. SERET (ENST, France)
M. SIMON (Montevideo Univ., Uruguay)
O. SPANIOL (Techn. Univ. Aachen, Germany)
P. SCHICKER (Consultant, Switzerland)
S. TOHME (ENST, Fance)
M. TONDRIAUX (Matra Marconi Space, France)
D. TRANCHIER (CNET Lannion, France)
P. TRAN-GIA (Würburg Univ., Germay)

V-W-Y-Z

P. VANKWIKELBERGE (Alcatel Bell, Belgium)
G. VENUTI (CSELT, Italy)
A. VOGEL (Montreal Univ., Canada)
P. VERMA (AT&T Bell Labs, USA)
G. WATSON (HP Labs, Bristol, UK)
S. WITTEVRONGEL (Gent Univ., Belgium)
A. WOLISZ (GMD-Fokus, Germany)
K. YAMAZAKI (KDD, Japan)
G. ZEMOR (ENST, France)
M. ZUKERMAN (Telecom Australia, Australia)

List of authors

A-B

M. AGHADAVOODI JOLFAEI (Aachen Univ., Germany)
A. ALONSO HERNAN (Telefonica Inv. y Des., Spain)
G. BAGNOLI (Fond. Ugo Bordoni, Roma, Italy)
T. BARKER (British Telecom, United Kingdom)
D. BEAUMONT (British Telecom, United Kingdom)
M. BECKER (INT, France)
A-L. BEYLOT (INT, France)
J. DE BIE (PTT Research, The Netherlands)
N. BJÖRKMAN(Telia Research, Sweden)
G. BOSVELD (PTT Research, The Netherlands)

C-D

P. CAMARDA (Politecnico di Bari, Italy)
J. CHEN(IBM Watson Research Center, USA)
H. CLAUSEN (Aarhus Univ., Denmark)
P. DELLO RUSSO (Politecnico di Bari, Italy)
G. DE ROBERTIS (Politecnico di Bari, Italy)
A. DOELMAN (PTT Research, The Netherlands)
A. DUPUIS (CNET, France)

F-G-H-I-J

L. FAGLIA (ITALTEL, Italy)
M. FRATER (Univ. of New South Wales, Australia)
M. GAGNAIRE (ENST, France)
R. GUERIN (IBM Watson Research Center, USA)
F. GUILLEMIN (CNET, France)
G. HEBUTERNE (CNET Lannion, France)
C. IKEDA (NEC Corporation, Japan)
A. JAJSZCZYK (Franco-Polish School, Poland)
J. JOHANSEN (Teledanmark, Denmark)
J-I. JUNG (Korea Telecom, Korea)

K-L

I. KOHLENBERG (INT, France)
P. KÜHN (Stuttgart Univ., Germany)
A. LATOUR-HENNER (Telia Research, Sweden)
K-Q. LIAO (Hull University, Canada)
M. LISTANTI (Roma Univ., Italy)

M

A. MACHT (DBP-Telekom, Germany)
L. MASON (INRS Telecommunications, Canada)
L-A. MERAYO FERNANDEZ (Telef. Inv. y Des., Spain)
T. MEUSER (Techn. Univ. Aachen, Germany)
V. MINAFRA (Politecnico di Bari, Italy)
S. MINZER (Bellcore, USA)
L. MOLA MORALES (Telefonica Inv. y Des., Spain)
J. MURPHY (Dublin Univ., Ireland)
L. MURPHY (Berkeley Univ., USA)

P-R

A. PAGLIALUNGA (CSELT, Italy)
U. QUERHEIM (Aachen Univ., Germany)
J. RODRIGUES (Telecom Portugal, Portugal)
I. RODRIGUEZ (Telefonica Inv. y Des., Spain)
L. RONCHETTI (Ericsson-Fatme, Italy)
O. ROSE (Würzburg Univ., Germany)
C. ROSENBERG (Ecole Polytech. Montréal, Canada)
J-V. RUYMBEKE (Belgacom, Belgium)
P. RYBERG JENSEN (Aarhus Univ., Denmark)

S

H. SALAS ASENCIOS (Telefonica Inv. y Des., Spain)
J. SCARR (PTT Research, The Netherlands)
D. SERET (ENST, France)
V. SIGNORE (Politecnico di Bari, Italy)
H. SUZUKI (NEC Corporation, Japan)
M. THEYS (PTT Research, The Netherlands)

T-U-W-Y

T. USO (GEC Alsthom, France)
M. WENAS (PTT Research, The Netherlands)
R. WINKLER (Fond. Ugo Bordoni, Roma, Italy)
J. WOLSKY (Techn. Univ. Aachen, Germany)
M. YOSHIDA (NEC Corporation, Japan)

APPLICATION AND SERVICES

Broadband Communications, II (C-24)
S. Tohmé and A. Casaca (Editors)
Elsevier Science B.V. (North-Holland)

Video Coding and Network Support for Multimedia Services over ATM Networks

David O. Beaumont

Visual Applications Division, BT Laboratories [MLB4/12], Ipswich IP5 7RE, UK

Tel: +44 473 642888
Fax: +44 473 643791
E-mail: d.o.beaumont@bt-web.bt.co.uk

Abstract
This paper gives an overview of the various digital video coding schemes that are employed in the provision of multimedia services, both currently and in the near future. The issues related to network support for these services are also discussed.

Multimedia is a term often used to describe a mix of video, audio and data technologies but it has always been difficult to offer a definitive definition of its applications. Computer, telecommunication and consumer industries all have different concepts of multimedia services but they are all increasingly reliant on video. Video, as a medium, usually places high demands on storage and transmission capacity; consequently there is much interest in how this demand can be reduced by video coding.

The advent of networks based on Asynchronous Transfer Mode (ATM) technology should offer greater possibilities for the support of truly integrated and seamless multimedia services. ATM networks may also provide the possibility of variable bit rate transmission for real-time services. In theory, variable bit rate transmission looks attractive for multimedia services employing video but in practice the principle seems fraught with problems. The use of variable bit rate transmission may increase the likelihood of transmission impairments such as cell loss. In short the use of variable bit rate transmission for real-time services over public networks opens up a 'Pandora's Box' of issues such as the network's quality of services, service requirements and tariffing which still need to be addressed.

Keyword Codes: **Supplementary Transaction C 5; I.4.2; I.4.0**
Keywords: **Application layer communication services, Multimedia services and systems; Image processing, Compression; General**

1. Introduction

The term multimedia applies to a mix of media components such as video, audio and data. This combination of technologies can currently be supported on a stand alone basis by computer platforms ranging from PCs to workstations which can be regarded as multimedia terminals, as shown in Figure 1. However it is the addition of a communications link which really increases the potential of a multimedia terminal, particularly in point-to-point and multipoint working.

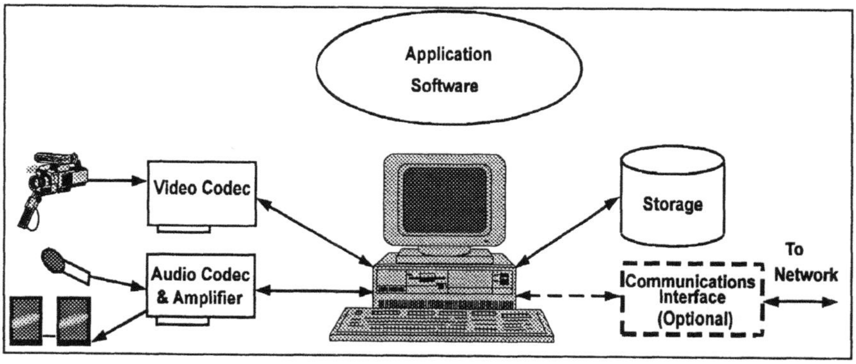

Figure 1: An example of a multimedia terminal

Digitised video transmission inherently requires high bit rate transmission which can be very costly in terms of connection charges. Consequently it is advantageous to minimise transmission costs by utilising some form of data compression. However the user has to decide between reasonably high terminal equipment costs with relatively low transmission charges or cheaper terminal equipment with higher transmission charges. For the multimedia customer who requires reasonably long connection times for point-to-point inter-working the usual decision is low connection charges and hence the need for high data compression.

Currently, the main digital network which offers the multimedia customer switched access is the Narrowband Integrated Services Digital Network (N-ISDN) at bit rates of 64-1920 kbit/s. The bit rates offered by N-ISDN will support a limited multimedia service but for a truly integrated multimedia service the future Broadband ISDN will provide greater potential. The future B-ISDN is based on Asynchronous Transfer Mode (ATM) technology which transports data packetized into cells. Unfortunately the B-ISDN also presents potential transmission problems for highly compressed datastreams such as coded video as cells may be 'lost' in the event of an attempted overload of the network.

2. Current multimedia support

There are currently three world standard video coding schemes suitable for use within multimedia terminals; CCITT H.261[1], ISO JPEG[2] and ISO MPEG-1[3]. The three coding schemes complement each other as they were designed to address the needs of specific services and applications. Whilst being complementary to one another they are not compatible and will not inter-work (i.e. an H.261 bitstream can't be decoded by a JPEG decoder). Another distinction between the coding schemes is that the CCITT H.261 recommendation was designed to operate over specific digital networks operating between 64 kbit/s and 1920 kbit/s whereas the ISO standards are just purely video coding design specifications and don't consider transmission requirements such as error protection. Further explanation of the schemes is given below:

2.1. CCITT H.261 coding

A limited form of multimedia service has been available for a number of years, namely videoconferencing and videotelephony. Videoconferencing and videotelephony are audio visual services based on terminals incorporating coders and decoders (CODECS) and existing digital networks like N-ISDN, T1 and E1. The CCITT H.261 recommendation was ratified in 1990 and along with other CCITT recommendations was incorporated into the CCITT recommendation H.320, for videotelephony terminals. The adoption of these standards by equipment manufacturers and the increasing penetration of N-ISDN has led to a new communication medium with a great deal of interoperability between conforming terminals.

The H.261 video codec is a hybrid design using a combination of spatial compression (INTRA coding) and temporal compression (INTER coding) with a discrete cosine transform (DCT) and motion compensation, as shown in Figure 2. A key feature of this coding scheme is that it codes a Common Intermediate Format (CIF) picture to which native TV standards (e.g. PAL, NTSC and SECAM) have to be converted. The CIF picture format is progressive (non-interlaced) and comprises of 360 horizontal picture elements (pels) by 288 lines at a maximum picture rate of 30Hz. The CIF format was derived as a compromise between the various TV standards. A subset of CIF called Quarter Common Intermediate Format (QCIF) is also defined which is half the horizontal and vertical resolution of CIF and is more suitable for videotelephony applications.

Work is now being undertaken by ITU-T SG15 (formerly CCITT SGXV) to develop new recommendations based on H.320 for use over Local Area Networks (LANs) and B-ISDN.

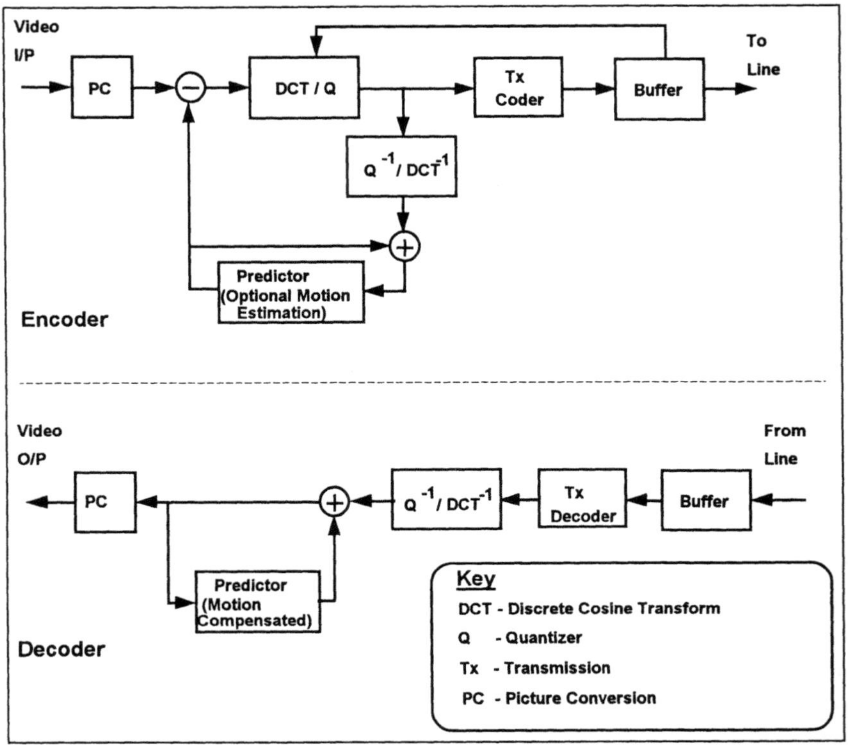

Figure 2: A simplified schematic of an H.261 codec

The prime objectives of the H.261 design were to obtain a high degree of data compression whilst incurring an acceptable processing delay so that it was usable for real-time interactive (conversational) services.

A number of attempts have been made to use this coding scheme in a true multimedia terminal. A European ESPRIT project called MIAS is one such example. This project developed a multimedia terminal which provided videotelephony, still image transfer, file transfers and group editing facilities whilst communicating over a basic rate N-ISDN (128 kbit/s) connection. A further important feature of the MIAS terminal was that it provided user friendly facilities such as chairpersonship for a multipoint connection between up to 5 MIAS terminals. The main limitation of this terminal was that its functionality was restricted to the limited capacity of the N-ISDN connection and therefore required momentary suspension of the real-time video coding whilst large file transfers were in progress. However this project did highlight the potential for interactive multimedia services.

2.2. ISO JPEG

The JPEG coding scheme was developed for coding high resolution still images for computer storage and retrieval applications. This scheme also uses a DCT but as the application is to code still images only spatial compression is used. The JPEG coding scheme does not limit itself to a range of fixed images formats as in H.261 but allows the user to define the image size and measure of compression.

2.3. ISO MPEG-1

The ISO MPEG-1 coding scheme's functionality has its roots in the H.261 coding scheme but has been significantly adapted to meet the needs for storage and retrieval applications of moving images. Processing delay was not a major concern of the MPEG-1 coding scheme as it was not intended for use in conversational services. Consequently it could utilise INTERPOLATED coding in addition to INTRA and INTER coding. This form of coding requires more processing storage with a resulting increase in delay, but when combined with INTRA and INTER coding techniques it provides a much higher quality coded picture than that obtained by H.261 at a similar bit rate. It should be noted that the use of INTER and INTERPOLATED coding techniques to increase compression inherently requires loss free recovery of the coded data. If errors are incurred in the recovered data used by a decoder, then corrupted reconstructed images will result and may persist for an unacceptable period of time (e.g. tens of seconds) dependant upon the transmission rate and the defensive measures employed by the decoder.

An example of a sequence of coded pictures used in the MPEG scheme is shown in Figure 3.

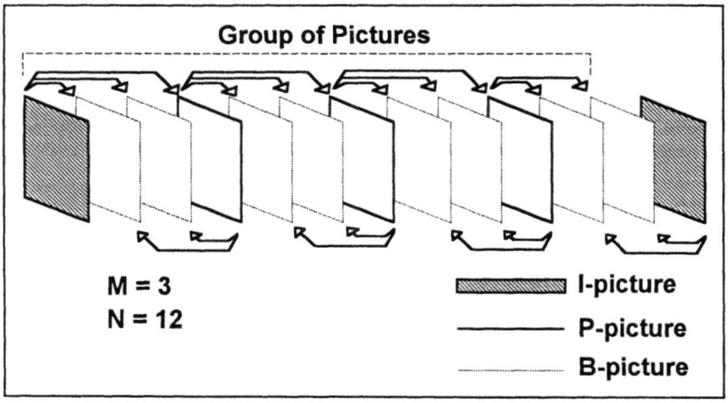

Figure 3: A example of a sequence of coded frames used by MPEG-1

where	I-picture	- Intra picture
	P-picture	- Predicted picture (Inter)
	B-picture	- Bi-directionally predicted picture (Interpolated)

N = 12	i.e. an I-picture every 12 pictures
M = 3	i.e. 2 B-pictures between every P picture

2.4. Current implementations

The high demand within the computer industry for the reduced storage requirements of both still and moving images has stimulated the microchip manufacturers to implement the ISO coding schemes. The telecommunications industry has also been keen to develop relatively low cost videotelephony terminals utilising very large scale integration (VLSI) 'chips'. Even though the above coding schemes are not compatible they do have some commonalty of functional elements (e.g. DCT, quantizers, etc.) and therefore a number of 'chip' manufacturers have made devices which incorporate a number of these coding schemes, selectable by the user. A further interesting development has been the use of so called 'moving' JPEG which is basically the JPEG algorithm being applied to moving image sources. When compared to H.261 or MPEG-1, the moving JPEG scheme requires a high bit rate and may be a costly alternative for conversational services but as it only uses spatial compression it does provide better subjective quality when transmitted over high error prone transmission circuits.

3. Future multimedia support

As previously mentioned, current attempts at truly integrated multimedia services using multimedia terminals and N-ISDN have been restrictive to the user. The restrictions fall into two main camps, the first being the limited picture quality of the moving image service and secondly the limited functionality imposed by N-ISDN. The ultimate goal of a truly integrated multimedia service must be the seamless manipulation of both moving and still images together with audio, voice and computer data files over an unrestricted network.

Developments within the video coding community are leaning towards a unified coding scheme to cover a wide range of applications and services. Both ITU-T SG15 and ISO MPEG are collaborating to produce new higher quality video coding standards called H.262 and MPEG-2 respectively. The H.262 recommendation effectively includes the common text of the MPEG-2 standard. Different PROFILES are being developed for the MPEG-2/H.262 coding scheme which are optimised for specific service applications and networks. Progress on the MPEG-2 algorithm, which is currently at the Committee Draft stage, has resulted in great advances in coding performance, with a target of CCIR 601 picture quality at between 4-10 Mbit/s. The importance of the MPEG-2 algorithm is increasing as it is now being considered for TV distribution and contribution applications and even High Definition TV [HDTV] applications. This view has been endorsed by a number of broadcasting bodies such as the European Broadcasting Union, World Broadcasting Union and the American 'Grand Alliance'.

The concept of multi-layered coding has been proposed as one method of offering differing scales of picture resolution (scalability), a key requirement for TV distribution and useful to multimedia. An example of a 4-layer coding scheme is shown in Figure 4. With such a scheme, the layers would be multiplexed together for transmission and terminals would extract only that part of the bitstream they are capable of displaying. The adoption of a common video coding standard will greatly assist in the development and introduction of truly integrated multimedia services of the future.

Currently coded video is transported over constant bit rate (CBR) networks. However the network standards bodies are striving to develop recommendations for future networks based on ATM technology. ATM networks are cell based; that is the respective datastreams

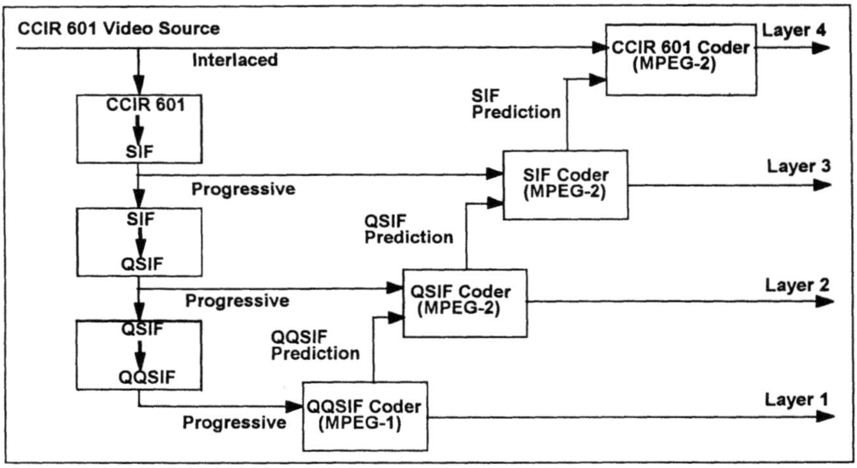

Figure 4: An example of multi-layered video encoding

are packetized into fixed length cells together with an ATM header for signalling and control purposes. There are currently two ATM network standards under development, one by the ATM FORUM (a group of companies interested in the development of ATM products) and the other by ITU-T SG13. The first network is an ATM Local Area Network (LAN) designed to couple PCs and computer workstations in a local environment such as an office or department. The second network structure is the future Broadband ISDN for public network operators offering large switched bit rates. It is the objective of both groups to design ATM networks which can support services that require bit rates which do not readily map onto the existing digital bit rate hierarchies. A further objective is the support for flexible allocation and de-allocation of associated data channels combined within a call connection. The advent of these types of network will greatly improve the support for multimedia services of the future where during a single call connection the demand for bit rate capacity will vary with time according to the particular multimedia requirements.

So far coding technologies and networks have been discussed but there is a third element needed for the successful introduction of interoperable multimedia services, namely the standardisation of how the multimedia objects or components are described and handled. An ISO body called Multimedia and Hypermedia information coding Expert Group (MHEG) is responsible for this function. Figure 5 shows a model of how the MHEG work combines with the MPEG video coding scheme.

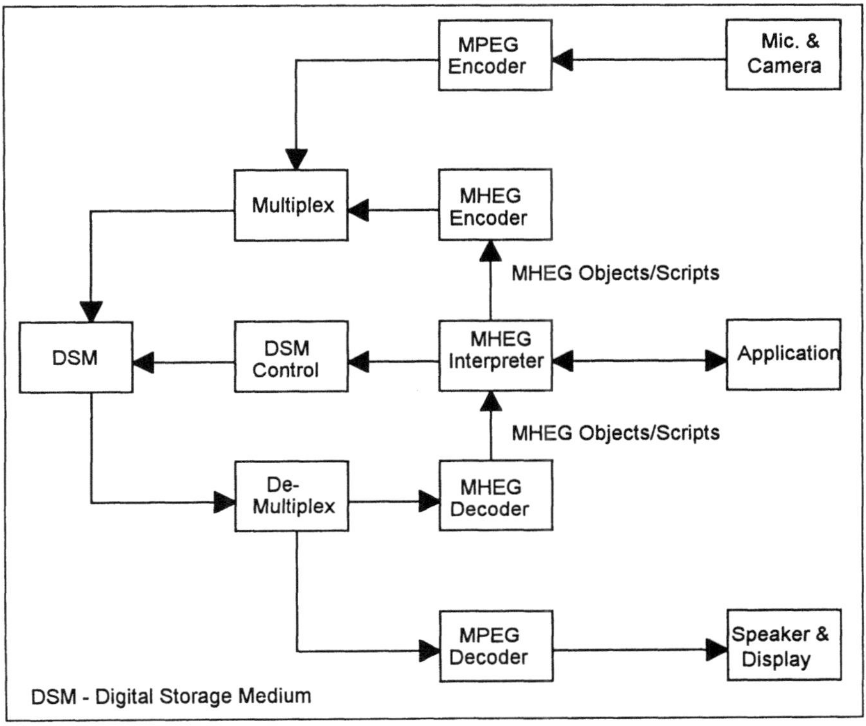

Figure 5: A schematic showing the MHEG and MPEG model

4. The impact of ATM on multimedia terminals

As previously mentioned, ATM network technology is a radical departure from current transmission systems and its introduction coincides with an increasing interest and demand for multimedia and related services. Unfortunately this situation has resulted in standards for both activities being developed in parallel, and there is some difficulty in tracking both activities which are in a constant state of flux. Recently, coded video has been seen as a key component for multimedia terminals with particular interest focusing on the MPEG standards. The potential transmission flexibility of ATM networks is the most attractive feature for future multimedia services, in particular offering the possible support for variable bit rate (VBR) transmission. However it is also important to remember that a multimedia connection may incorporate other network technologies in addition to ATM and therefore the terminal should maintain some independence from the network and not solely rely on specific network features.

VBR transmission is just one such topic that is causing concern for terminal designers, network operators and standards bodies. From a service operators point of view VBR transmission seems the solution to existing service constraints, by offering the possibility of:

- reduced processing delay in codecs,
- near constant picture quality, and
- more efficient use of transmission capacity.

However the use of VBR transmission may also cause:

- a higher probability of cell loss,
- greater cell delay variation, and
- higher network management costs.

Cell loss is major problem for highly compressed datastreams like coded video. Video decoders can take defensive measures to maintain a minimum picture quality in the event of cell loss but the decoder must be able to detect that cell loss has occurred. This criteria has been addressed for constant bit rate video services by using the Sequence Number technique defined for ATM Adaptation Layer (AAL) Type 1. However VBR real-time services may require an AAL to provide additional features than those offered by AALs Type 1 and Type 5. The development of an appropriate AAL suitable for MPEG-2 bitstreams is currently an issue within the networks and video coding standards bodies.

One approach that video codec designers[4,5] have taken to initially overcome the problem of cell loss with VBR transmission has been the use of a 2-layer CBR/VBR coding scheme, as shown in Figure 6. This scheme relies on a base layer picture coded at CBR with guaranteed transmission and an enhancement layer coded at VBR. When the two datastreams are decoded and combined in the decoder the result is an image with near constant picture quality. In the event of cell loss occurring in the VBR transmission then the decoder displays only the base layer picture for the affected area of the image.

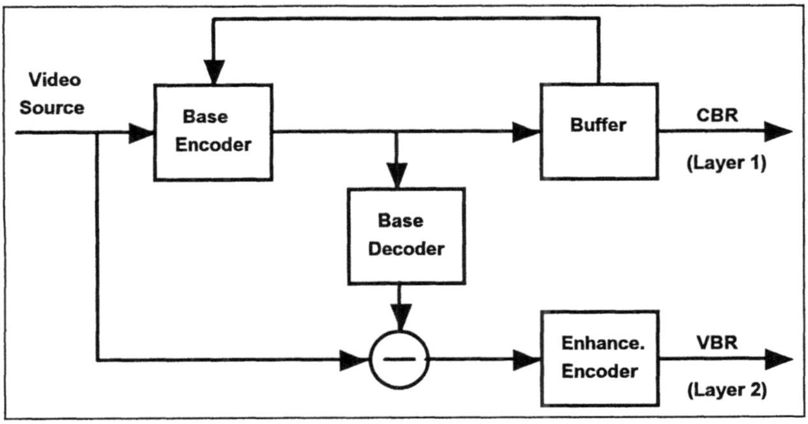

Figure 6: Schematic of a 2-layer CBR/VBR encoder

A better approach for VBR transmission from a service providers point of view would be to have guaranteed transmission similar to CBR transmission. Current developments in ATM standards suggest that this might be possible by implementing Generic Flow Control (GFC) for controlled terminals, as shown in Figure 7. This claim is based on the premise that the main cause of cell loss will be at the ingress to the network and, if this activity was handled in a controlled manner by sending feedback information to the terminals connected to an ATM node, then transmission can be guaranteed as the bit rate entering the node will not exceed the node's output capacity.

The use of GFC may also provide multimedia terminals with an added benefit by providing the necessary information they need to control the priority of data flow across a multimedia connection. This aspect is particularly important when a terminal has to control a multimedia connection consisting of a number of Virtual Channels (VCs), some with real-time data such as coded video and other channels for data such as file transfers. In such a scenario, when GFC indicates that the terminal must reduce its aggregate bit rate the terminal can use its discretion as to how to implement the change in a controlled manner, favouring the real-time video component over the file transfer.

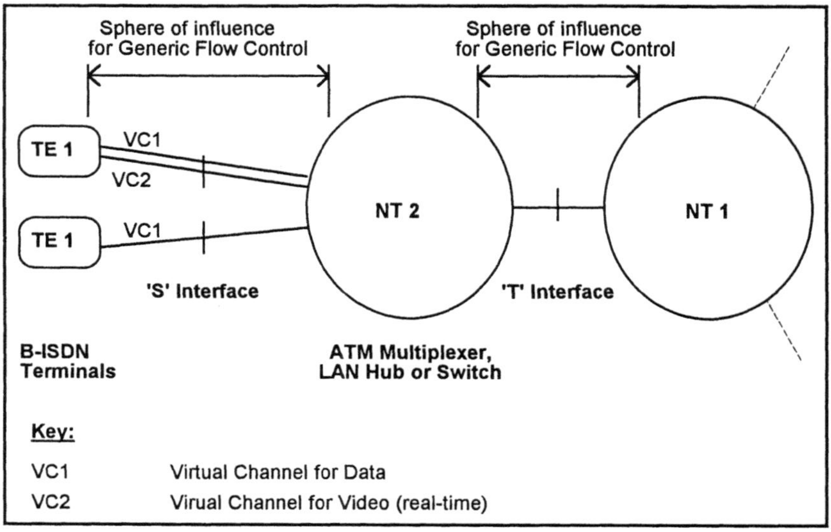

Figure 7: A schematic showing operating environment for GFC

5. Future services

The implementation of new video coding schemes such as H.262/MPEG-2 combined with their use over future ATM networks will greatly assist in improving existing multimedia services and encourage the development of new services.

The support for multi-layered coding will improve a multipoint service by allowing all terminals to display a received picture to the limit of their processing capability, except when receiving pictures from a lower resolution source. This compares with current arrangements for multipoint services where all users have to abide by the highest common picture resolution. This is illustrated in Figure 8 where all terminals will display videotelephony type picture resolution (QCIF).

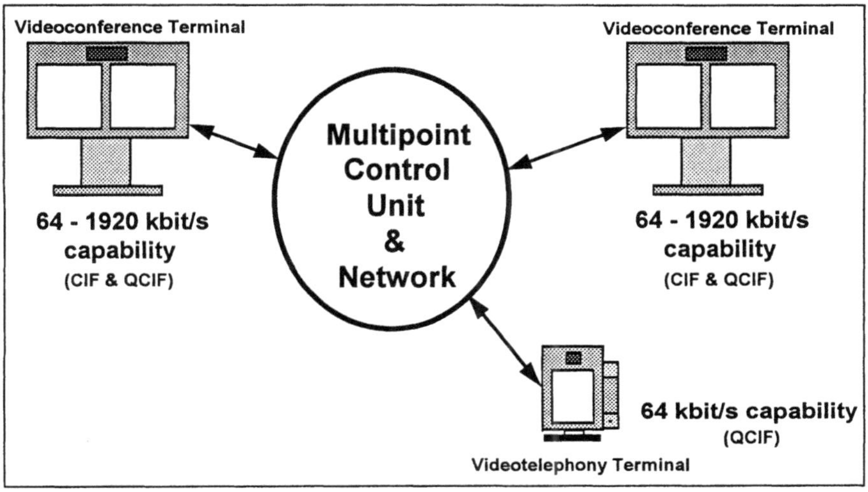

Figure 8: An example of multipoint working

Multi-layered coding also offers the possibility of compatibility and interoperability with existing video services as shown in Figure 9. In this example an advanced videoconferencing terminal using H.262 can decode the bitstream from the videotelephony terminal using H.261 by only using layer 1. The same is true for the terminals that use a combination of MPEG coding schemes.

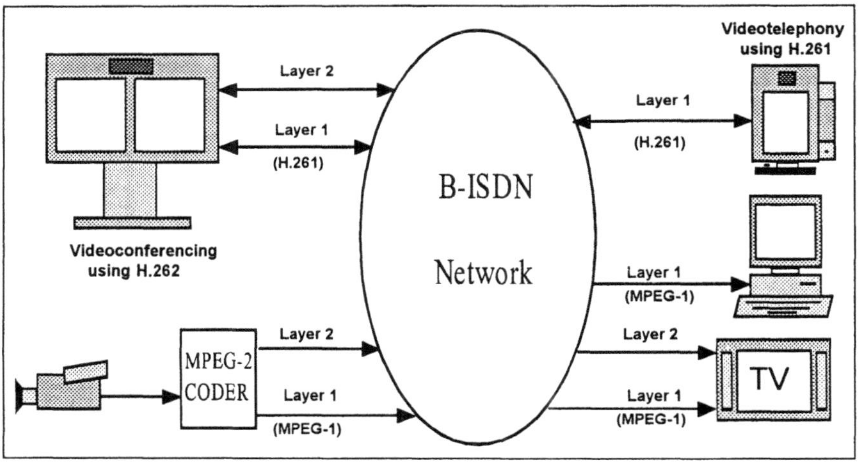

Figure 9: An example of compatible coding

To illustrate the possible uses of the various coding schemes Table 1 presents a summary of candidate applications and services, the appropriate video coding schemes employed and possible bit rates.

Table 1

A summary of candidate applications and services

Application/ Service	Video Coding Scheme	Service Requirements	Bit Rate (kbit/s)	Desirable Features	Comments
Videotelephony/ Videoconferencing	H.261	Conversational real-time	48 - 1856	Low delay, inter-regional operation	Single layer
Storage & Retrieval	JPEG	Still image	Undefined	Good picture quality	Single layer
	MPEG-1	Moving image	Possibly... 1,000-5,000	Similar to S-VHS picture quality or better	Single layer
Improved Videoconferencing	H.262	Conversational real-time	Possibly... 48 - 10,000+	Low delay, inter-regional operation, compatibility with H.261 , higher picture quality than H.261	Possibly ... Multi-layer
Improved Storage & Retrieval, Multimedia, TV distribution	JPEG H.262 MPEG-2	Still image Conversational real-time Moving image	Possibly... 48-10,000+ [HDTV 15,000 to 25,000]	Scalability of picture resolution. video/voice communication coupled with additional services, high picture quality	Single layer or Multi-layer

6. Conclusions

A summary of the various digital video coding schemes suitable for multimedia services has been presented together with their individual characteristics for specific video applications. The developing video coding standards of ISO MPEG and ITU-T SG15 are gaining increasing significance as the work is progressing towards a unified video coding scheme which will assist in the manipulation of video within multimedia applications and services.

Whilst video coding schemes such as ISO MPEG provide the vehicle for acceptable data compression it is important that the higher functions of multimedia applications are also standardised. The work within ISO MHEG is valuable as it should assist in the seamless usage of multimedia components and provide interoperability at the application level. A growing market already exists for audio-visual services and it is important that customers can identify a clear co-existence with new multimedia services. It is foreseen that interoperability and compatibility will become important marketing issues for multimedia.

The advent of new networks based on the use of ATM will offer the transmission flexibility to enable users to make best use of multimedia applications without the limitations imposed by present day fixed bit rate networks. However the use of VBR transmission will require more stringent network management controls and an a radical approach to connection tariffing.

As computer processing power becomes cheaper and more widely available, and video coding offers higher compression with high picture quality, new markets for multimedia applications will flourish. Service industries such as medical and advertising will see greater potential for multimedia particularly when inter-working is required and the flexibility of ATM networks exist. Broadcasters and TV production units may be able to exploit their wealth of archive source material for multimedia users, provided of course that the copyright issues can be solved.

References

1. CCITT Recommendation H.261: Video codec for audio-visual services at p x 64 kbit/s.

2. ISO/IEC IS 10918-1: Information technology - Digital compression and coding of continuous-tone still images

3. ISO/IEC IS 11172: Coding of moving pictures and associated audio for digital storage media at up to about 1.5 Mbit/s.

4. M. Ghanbari, "An adaptive video codec for ATM networks": Third International Workshop on Packet Video, Morristown, U.S.A. 1990

5. D. O. Beaumont & D. G. Morrison, "A prototype VBR video codec for use on an experimental ATM network": Third International Workshop on Packet Video, Morristown, U.S.A. 1990

Broadband Communications, II (C-24)
S. Tohmé and A. Casaca (Editors)
Elsevier Science B.V. (North-Holland)

Provision of the Broadband Connectionless Data Bearer Service in N-ISDN

Maria Wenas and Mark Theys

Royal PTT Nederland NV, PTT Research, Dr. Neher Laboratories, P.O. Box 421, 2260 AK Leidschendam, The Netherlands

ABSTRACT

This paper investigates how BCLB can be provided in N-ISDN[1]. Firstly a brief description of BCLB is given which indicates some open issues. Next the paper reflects several network architectures that give possibilities to provide BCLB to N-ISDN users. It turns out that N-ISDN preferably should be used as a transparent access network to BCLB only. To achieve this, BCLB can be provided over several types of N-ISDN bearer services (Circuit Mode, Frame Mode or Packet Mode). Each type has its own advantages and disadvantages. However, all types require the introduction of a convergence layer. The functions to be supported by the convergence layer depend on the type of N-ISDN bearer service used.

Since N-ISDN connections have limited bandwidth and since N-ISDN has a higher bit error probability than B-ISDN, BCLB can be provided only to N-ISDN users for QOS classes that allow a relatively large packet delay and packet loss probability. The Quality of Service provided to narrowband BCLB users may be improved by means of extra functions in the convergence layer. Packet delay can be decreased by allowing pipelining. It turns out that the decrease in delay is only significant if a packet is transferred from a narrowband BCLB user to another narrowband BCLB user. The packet loss probability can be traded for additional delay. Therefore a retransmission function must be adopted in the convergence layer.

Keyword Codes: Communication Systems
Keywords: ISDN (Integrated Services Digital Network), ATM (Asynchronous Transfer Mode) networks, Internetworking

1. INTRODUCTION

At the moment, ITU-T is standardizing B-ISDN. B-ISDN can be interpreted in many aspects as the successor of the traditional or *narrowband* ISDN (N-ISDN). Complete

---

[1] Initial work has been done in ITU-T to standardize access to specific B-ISDN services from a 64 kbit/s based ISDN (N-ISDN). One of the B-ISDN services is BCLB (annex 7 of I.580, July 1993 version).

replacement of N-ISDN by B-ISDN is generally seen as a long term objective.

With the introduction of B-ISDNs, existing N-ISDNs will not be taken out of operation immediately. For economical reasons, B-ISDNs will be introduced piece by piece. As a result, there will be a period during which B-ISDNs and N-ISDNs coexist. In such an evolutionary phase, N-ISDN users (N-users or narrowband users) must of course be able to communicate with B-ISDN users (B-users or broadband users).

For communication between a narrowband user and a broadband user two cases can be distinguished. In the first case, communication takes place via an N-ISDN bearer service. This usually will be possible, since it has been agreed that B-ISDN supports all N-ISDN bearer services. In the second case, communication takes place via an exclusive B-ISDN bearer service (i.e. via a bearer service that is not an N-ISDN bearer service). This usually will be impossible, since B-ISDN allows much higher bit rates than N-ISDN. Yet, there are some B-ISDN bearer services that can be provided to narrowband users with a lowered Quality of Service (QOS). One of these services is the Broadband Connectionless Data Bearer Service (BCLB).

This paper starts with a brief description of BCLB. It then shows how BCLB can be provided in N-ISDN. Therefore, it focuses on network architecture and protocol architecture. Finally the quality of the service provided in N-ISDN is discussed and conclusions are drawn.

2. DESCRIPTION OF BCLB

The Broadband Connectionless Data Bearer Service (BCLB) is a B-ISDN bearer service. It is described in [F.812], [I.211] and [I.364]. Only the main characteristics of BCLB are described here.

BCLB is a connectionless service. That means that BCLB users can send packets to each other without prior *end-to-end* call establishment procedures. Each packet contains among others the address of the destination user. The number plan used by BCLB is E.164. This is the same number plan as is used in B-ISDN and N-ISDN.

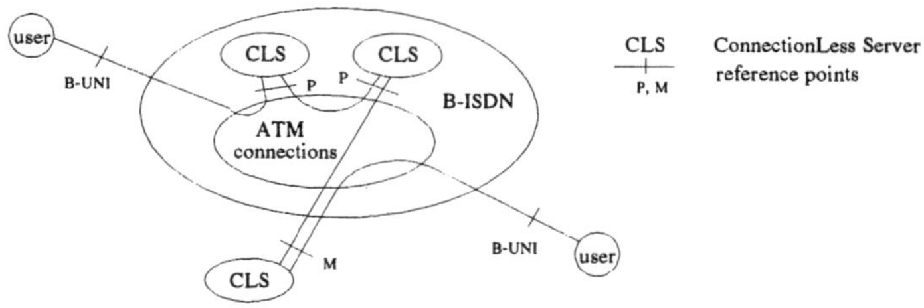

Figure 1: Standardized architecture for the provision of BCLB in B-ISDN

BCLB is provided via a network of interconnected connectionless servers (CLSs; see figure 1). According to [I.211], a CLS resides either within or outside the B-ISDN. In the former case, the CLS is connected to the ATM network via a P reference point. In the latter case, the server is connected to the B-ISDN via an M reference point. A CLS that is connected to the B-ISDN via an M reference point, may be administered by another organization than the B-ISDN provider.

CLSs are connected to each other via ATM connections. Also, broadband users that want to use BCLB, must have an ATM connection to a connectionless server. It is intended that both types of ATM connections can be established both on demand and semi-permanently.

As soon as a broadband user is connected to a CLS via an ATM connection, he can send any number of connectionless packets to the CLS. Each connectionless packet may be destined for a single or for multiple destination users. A CLS that receives a packet forwards the packet to another CLS or to one or more destination users. In routing the packet, the CLS uses the destination address supplied in the packet.

In considering the actual transfer of a connectionless packet, two cases must be distinguished: the transfer of the packet between a broadband user and a CLS and the transfer of the packet between two CLSs. The protocol stacks for both cases are depicted in figure 2.

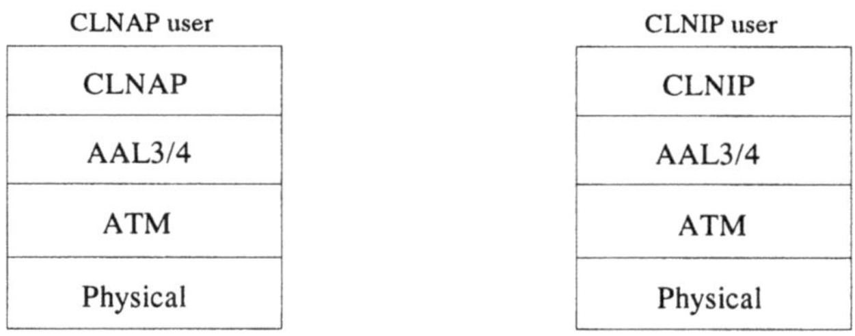

stack for user-CLS communication   stack for CLS-CLS communication

Figure 2: BCLB protocol architectures as defined in I.364

Between a broadband user and a CLS, a connectionless packet is transferred via the left protocol stack in figure 2. This protocol stack is defined in [I.364]. The main functions of the layer above the ATM Adaptation Layer (AAL) are the creating, sending and receiving of connectionless packets. These functions are performed by the Connectionless Network Access Protocol (CLNAP). The main function of the AAL is to segment connectionless packets created by the CLNAP, so these packets can be transferred in ATM cells.

For the transfer of a connectionless packet between two CLSs, ITU-T has just defined the right protocol stack in figure 2. The Connectionless Network Interface Protocol (CLNIP) performs similar functions as CLNAP. In addition, it performs functions that are exclusively related to CLS-CLS communication. An example of such a function is the reduction of destination address resolution in packets whose original destination address was a group address.

A possible overall protocol architecture for the transfer of a connectionless packet between two broadband users is depicted in figure 3. The mapping entity may for instance perform the routing function.

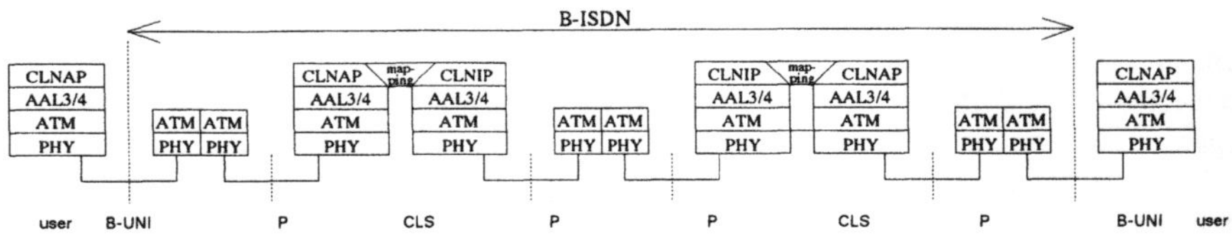

Figure 3: Example of BCLB protocol architecture between two broadband users

So far, the concept of BCLB is rather clear. However, the standardization of BCLB has not yet been finished and a number of open issues still need to be cleared. Open issues that are relevant for this paper are:
- routing architecture;
- login possibilities;
- signalling.

Routing architecture

The main function of a CLS is the forwarding of connectionless packets to other CLSs or to destination users. In order to forward a packet in the correct direction, CLSs must agree some routing protocol. So far, no proposals have been submitted for a routing protocol for BCLB.

The protocol that finally will be adopted may imply a (partly) hierarchical routing structure. An example of such a routing structure is given in figure 4.

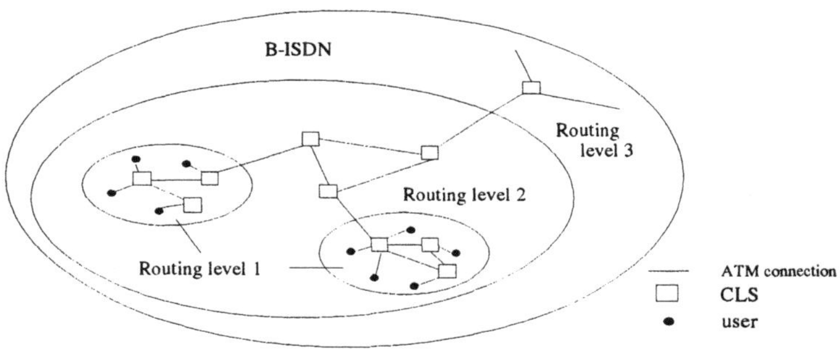

Figure 4: Possible BCLB routing structure (hierarchical)

In the figure, routing areas are introduced. Each routing area is assigned a routing level *p*, where *p* is an integer that is equal to or larger than one. Each routing area of level *n* (n> =2) consists of a number of routing areas of level *n*-1. Each E.164 address uniquely belongs to a level 1 area. Within an area, packets are routed based on some specific part of the destination's E.164 address. The specific part of the address is determined by the area's routing level. The higher the level, the more significant the specific part within the E.164 address. A special case form the level 1 areas. In these areas routing is always based on the least significant part of the destination's address. Routing takes place in such a way that a packet that is transferred between two BCLB users belonging to some common area, never leaves the common area (regardless of the routing level of that common area).

Note that in the routing architecture of figure 4, users can only have an ATM connection to a CLS that belongs to a level 1 area. Further, not all CLSs may be connected with each other (a CLS that resides in a level 1 area may for instance not be connected to a CLS residing in a level 3 area).

Login possibilities

A BCLB user must have an ATM connection to a CLS (i.e. the user must login into a CLS). So far, it is not clear to what CLSs a user can be connected. As was shown above, the BCLB routing protocol can influence a user's login possibilities.

Another aspect related to login is addressing. Since a BCLB user is a B-ISDN user, the E.164 address used for routing BCLB packets can be the same as the E.164 address of the user's UNI. As an alternative, these addresses may be different. In this case, the CLS to which the user is connected must associate the two addresses (for instance in a table). The assignment of two E.164 addresses to a single user may be convenient for the provision of BCLB via external administrations. On the other hand, such an assignment will waste address space.

Signalling

ITU-T has stated that it must be possible to establish any ATM connection used for BCLB both on demand and semi-permanently. However, so far signalling for BCLB has not yet been considered by ITU-T. The reason for this is that signalling for B-ISDN is standardized in three releases. Signalling for BCLB is scheduled for release 2. At the moment, release 1 is being completed. As a consequence, in the near future BCLB only can be provided via semi-permanent ATM connections.

3. NETWORK ARCHITECTURE FOR THE PROVISION OF BCLB TO NARROWBAND USERS

For the provision of BCLB to N-ISDN users, several network architectures are possible. This paper discusses an architecture by considering the following three aspects:
- the introduction of narrowband connectionless servers;
- a network adapting functionality;
- direct or indirect provision of BCLB to N-ISDN users.

I. Introduction of narrowband connectionless servers

If a narrowband user wants to use BCLB, he must be connected to a CLS. A solution is to connect the narrowband user to a B-ISDN CLS. Note that in this case the problem that the narrowband user supports N-ISDN connections, whereas the B-ISDN CLS supports ATM connections, has to be solved. As an alternative, the narrowband user could be connected to newly introduced CLSs that support N-ISDN connections only. In this paper, these new CLSs will be called N-CLSs (narrowband CLSs).

II. Network adapting functionality

For the provision of BCLB to N-ISDN users, B-ISDN and N-ISDN must be adapted and connected to each other. This can be accomplished via a separate unit, an NA (Network Adaptor). Alternatively, the network adapting functions can be integrated in a CLS. In this paper, this latter CLS will be called an NA-CLS (Network Adapting Connectionless Server). The difference between a B-ISDN CLS, an N-CLS and an NA-CLS is that a B-ISDN CLS and an N-CLS each support one type of connection only (namely ATM and narrowband connections respectively). In contrast, an NA-CLS supports ATM connections *and* narrowband connections.

III. Direct or indirect provision

As in B-ISDN, BCLB can be provided in N-ISDN either directly or indirectly.

If BCLB is provided directly in N-ISDN, the N-ISDN 'knows' the service BCLB. In fact, BCLB is added to the N-ISDN bearer services. In signalling protocols across the UNI, narrowband users can explicitly ask for the service BCLB. If N-CLSs exist, they functionally reside inside the N-ISDN.

Indirect provision of BCLB here is defined as provision of BCLB on top of an existing N-ISDN bearer service. The N-ISDN does not know the service BCLB. It only provides a number of bearer connections. If N-CLSs exist, they reside outside N-ISDN. N-CLSs, NAs and NA-CLSs are connected to N-ISDN via ordinary UNIs.

An example of a network architecture is depicted in figure 5. This architecture was derived by copying the architecture used in B-ISDN into N-ISDN. The architecture contains N-CLSs that reside inside N-ISDN. B-ISDN and N-ISDN are connected to each other via NA-CLSs. The P_N and M_N reference points are reference points introduced by this paper. They are similar to the B-ISDN P and M reference points respectively.

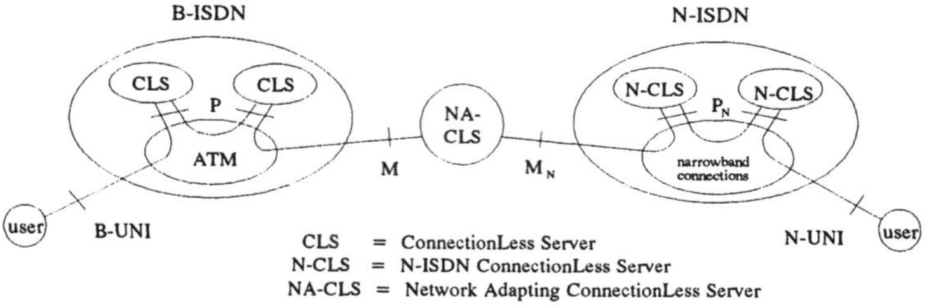

Figure 5: Direct provision of BCLB in N-ISDN by copying the B-ISDN architecture

In the architecture in figure 5, N-CLSs reside inside N-ISDN. Therefore, BCLB is provided directly to N-ISDN users. In general, direct provision of BCLB has a major disadvantage: it requires changes in existing N-ISDN signalling protocols. As a consequence, existing N-ISDN equipment will need to be modified or replaced. Obviously this is very uneconomical. Therefore, this paper recommends to provide BCLB indirectly in N-ISDN.

A second network architecture is depicted in figure 6. The architecture again connects B-ISDN and N-ISDN via NA-CLSs and also uses N-CLSs. This time however, the N-CLSs reside outside N-ISDN. As the N-CLSs are connected to the N-ISDN via ordinary UNIs, N-ISDN is used indirectly (or *transparently*).

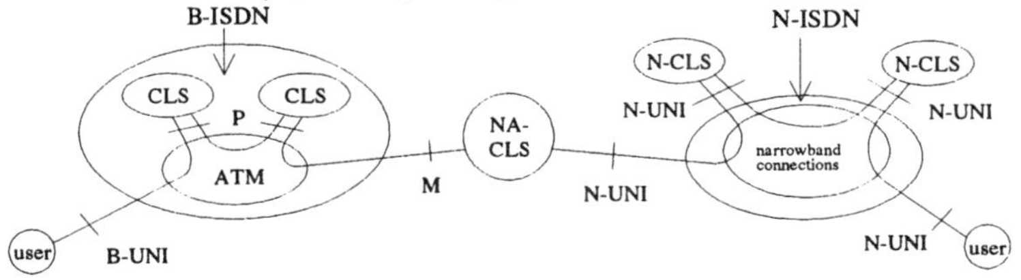

Figure 6: Indirect provision of BCLB in N-ISDN using N-CLSs

In the architecture in figure 6, no changes to N-ISDN are needed. However, the architecture leaves two other drawbacks. First of all, three types of CLSs exist: B-ISDN CLSs, NA-CLSs and N-CLSs. The development of NA-CLSs and N-CLSs will not be economical if only few N-ISDN users want to use BCLB. Secondly, the architecture has consequences for the development of a routing protocol for BCLB. To illustrate this, note that in a pure broadband ISDN a routing protocol can be adopted that requires all CLSs to have full connectivity. This routing protocol cannot be used in the architecture of figure 6, since this architecture does not allow an N-CLS to be connected to a B-ISDN CLS.

As another example of the possible influence of the BCLB routing protocol, assume that in B-ISDN the hierarchical routing architecture of figure 4 is adopted. Further suppose that a BCLB user uses the E.164 address of his B-UNI for sending and receiving connectionless packets. If the provision of BCLB is extended to N-ISDN using the architecture of figure 6, each level 1 area that has a narrowband BCLB user must have an N-CLS. Clearly this can be very unattractive. For, if the level 1 areas are small, many N-CLSs will be needed. Also, it may be possible that in a certain level 1 area an N-CLS is needed for a single narrowband user only.

From the above it follows that the architecture of figure 6 puts constraints on the development of a routing protocol for BCLB. Since this is inconvenient for further standardization, one may try to find a network architecture that removes these constraints. A possible solution to this problem is to eliminate any NA-CLS and N-CLS. An indirect network architecture that contains neither NA-CLSs nor N-CLSs is depicted in figure 7.

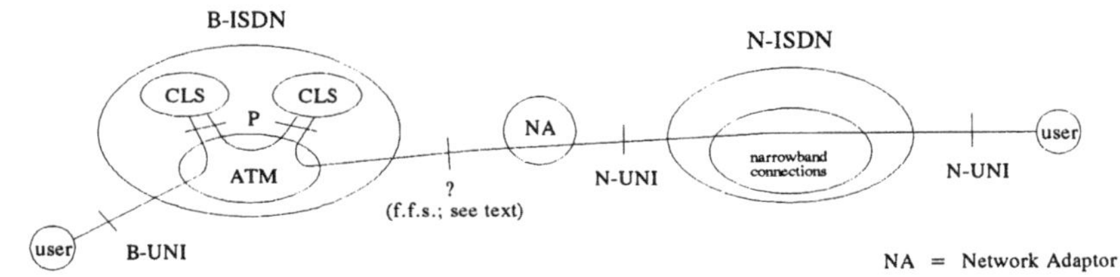

Figure 7: Indirect provision of BCLB in N-ISDN using NAs only

In the architecture of figure 7, a narrowband user directly is connected to a B-ISDN CLS. This is accomplished via a *hybrid* connection. This hybrid connection is partly an ATM connection and partly an N-ISDN bearer connection. Both parts are connected to each other in an NA. Since BCLB is provided indirectly, the NA is connected to N-ISDN via an ordinary UNI. The reference point for connecting an NA to B-ISDN is left for further study. The reason for this is that none of the reference points identified so far (B-UNI, P, M) seem obvious for connecting an NA to B-ISDN. Please note however, that it may be convenient (or even necessary) to define a new reference point. Such a reference point could exclusively be used for interworking between B-ISDN and N-ISDN. The reference point could be used for multiple services (thus not only for BCLB). An interesting aspect is that interworking units connected to the reference point may be able to shield a calling user from the question whether the called party has a broadband or narrowband UNI by cooperating with the control plane.

Since the network architecture of figure 7 is superior to both other architectures, this paper recommends to adopt this architecture for the provision of BCLB to narrowband users. The superiority of the architecture can be summarized by:

- the architecture does not require existing N-ISDN equipment to be modified or replaced;
- the architecture only introduces one new piece of equipment;
- the architecture does not put constraints on the development of a routing protocol for BCLB.

4. PROTOCOL ARCHITECTURE FOR THE PROVISION OF BCLB TO N-ISDN USERS

The goal of this section is to present a protocol architecture for the network architecture depicted in figure 7. This network architecture was recommended in the previous section.

In figure 7, it was assumed that BCLB was provided in N-ISDN on top of an existing N-ISDN bearer service. In this section, only the bearer services circuit switching (CS) and frame relaying (FR) are considered. The protocol architectures of these bearer services are depicted in figure 8.

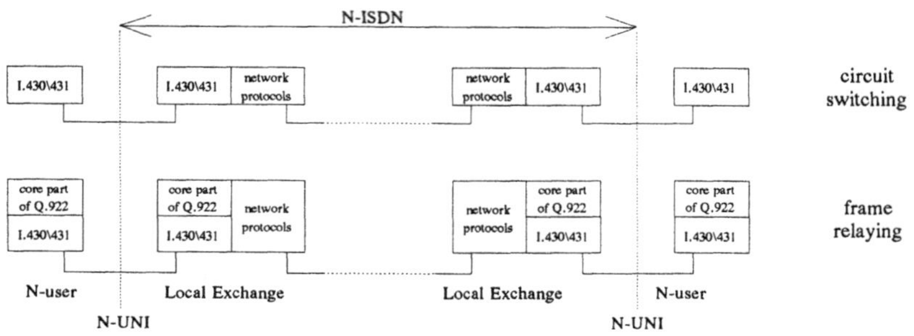

Figure 8: Protocol architectures of the N-ISDN CS and FR bearer services

To arrive at a protocol architecture for the provision of BCLB in N-ISDN, the protocol architecture for BCLB in B-ISDN has to be merged with the protocol architecture of CS or FR. In figure 9, a possible solution is depicted for a bearer connection of each type. In both cases, the actual merging has been accomplished by the introduction of a convergence layer (CONV).

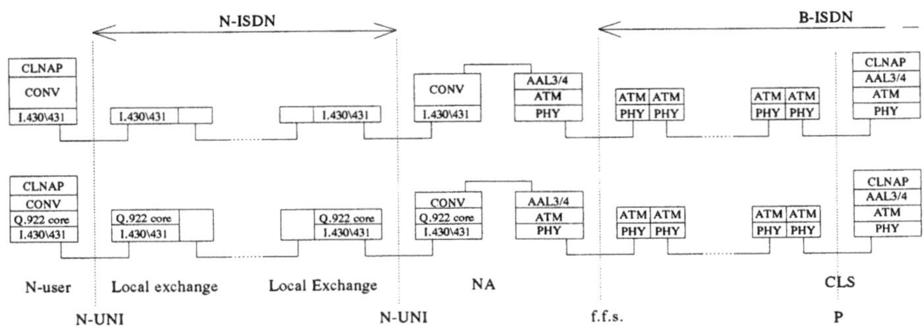

Figure 9: Proposed protocol architecture for the provision of BCLB in N-ISDN

The functions that must be or could be supported by the convergence layer depend on the type of bearer connection that is used in N-ISDN (i.e. a circuit switched or a frame relayed bearer connection). In the following sub-sections, it is investigated per bearer connection type what functions the convergence layer must or may support.

4.1. Functions of the convergence layer for a circuit switched bearer connection

For a circuit switched bearer connection, the convergence layer *must* support the following functions:

Error detection

Since bit errors can occur at the physical layer, this function is necessary to detect if a connectionless packet is corrupted. Note that a similar function is performed by the AAL in B-ISDN.

Delimiting of connectionless packets

This function is necessary to delimit a packet at the physical layer. The function allows a receiving convergence layer to recognize connectionless packets within a continuous stream of bits from the physical layer.

Functions that *may* be supported by the convergence layer are:

Segmentation and Reassembling

In B-ISDN, connectionless packets are transferred in ATM cells. As a consequence, a segmentation and reassembling function cannot be avoided. In N-ISDN, a packet may be transferred in the consecutive octets of a circuit switched connection. As a result, a segmentation and reassembling function is not absolutely necessary in the convergence layer. Yet, it may be convenient to introduce such a function.

Compared to B-ISDN, N-ISDN only supports connections with a relatively low bandwidth. As a consequence, connectionless packets will experience a much larger delay in N-ISDN than in B-ISDN. By introducing a segmentation and reassembling function in the convergence layer, pipelining becomes possible in N-ISDN (i.e. an NA can forward segments of a packet even if it has not yet received the entire packet). As a result, delay may decrease. One should note however that this decrease will be on the cost of additional delay and overhead. For a closer look at pipelining, the reader is referred to the next section.

Another advantage of a segmentation and reassembling function is that an NA may be simplified. For, in an implementation it may be possible to combine the segmentation function of the AAL with the reassembling function of the convergence layer and vice versa.

Multiplexing

If a segmentation and reassembling function is introduced in the convergence layer, one may consider to allow multiplexing of connectionless packets. In effect, the segments of several connectionless packets can be interleaved.

Retransmission

In B-ISDN, the cell loss rate of ATM cells is expected to be very low. A typical value may be 10^{-8} (see [PRYC92] page 215). Assuming this value, the probability that a maximum sized connectionless packet of 9232 octets is lost, is $1\text{-}(1\text{-}10^{-8})^{210}=2.1*10^{-6}$. (The number 210 represents the number of ATM cells that is needed to carry the packet. The number can be deduced from the definition of AAL3/4 in [I.363]). Although N-ISDN and B-ISDN may use the same optical fibres within the network, the packet loss probability in N-ISDN will be much higher than in B-ISDN. This is caused by the physical medium that is used across the UNI.

For a 64 kbit/s circuit switched connection in N-ISDN, bit error rates will be in the order of 10^{-6}. With this error rate, a maximum sized connectionless packet will experience a loss probability of $1\text{-}(1\text{-}10^{-6})^{8*(9232+10)}=7.1*10^{-2}$ (it is assumed that the convergence layer adds an overhead of 10 octets). Compared to B-ISDN, this is very high.

The high packet loss probability in N-ISDN may be taken for granted. As an alternative connectionless packets in N-ISDN can be assigned a lower maximum length. A drawback of this alternative is that the service provided to a narrowband user is not consistent with the service provided to a broadband user. A better alternative may be to align the loss probability in N-ISDN more with the loss probability in B-ISDN. This can be achieved by the introduction of a retransmission function in the convergence layer. Note that such a function not only makes the convergence layer more complex, but that it increases the average packet delay as well.

4.2. Functions of the convergence layer for a frame switched bearer connection

The functions in the previous sub-section were classified as necessary or optional. If BCLB is provided in N-ISDN over a frame relayed bearer connection, the classification changes:

Delimiting of packets

This function is not needed for provision of BCLB over frame relayed bearer connections. The reason for this is that the Q.922 core protocol already performs this function.

Error detection

This function may be needed in the convergence layer. The function is also performed by the Q.922 core protocol. However, for long frames this error detection is insufficient. Therefore, additional error detection in the convergence layer is necessary if a Q.922 core frame carries a considerably long (part of a) connectionless packet.

Segmentation and Reassembling

Whether this function is necessary or optional depends on the maximum frame size the N-ISDN supports. If this size is large enough to carry a maximum sized connectionless packet, no segmentation and reassembling is necessary. In general however, the network will not support such frame sizes (a connectionless packet can consist of more than 9000 octets; for frame relaying ITU-T recommends the support of frame sizes in the order of 1600 octets). In this case, a segmentation and reassembling function cannot be avoided.

Multiplexing

For this function the same holds as for the circuit switched case.

Retransmission

For this function, the same text as for the circuit switched case holds. As an extension, note that for FR it is possible to put a retransmission in the datalink layer instead of the convergence layer. This is accomplished by replacing the Q.922 core protocols in NAs and in N-ISDN terminals by the *full* Q.922 protocol. In this case retransmission can be performed by the Q.922 protocol. Note that the retransmission takes place end-to-end between the NA and the N-ISDN terminal. Therefore, the N-ISDN still provides an FR connection (see figure 10).

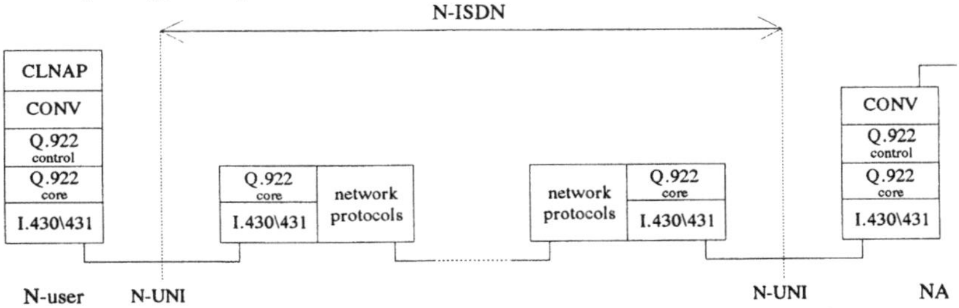

Figure 10: Retransmission via Q.922 on top of FR bearer service

4.3. Use of other bearer connections

Besides CS and FR, the bearer services Frame Switching (FS) and Packet Switching (PS) are defined in N-ISDN. For these services, similar stacks as in figure 9 can be constructed. However, the use of these services for the provision of BCLB to a narrowband user is not recommendable. Since *internal* network links usually are very reliable, link-by-link error correction as in both FS and PS has no advantages over end-to-end error correction as for instance in figure 10. Link-by-link error correction within N-ISDN rather should be prevented, since it unnecessarily increases a connectionless packet's delay.

5. QUALITY OF SERVICE AND INFLUENCE OF PIPELINING ON QUALITY OF SERVICE

From the previous section, it follows that provision of BCLB in N-ISDN is only possible if the narrowband user selects a QOS class (Quality of Service class) that allows:
- a relatively large packet loss probability;
- a relatively large delay.

The first requirement results from the relatively high bit error probability in N-ISDN. This error probability at its turn is mainly caused by the use of existing copper cables across the UNI. Via a retransmission function, this drawback can be traded for additional delay.

The second requirement, the relatively large delay, is caused by the possible bandwidths for connections within N-ISDN. These bandwidths are relatively low (e.g. 64 kbit/s). In the remainder of this section it is shown that delay may be decreased slightly by allowing pipelining in N-ISDN.

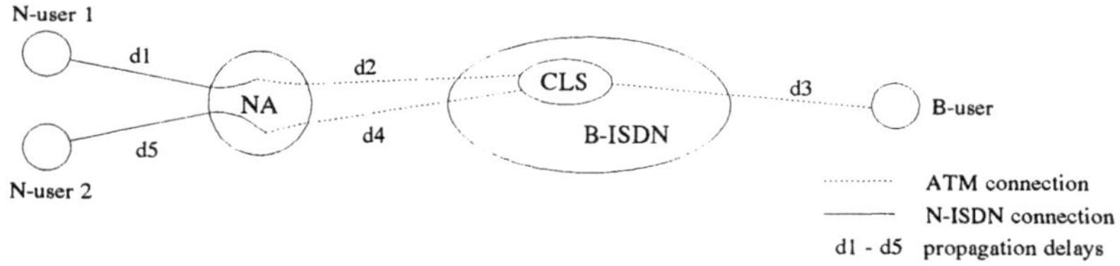

Figure 11: Reference configuration to illustrate the influence of pipelining

To illustrate the influence of pipelining on delay, a reference configuration has been depicted in figure 11. The figure shows three BCLB users, one NA and one CLS. The parameters $d1$ to $d5$ represent the propagation delays of the links that interconnect the users, NA and CLS. It is assumed that links within N-ISDN can transport information at a maximum rate of y bit/s; the ATM links have a maximum transmission speed of x bit/s ($x >= y$). Any connectionless packet is assumed to have a length of l bits. Throughout the remainder of this section, it further is supposed that all ATM and narrowband connections are established semi-permanently and that no retransmission is applied in N-ISDN.

Consider figure 11. Assume that N-user 1 transmits a packet to the B-user. Assume further that the CLS pipelines the individual ATM cells belonging to the packet (i.e. the CLS acts as an ATM switch). For convenience, suppose that any pipelining takes place ideally (i.e. a device that pipelines is able to forward any bit it receives with zero delay). Now two cases can be distinguished.

In the first case, the NA does not support pipelining. The overall delay (the time that passes between the N-user starting to transmit the packet and the B-user having received the packet completely) in this case is $d1 + l/y + d2 + d3 + l/x$ seconds. Here $d1$, $d2$ and $d3$ represent propagation delays. The term l/y is the reception time in the NA (the time that passes between the NA receiving the first bit and the NA receiving the final bit of the packet). Likewise, l/x is the reception time of the B-user.

In the second case, the NA supports (ideal) pipelining. In this case the total delay becomes $d1 + d2 + d3 + l/y$ seconds. Again $d1$, $d2$ and $d3$ represent the propagation delays of all links involved. The term l/y is the reception time of the B-user. Note that the B-user's reception time has increased while the overall delay has decreased.

In figure 12, the delays are shown graphically. As can be seen, pipelining in the NA reduces overall delay by l/x seconds. It should be noted however, that in general this decrease will be little in relative terms. For, a typical value may be 0.6%. (This reduction is achieved if y=64 kbit/s, x=10 Mbit/s, l=32 kbit and if all propagation delays are negligible). Also, in practice the profit of pipelining will be less than l/x seconds. The reason for this is that pipelining requires additional overhead (and thus an increased transmission time) in the convergence layer. In some specific implementations, pipelining even may *increase* rather than decrease the overall delay.

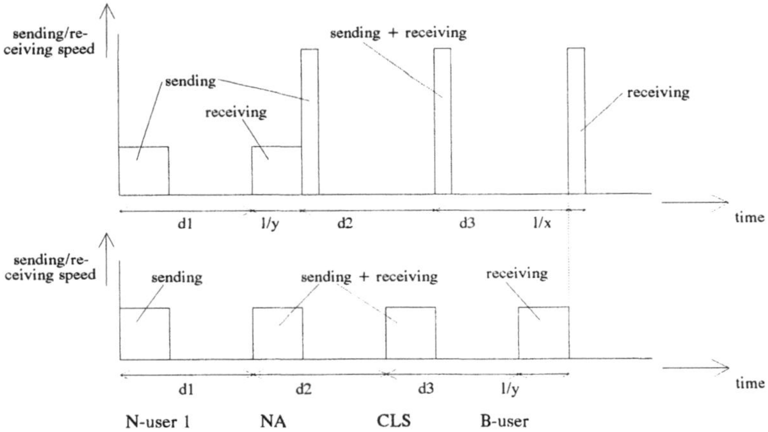

Figure 12: Example of overall delays (top: non-pipelining NA; bottom: pipelining NA)

Table 1 gives overall delays for packet transfer in different situations. Each situation is characterized by the user that sends the packet, the intended recipient and the question whether the CLS allows pipelining. From the figures can be concluded that pipelining in an NA only leads to a significant decrease in overall delay if an N-user sends a packet to another N-user *and* if the CLS supports pipelining. As a consequence, a pipelining option should only be introduced in N-ISDN if this can be accomplished by little additional costs.

Table 1: Influence of pipelining (in NA) on overall delay

source	dest.	operation mode CLS	overall delay if NA cannot pipeline (in s)	gain if NA pipelines (in s)	typical gain (in %)
N-user 1	B-user	pipelining	d1 + l/y + d2 + d3 + l/x	l/x	0.64
N-user 1	N-user 2	pipelining	d1 + l/y + d2 + d4 + l/x + d5 + l/y	l/x + l/y	50
B-user	N-user 1	pipelining	d3 + d2 + l/x + d1 + l/y	l/x	0.64
N-user 1	B-user	non-p.lining	d1 + l/y + d2 + l/x + d3 + l/x	l/x	0.64
N-user 1	N-user 2	non-p.lining	d1 + l/y + d2 + l/x + d4 + l/x + d5 + l/y	2*(l/x)	0.64
B-user	N-user 1	non-p.lining	d3 + l/x + d2 + l/x + d1 + l/y	l/x	0.64

Note 1: for definitions see text and figures 11/12
Note 2: for the typical gain x, y and l are assumed 10 Mbit/s, 64 kbit/s and 32 kbit. In addition all delays are assumed negligible

6. CONCLUSIONS

This paper described how the B-ISDN bearer service BCLB can be provided to N-ISDN users. It first described BCLB. It appeared that the concept of BCLB is rather clear and that standardization should focus now on routing, login possibilities and signalling.

Next the paper investigated how the provision of BCLB can be extended to narrowband users. To do so, the paper took a look at possible network and protocol architectures.

Regarding the network architecture, it was recommended to introduce a general reference point for interworking between B-ISDN and N-ISDN. Further, it turned out that N-ISDN should be used as a *transparent* network for *accessing* BCLB only. N-ISDN should be used transparently in order to avoid the need for any changes in current N-ISDN standards. N-ISDN should be used as an access network (i.e. not containing any CLSs) to prevent that the provision of BCLB in N-ISDN puts any constraints on the development of a routing protocol for BCLB.

For the protocol architecture, it was recommended to introduce a convergence layer. The functions to be performed by this convergence layer depend on the N-ISDN bearer service used.

Finally, the quality of service provided to N-ISDN users was investigated. It was shown that BCLB can only be provided to N-ISDN for QOS classes that allow a relatively high packet loss probability and a relatively large packet delay. The packet loss probability can be eliminated on the cost of additional delay and complexity. Delay can be decreased by letting an NA operate in a pipeline mode. However, the decrease in delay is significant in one specific case only.

REFERENCES

[E.164] ITU-T Recommendation E.164
Numbering plan for the ISDN era

[F.812] ITU-T Recommendation F.812
Broadband Connectionless Data Bearer Service

[I.211] ITU-T Recommendation I.211
B-ISDN Service Aspects

[I.363] ITU-T Recommendation I.363
B-ISDN ATM Adaptation Layer (AAL) Specification

[I.364] ITU-T Recommendation I.364
Support of Broadband Connectionless Data Bearer Service on B-ISDN

[PRYC92] Prycker, M de
Asynchronous Transfer Mode: Solution for Broadband ISDN, 2nd ed.
London: Ellis Horwood Limited, 1992

Broadband Communications, II (C-24)
S. Tohmé and A. Casaca (Editors)
Elsevier Science B.V. (North-Holland)

LANs Interconnection via ATM: An Experimental Prototype

P. Camarda, P. Dello Russo, G. De Robertis, V. Minafra, V. Signore

Politecnico di Bari - Dip. di Elettrotecnica ed Elettronica
Via E. Orabona, 4 - 70125 Bari (Italy)

Abstract

An experimental prototype of an IP router, interconnecting standard Ethernet LANs by means of ATM networks, is presented. The suite of ATM protocols up to the SAR sublayer has been implemented partially in hardware and partially in software. In particular, the physical layer is implemented in hardware with a data rate of 155.520 Mbps while the other functions are software based. The global throughput for this implementation can be up to 6-8 Mbps, fast enough for the considered context. An upgrade of the design is presented where most of the functions up to the AAL protocol are hardware based. The throughput of this upgraded architecture is expected to improve considerably up to several tens of Mbps.

Keyword Codes: B.4.1; C.2.1; C.2.5
Keywords: Data Communications Devices; Network Architecture and Design; Local Networks

1. Introduction

ATM has been proposed as the switching and multiplexing technique for BISDN. Due to its great flexibility in managing network resources, it will be able to support new multimedia services, resulting from a mix of the basic service components (audio, video, data) and from complex interconnection topologies. At the same time ATM should improve the performances of existing high-demand services, such as high-speed data transfer, distributed processing, distributed data-base, LAN/MAN interconnection. Dealing with LAN-like services two kind of problems have to be faced:
- most of existing LAN protocols are connectionless, while ATM operates in a connection-oriented mode. As a consequence, if an ATM network has to be used as a backbone for LAN interconnection, an interface between LAN and ATM, namely a gateway, must be provided, in order to create a mapping between source and destination addresses, which are the basic routing informations for connectionless protocols, and the connection IDs (VPI/VCI for ATM protocols) used by connection-oriented networks.

- due to the bursty nature of data transfer services, a flexible bandwidth reservation technique and a fair congestion control mechanism have to be considered [1,2,3], in order to satisfy QoS requirements of both LAN-to-LAN connections and other connections sharing the same ATM links.

In this paper the first problem is addressed: in the first part, the techniques of LANs interconnection are described and compared; in the second part the implementation and the performances of a gateway prototype is reported (part of this apparatus is being integrated in an ATM test bed for the Telecommunication Project of the CNR - National Research Council [4]); in the last part some key indications for a possible upgrade are given.

2. Overview of LAN interconnection techniques

A classification of telecommunication services can be given, based on three attributes [5].

i) Time constraints between source and destination
ii) Bit rate: constant or variable
iii) Connection mode: connectionless or connection-oriented.

According to this classification four service classes have been pointed out. LAN interconnection (absence of time constraints, variable bit rate, connectionless) can be considered a class D service, consequently AAL 3/4 or recently introduced AAL 5 protocols can be adopted, in order to let ATM subnetwork support it.

Discussing LAN interconnection, the following LAN configuration is often held as an example: an Ethernet LAN with the TCP/IP protocol suite running over it, quite common both in business companies and in research communities. At the top of the protocol stack, applications such as FTP or TELNET use operating system primitive functions in order to access TCP/IP services. However the interconnection principles are general, and they can be applied to any kind of network.

Interconnecting LAN, an ATM network should provide adaptation entities in order to support the connectionless service. Depending on the location of these entities, two interconnection criteria, namely, direct and indirect method, have been suggested [6,7,8].

2.1. Indirect method

Adaptation functions are performed externally to ATM domain and connectionless protocols are supported in a transparent way. This is possible connecting each pair of LAN-ATM gateways through a semi-permanent ATM path (fig. 1), i.e. a Virtual Path connection, established whenever a LAN is booted, using normal ATM signalling procedures. The initializing gateway, during these procedures, specifies the receiving gateways BISDN addresses (using for example the E.164 format).

The control processor of each ATM switch crossed by the signalling messages, uses these addresses as routing information toward the destination gateways and furthermore reserves to the chosen physical output link a VPI, with a local significance. On the contrary the VCIs have an end-to-end significance and are handled by the gateways in order to identify, for example, each pair of hosts involved in exchanging datagrams. In respect of VP connections, each ATM switch acts as a VP switch, swapping only the VPI value and leaving unmodified the VCI value of ATM cells belonging to those connections [5,9].

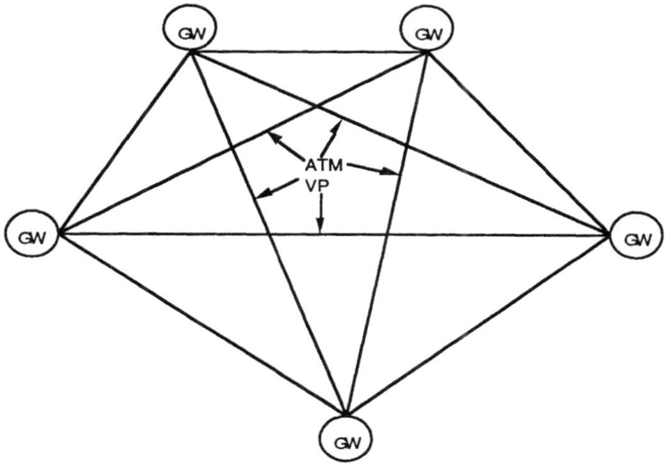

Fig. 1 - LAN interconnection through ATM: indirect method

2.2. Direct method

Some of ATM switches, namely ConnectionLess Servers (CLS), connected each other via ATM VPs (fig. 2), are explicitly conceived to support connectionless services. ATM cells deriving from segmentation of IP datagrams for example, pass trasparently through normal ATM switches, while are processed by the CLSs, which can perform CL-AAL functions besides ordinary PHY-level and ATM-level functions, allowing CLSs behave, also, as normal switches. Of course, connectionless cells should be distinguished from ordinary cells using a limited set of VCIs. CL-AAL functions consist in routing connectionless cells toward the destination gateway, whose address or another related information must be inserted in the payload. As the routing algorithm is executed more than once and not only during the signalling procedure, there is not a predefined path between source and destination gateways.

Two solutions have been proposed for the basic routing information [6].

i) Direct method, cell based: each connectionless cells carries in its payload the destination gateway BISDN address. The routing algorithm is executed cell by cell, thus different cells belonging to the same message might reach the destination gateway through different paths.

ii) Direct method, message based: the destination BISDN address is inserted only in the BOM cells payload, together with a Message IDentifier (MID), while all subsequent cells carry only the MID. When a CLS receives a connectionless BOM cell, it creates a mapping between the BISDN address and the MID and executes the routing algorithm in order to determine a path toward the destination gateway. All following cells carrying the same MID are routed through the same path. However, as the MID, ST, LI payload subfields are needed, AAL 5 protocol cannot be used.

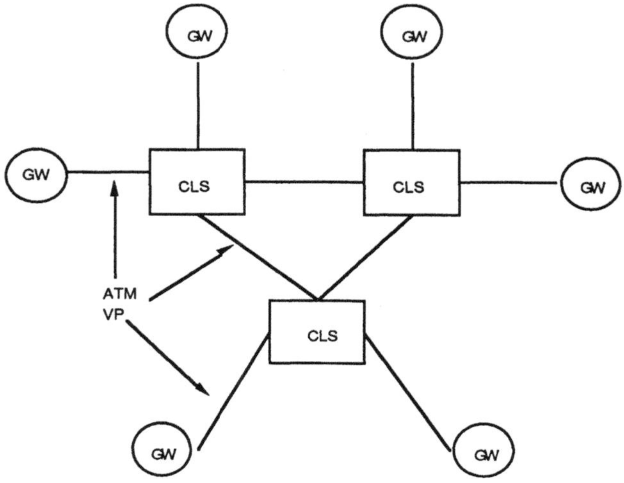

Fig. 2 - LAN interconnection through ATM: direct method

Comparing the cell based and message based options, the following considerations can be made.

- Routing performed cell by cell, typical of the cell based option, better fits suddenly variable traffic conditions than the one performed message by message. However, the routing algorithm is repeatedly applied, and heavy processing speed requirements are imposed to the CLSs. Moreover, due to the destination address carried cell by cell with the cell based option, a greater overhead outcomes than in the message based one, as less space is left in the cell payload for user information.
- For the message based solution the AAL 4 payload structure can be used, while the one for cell based is out of current standards.

Given the pros and the cons of the two options, the message based solution is preferable, unless the amount of connectionless traffic is considerably superior and then a dynamic routing procedure is needed.

2.3. Gateway functions

Gateway functions depend on whether the indirect method or the direct one is adopted. With the indirect method the transmitting gateway intercepts IP datagrams destined to another LAN. The destination LAN IP address is used as an entry point for a static table reporting the corresponding gateway BISDN address. A dynamic lookup table, updated during the signalling procedure, maps this address on the VPI identifying the virtual path which must be used by ATM cells in order to reach the desired gateway.

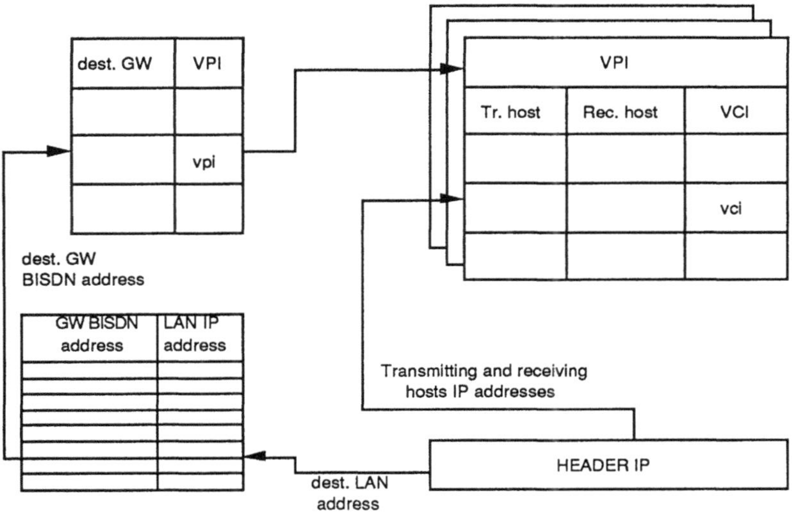

Fig. 3 - Procedure executed by the transmitting gateway for VPI/VCI evaluation: indirect method

Using this VPI value, and the IP addresses of the source and destination hosts extracted from the IP header, a set of dynamic tables is entered in order to determine the associated VCI value. The procedure adopted by the transmitting gateway in order to get VPI and VCI values from IP address is represented in fig. 3. A multilayered representation of gateway functions is represented in fig. 4. Unless a high-level multiplexing, i.e. interleaving of ATM cells carrying the same VPI/VCI values and belonging to different messages exchanged between the same pair of hosts, is needed, the simple and efficient AAL 5 protocol can be used. The described procedure is meaningful only if the source and destination LANs use the same internetworking protocol (not necessarily IP), and then the gateway can be configured as a router. Otherwise the gateway should be configured as a MAC bridge and a higher adaptation level is requested, as evidenced in fig. 5. MACADP layer handles MAC frame, reading the destination host address and deriving from a static data-base the BISDN address of the gateway which connects the destination LAN. Then the former procedure for VPI/VCI determination can be executed. At the receiving side the gateway reassembles IP datagrams or MAC frame, whether it behaves as a router or a bridge, linking the user data carried by all ATM cells identified by the same VPI/VCI value.

With the direct method (message based option), the transmitting gateway performs a limited set of functions. When a packet destined to a different LAN is caught, it associates an unused MID value to the destination BISDN address (determined in the same way as in the indirect method). The same MID is used for all cells belonging to the same datagram or frame. Furthermore the address is inserted in the BOM cell. The header of ATM cells is filled with a VCI value identifying connectionless cells, and the VPI value of the virtual path connecting the gateway to the closest CLS.

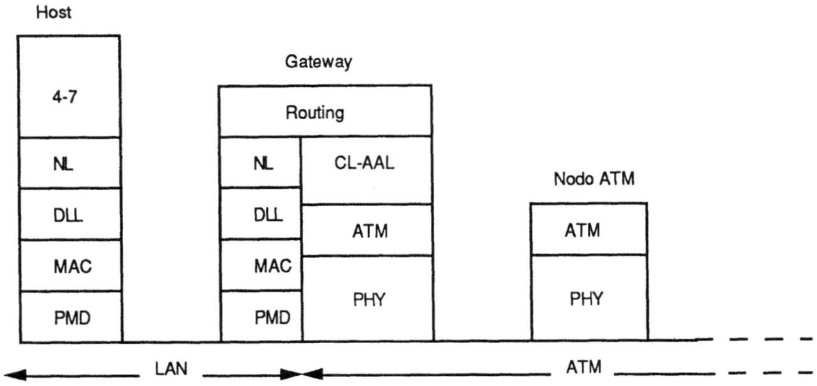

Fig. 4 - Layered structure of the functionalities needed in LAN interconnection via ATM, using the indirect method and a gateway configured as a router.

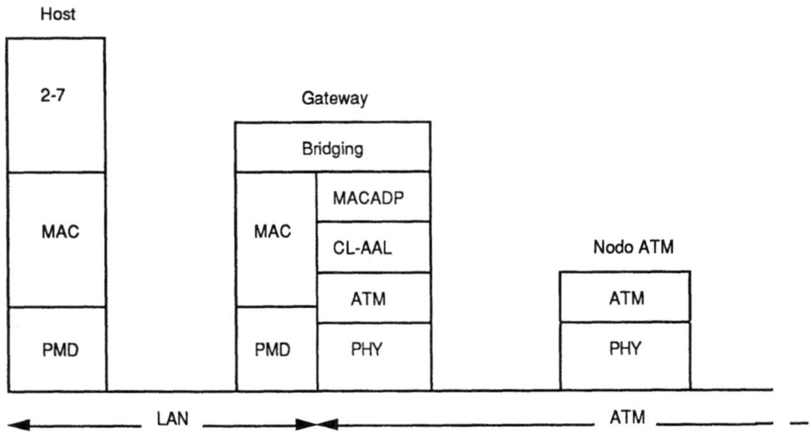

Fig. 5 - Layered structure of the functionalities needed for LAN interconnection via ATM, using the indirect method and a gateway configured as a MAC bridge

This procedure is represented in fig. 6, while the multilayered representation is in fig. 7 for the router and in fig. 8 for the MAC bridge. At the receiving side the gateway reassembles the original packets linking user data of cells carrying the same MID value [10].

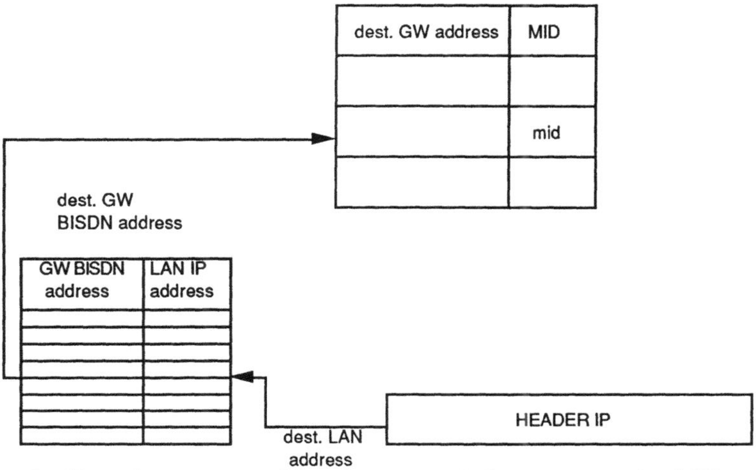

Fig. 6 - Procedure executed by the transmitting gateway for MID evaluation: direct method

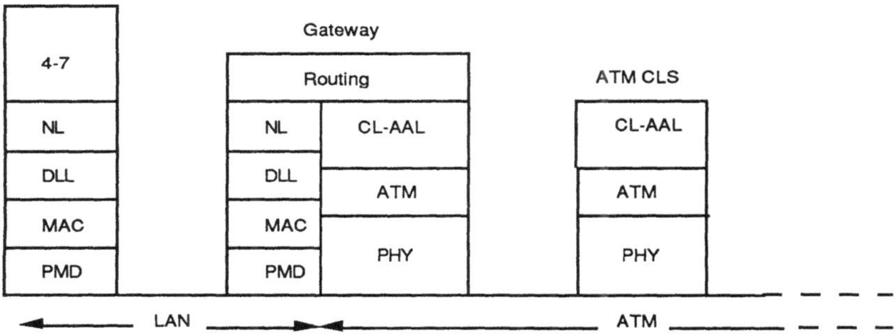

Fig. 7 - Layered structure of the functionalities needed for LAN interconnection via ATM, using the direct method and a gateway configured as a router

2.4. Comparison between indirect and direct method

The main advantage of the indirect method resides in the location of the connectionless adaptation entities. In fact ATM switches do not perform specific functions as the CLSs do in the direct method, and the required procedures are distributed at the edge of ATM network, being executed by the gateway, that has to guarantee lower throughput than an ATM switch. Indeed, the latter deals with multiple high-capacity ATM links, while the former interfaces with a lower speed LAN. Furthermore, for the direct method the AAL 4 protocol must be used, while for the indirect one the AAL 5, simpler to be implemented in hardware, can be exploited.

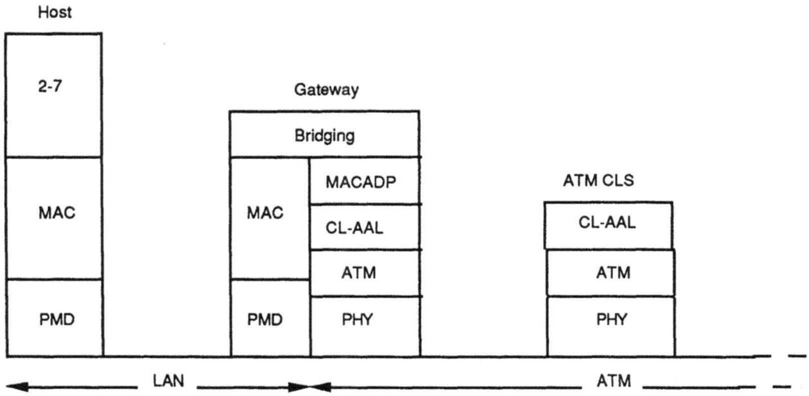

Fig. 8 - Layered structure of the functionalities needed for LAN interconnection via ATM, using the direct method and a gateway configures as a router

On the other hand if n is the number of gateways and m the number of CLSs (generally m<<n) the number of unidirectional VP connections to be established is

$$v = n(n-1)$$

for the indirect method, and

$$v = n + m(m-1)$$

for the direct one. It can be said that for n>10, v increases approximately as the square of n for the former case, and linearly for the latter. As network resources are reserved in a semi-permanent way to each vitual path, using the indirect method might become a source of ineffectiveness if n is very large.

In conclusion, the indirect method is more straightforward and efficient for a small number of LAN-ATM gateways, otherwise the direct one better suits the ATM concept of a flexible management of network resources.

3. Experimental prototype

A VME based SUN workstation, to which have been added two VME modules, has been used in a laboratory experiment to implement an IP router with the complete suite of ATM protocols, following the indirect criteria. The first board (DPV30D - manifactured by Loughborough Sound Images Ltd) implements the ATM layer and the SAR sublayer of the protocols (fig. 9), while the second custom board (designed specifically for this experiment) implements the physical layer (fig. 10). All other needed protocol modules are implemented by the SUN workstation. In particular, the LSI board has a complex architecture which has been only partially exploited in this experiment. The peculiar characteristics of this board well suited to our experiment are the presence of two high speed independent processors (TMS320C30) and two 32 bit wide parallel I/O, which globally allows us to implement a simple ATM interface (most of the functions are software based) fast enough to be exploited for the interconnection of standard Ethernet LANs.

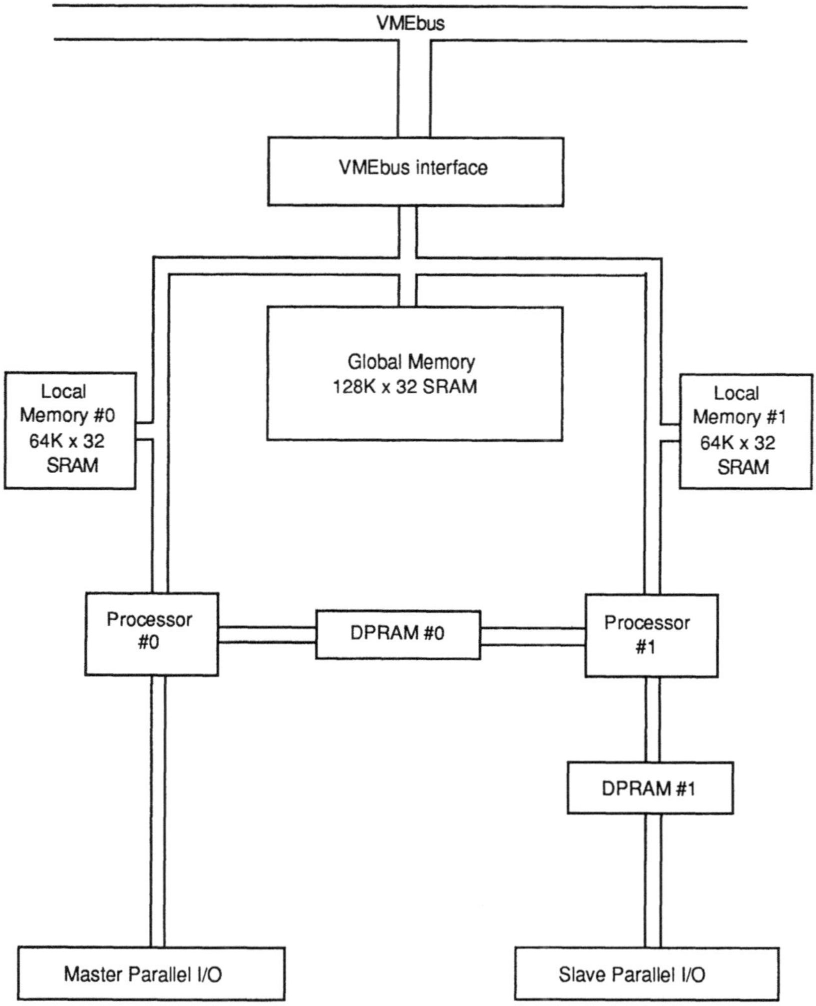

Fig. 9 - Simplified functional diagram of LSI DPV30 DSP board.

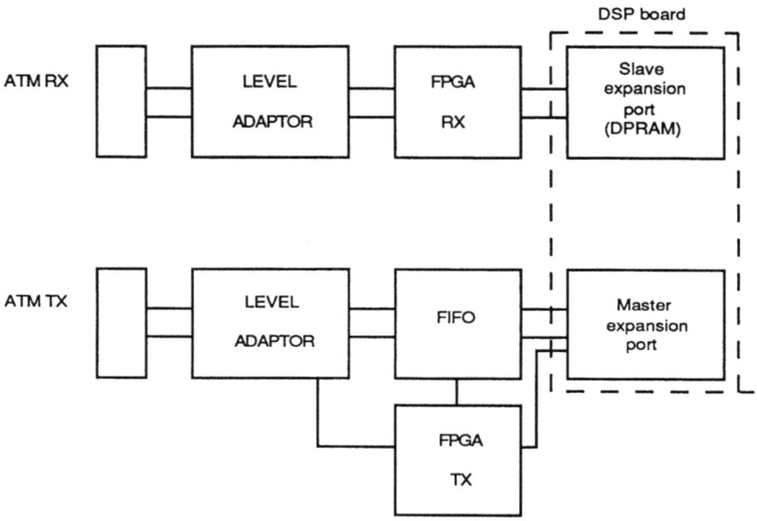

Fig. 10 - Functional diagram of ATM physical board.

The second board implements the physical layer at a data rate of 155.520 Mbit per second. It is composed by two independent sections, one for transmission, the other for reception (fig. 11). In this last section, all cells are checked out for CRC correctness, discarding the faulty cells (in this first version the single bit error correction has not been implemented, although it can be added to the Field Programmable Gate Array - FPGA used to implement these functions). Empty cells are also discarded, transferring to the ATM layer board only full cells. The two sections of this physical board are connected to the LSI board by two high speed parallel I/O ports which present a transfer rate of 20 MBytes per second (i.e., far greater than the physical data rate) allowing all flux of informations coming from the broadband network to be stored into a DPRAM buffer of the ATM board. The two Kwords buffer provided by the ATM board, can store up to 128 cells to be analysed by the ATM and upper layer protocols. In the transmission section of the physical layer board, a TX FPGA enqueues the cells in a FIFO buffer from where are transmitted to the broadband network. At most 8 cells can be present in the FIFO, although this maximum is never reached in our experimental environment due to the faster processing of the physical board. The communication between upper layer and ATM layer in the transmitting side, is accomplished through a circular queue of 512 cells. It is to mention that in our experiment, and usually in general, this queue is almost never full, due the faster rate of ATM layer. A possible exception comes out from a possible form of flow control at ATM layer.

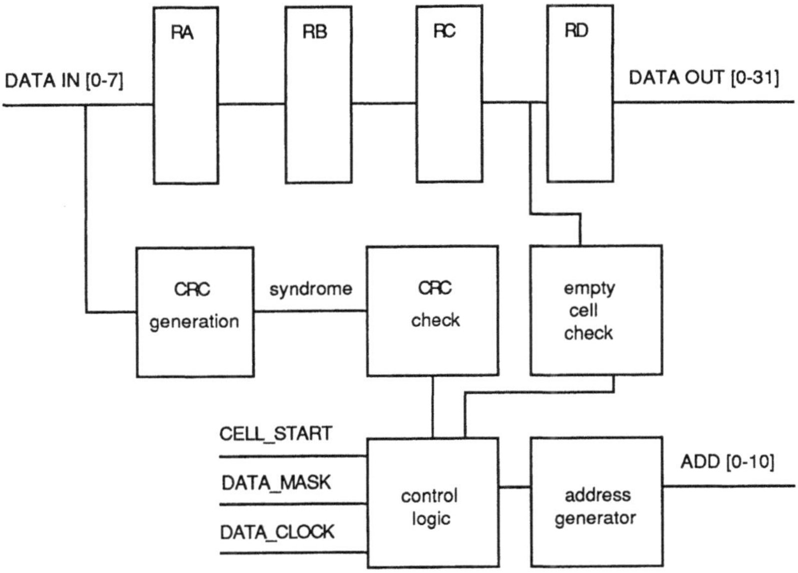

Fig. 11 - Functional diagram of RX FPGA

The scheme presented in fig.12 can help to evaluate the global performance of the router. The physical layer transmitter operates at line speed on the FIFO buffer and the time to transmit a cell is approximately 2.7μs. At the receiver side, where a cut-through technique is implemented, only the header of the cell is stored and it is analysed in a fraction of cell duration. Empty and faulty cells are discarded, while the others are stored in the DPRAM#1. The receiver ATM layer operates independently on processor #1 and carries out the VCI demultiplex and other minor functions, in a measured time of approximately 30 μs. Similar considerations can be developed for the transmission side of ATM. The system performance up to the ATM layer is determined by the software based module developed for this layer, which gives a sustained throughput of approximately 20 Mbit per second. Considering the SAR sublayer, that in this context must be software based, the total throughput decreases up to a few Mbit per second. In fig.13 is reported the throughput obtained by the segmentation module, as a function of the frame dimension, for various active connections. As can be verified the throughput for this first version of the software, written in plain C, is in the range 2.0-2.6 Mbps. An optimization step is being carried out using the sophisticated features provided by the considered architecture (i.e., DMA controller, interrupt controller, etc.) and determining the most time consuming software modules with the target of translating them into assembler. The global throughput after this optimization phase is expected to be in the range of 6-8 Mbps, which is adequate for Ethernet interconnections.

In fig.14 a picture of the two boards employed in the experiment is reported.

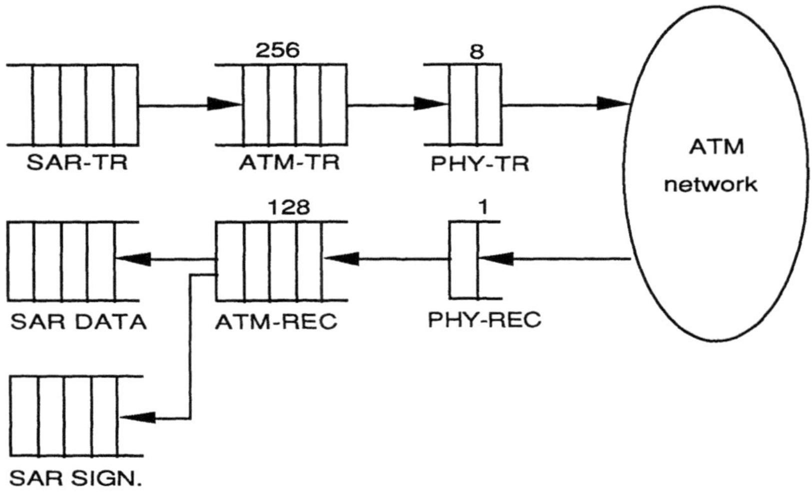

Fig. 12 - Layered buffering in the gateway protoype.

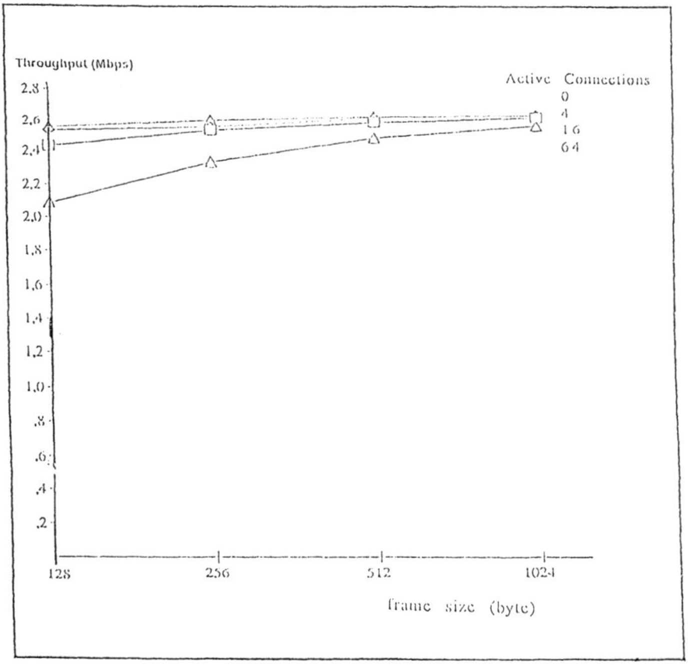

Fig. 13 - Throughput of the segmentation sublayer vs frame size for various active connections

Fig. 14 - Picture of the two employed boards

4. Upgraded architecture

The results obtained by the experimental prototype show that to interconnect high speed networks (FDDI, DQDB, CRMA, etc.) a more powerful architecture is needed. In particular, as shown in the previous section, the most time consuming software module is the SAR sublayer. On the other hand, the simplification introduced by AAL 5 makes relatively easy an hardware implementation of the protocols suite up to the SAR sublayer. A preliminary design of an upgraded architecture is reported in fig.15. As can be verified the upgrade is obtained by introducing two custom integrated circuits implementing the ATM and SAR sublayer and by a new design of the Rx FPGA (fig.16), which comprehends a hardware VCI check to discard the cells not addressed to the local station. Moreover, in the ATM-SAR logic adequate memory for data buffering is provided, in order to improve the efficiency of the system. The receiving section integrates the reassembling function exploiting a buffer structure as reported in fig.17. The host workstation, in this new context, implements by software the CS sublayer and the other ATM protocols not considered in the hardware implementation. Alternatively, another intelligent card could be used to implement this function. In this case the host workstation is not involved in ATM protocols processing and the only interaction is given by the exchange of IP datagrams through the VME bus. The global performance of the system is estimated by considering that up to the SAR sublayer the line data rate is kept, while the software based protocols act on long packets and a sustained throughput of several tens of Mbps is guaranteed.

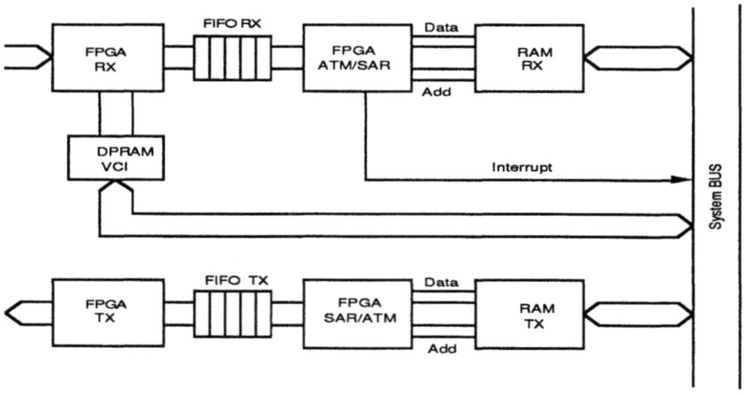

Fig. 15 - Upgraded architecture with a custom board implementing ATM protocol stack up to SAR sublayer

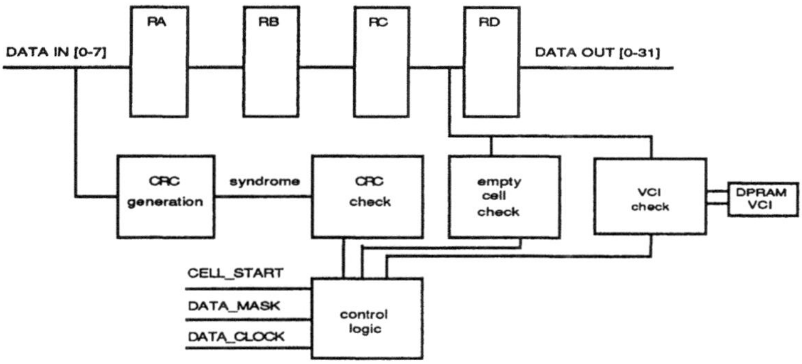

Fig. 16 - Structure of the RX FPGA in the upgraded architecture.

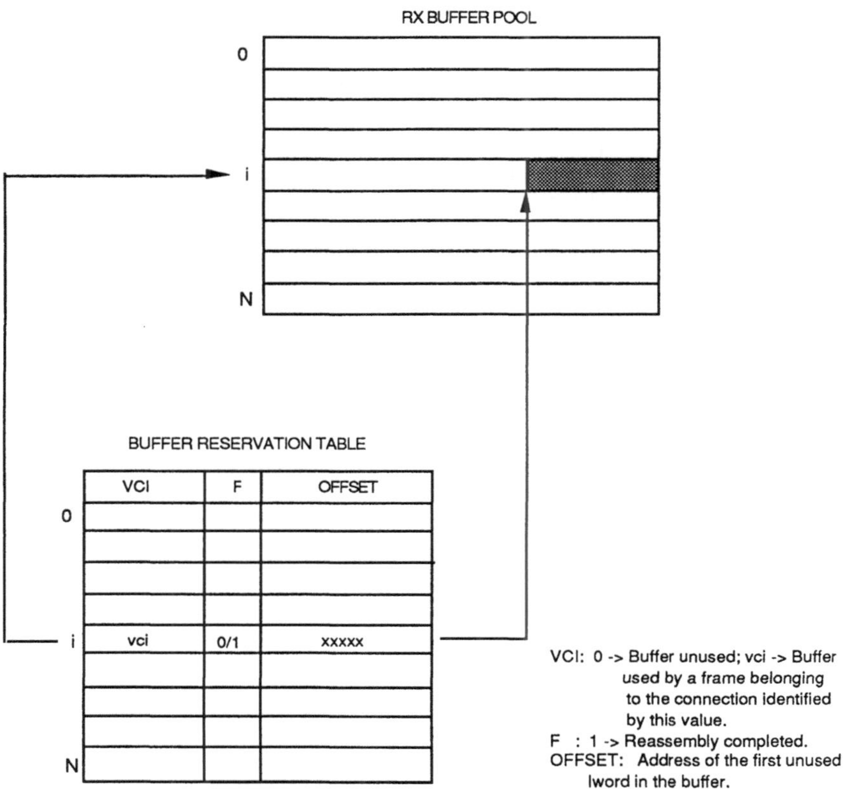

Fig. 17 - Buffers management in the receiving RAM.

Acknowledgment
This work has been supported, in part, by the Italian Research Council (CNR), in the frame of the Telecommunication Project.

References

[1] M. Gerla, T.Y.C. Tai, G. Gallassi, "Internetting LAN's and MAN's to B-ISDN's for Connectionless Traffic Support" **IEEE Journal on Selected Areas in Communications**, Vol.11, No.8, Oct.1993.

[2] J. A. S. Monteiro, M. Gerla, L. Fratta, "Leaky bucket input rate control in ATM network". **ICCC'1990**, New Delhi (India).

[3] G. Gallassi, G. Rigoglio, L. Fratta, "Bandwidth assignment and bandwidth enforcement policies". **IEEE Proceedings Globecom'89**, pp. 1788-1793.

[4] G. P. Balboni, W. Chiarottino, C. Demichelis, R. Melen, M. Pelizza, M. Ravera, "An ATM switching testbed". **European Transactions on Telecommunications**, Vol. 2, No.4, 1991.

[5] M. de Prycker, " ATM, a solution for Broadband ISDN", **Ellis Horwood**, 1991.

[6] S. Iisaku, M. Ishikura, "ATM network architecture for supporting the connectionless service". **IEEE Proceedings**, 1990.

[7] M. Gerla, T. Y. Tai, J. A. S. Monteiro, G. Gallassi, "Interconnecting LANs and MANs to ATM" **IEEE 16th Conf. on Local Computer Networks**, Minneapolis, Oct. 1991.

[8] M. Gerla, T. Y. Tai, G. Gallassi, "LAN/MAN interconnection: a simulation study" **IEEE INFOCOM '92**, Florence (Italy).

[9] R. Bubenik, M. Gaddis, J. DeHart, "Communicating with virtual paths and virtual channels" **IEEE INFOCOM '92**, Florence (Italy).

[10] E. Cooper, O. Menzicioglu, R. Sansom, F. Bitz, "Host interface design for ATM LANs" **16th Conference on Local Computer Networks**, Minneapolis Oct. 14-17 1991.

Broadband Communications, II (C-24)
S. Tohmé and A. Casaca (Editors)
Elsevier Science B.V. (North-Holland)

Translation of user's QoS parameters into ATM performance requirements

Dominique SERET[a] and Jae-il JUNG[b]

[a]Network Department, ENST, 46 rue Barrault, 75634 Paris Cedex 13, France
e-mail : seret@res.enst.fr

[b] Korea Telecom, Telecommunications Network Research Lab., Socho-Ku, Umyun-dong, 17, 137-140 Seoul, Korea
e-mail : jijung@ktman.kotel.co.kr

Abstract

The paper deals with the Quality of Service concept in broadband ISDN . Its main goal is to present the translation of user's QoS parameters between layers. We concentrate on the translation from the AAL to the ATM layer because the AAL is designed to be service dependent and specific. The QoS parameters at the ATM layer resulting from the translation will be the performance requirements on a connection basis for the ATM network. Finally, we discuss the impact of QoS parameter translation on guaranteeing QoS in the ATM network.

Keyword codes : C.2.0, C.2.3.

Keywords : Quality of service, Broadband Communications, Network Management

1. INTRODUCTION

ATM is a very flexible transfer mode which does not impose a fixed rate of transmission to the sources, therefore being adequate to support any kind of service, either at a constant or variable bit rate. This ensures that ATM can be used not only for existing services but also for new services and applications [1][2]. A major problem, however, in the B-ISDN is how to guarantee the QoS to each service, especially given the wide range of services that it will be supporting and the higher degree of ATM resource sharing. To provide QoS guarantees requires on the one hand the knowledge of QoS requirements and on the other hand the construction of an ATM core network that can guarantee this QoS .

Little is known about the ATM performance requirements of future broadband services, especially multimedia applications. Even the requirements of foreseeable broadband services (i.e., N-ISDN services) have not been thoroughly analyzed in terms of ATM network performance parameters ; only some (not exact) definitions

[3][4] are known. This is because the performance requirements depend not only on the diverse user's QoS requirements, but also on the layer processing and workload in host systems. So, in order to guarantee QoS to users, it is necessary to introduce a set of QoS parameters whose properties indicate the nature and the requirements in a layered protocol stack. These parameters are defined for each layer such that each layer can guarantee the demanded QoS to its next-higher layer and demands a possibly different QoS from its next-lower layer [5]. So, it is important to translate user's requirements into ATM layer QoS parameters.

The QoS expected from the network by various services varies over a wide range : some are sensitive to delays, others are sensitive to loss rates, while yet others are sensitive to delay variation. Therefore, QoS control as well as traffic control must be considered in resource allocation and management to guarantee the QoS. Several papers have dealt with certain aspects of how to guarantee QoS in high-speed networks [7][8][9][10]. This paper addresses the selection of a set of QoS parameters that can adequately describe user's requirements as well as the mechanism to translate them into ATM performance constraints. We concentrate on the translation from the AAL to the ATM layer, as the AAL is designed to be service dependent and specific. Our study shows that the QoS translation is a good approach in QoS guarantees for an ATM network.

In Section 2, we define QoS parameters in the AAL and ATM layer according to the QoS framework we have presented in [11],[12]. Section 3 presents several case studies in which the translation method is applied to video and data services. We also discuss the impact of QoS parameter translation on guaranteeing QoS in the ATM network. Conclusions are presented in Section 4.

II. USER's QoS PARAMETERS AND ATM PERFORMANCE

The aim of ATM is to achieve a simple and flexible B-ISDN. The functions of the ATM layer satisfy both the requirements of simplicity and flexibility, as the ATM layer provides a layer service which is independent of the information unit structure and bit rate of the telecommunication services supported. These characteristics, however, lead to some complications at the level of the ATM layer as shown in [13]. It is therefore necessary to guarantee for each service that the respective QoS is not degraded. It is up to the AAL to adapt the services provided by the ATM layer to the requirements of the higher layer [14].

The AAL is subdivided into the Segmentation and Reassembly (SAR) sublayer and the Convergence sublayer (CS). The essential functions of the SAR sublayer are, at the transmitting side, the segmentation of higher layer PDUs into a suitable size for the information field of the ATM cell and, at the receiving side, the reassembly of the information fields into higher layer PDUs. The CS is service dependent and provides the AAL service at the AAL-SAP [15].

2.1. A layered performance model

Figure 1 illustrates various causes of errors or degradations with a layered model for an ATM network and proposes possible recovery methods [11]. The QoS parameter translation in Section III consists of analyzing the AAL performance, according to a layered model. We can identify five classes of errors or degradations : cell error, cell loss, cell misinsertion, cell delay and jitter. Each class of error or degradation has its own origin (in the medium, in the physical layer or in the ATM and/or AAL layer) and can be totally or partially recovered in the physical layer, the ATM, the AAL layer or even in higher layers.

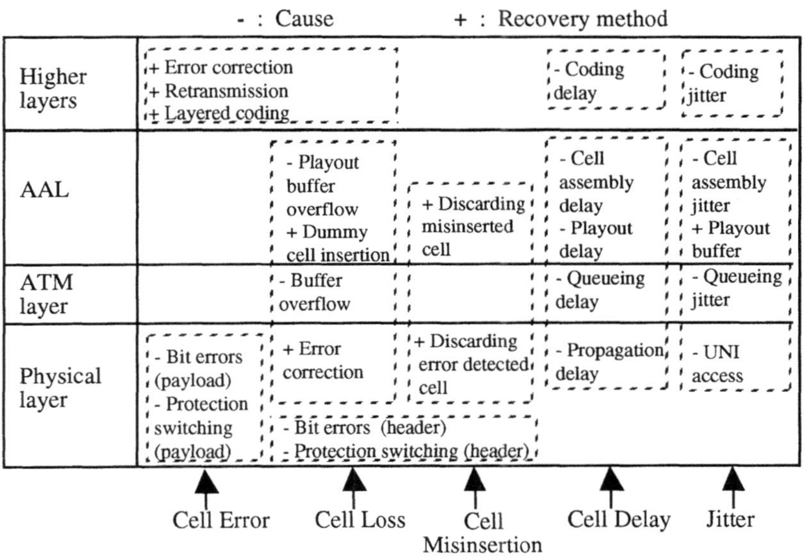

Fig. 1. Layered Performance Model

Because each layer receives the SDU (service data unit) from a higher layer and uses the corresponding PDU (protocol data unit) to communicate with peer layers, the QoS parameters have to be based on the SDU. The user's AAL QoS parameters are a factor of the higher layers in Figure 1 and are based on the AAL SDU. These parameters are accessed through the SAP (Service Access Point) between layers and can be globally viewed as managed objects. The result of parameter translation in AAL will be the ATM performance requirements (buffer overflow, queueing delay and queueing jitter ...). The user's QoS requirements are specified in terms of bounds on QoS parameters. These bounds can be either deterministic or statistical :

- deterministic bound

  QoS_parameter $\leq$ upper_bound or QoS_parameter $\geq$ lower_bound.

- statistical bound

  Prob (QoS_parameter $\leq$ upper_bound) $\geq$ prob_bound.

A deterministic bound is a rigorous constraint ; a statistical bound defines some tolerance. QoS requirements are usually specified as deterministic bounds from the user's viewpoint, while the ATM network can hardly guarantee deterministic bounds [10]. In this paper we consider user's requirements dealing only with the data transfer phase and we assume that their bounds are deterministic. We define the QoS parameters of class A and B services, namely CBR or VBR services with timing constraints and connection oriented mode. QoS parameters in class C and D for data services [14][15] will be a subset of these parameters.

2.2. QoS parameters in the AAL

The user's requirements are represented by bounds on QoS parameters which are attributes of the needs of the higher layers and must therefore be independent of the ATM network.

• Frame error ratio (*FER*) bound : FER_M

A frame is the service data unit received from the higher layers.*FER* means the percentage of frames in a given connection that may be errored. Let *total_frames* denote the total number of frames received from the higher layers and *errored_frames* the number of frames containing errors that are not corrected by the AAL.

$$FER = \frac{errored\_frames}{total\_frames} .$$

The bound FER_M is defined as $FER(t) \leq FER_M$ for all t, where $FER(t)$ is the errored frame ratio actually supported by the AAL, and FER_M is an upper bound specified by the higher layers.

The use of the *FER* might be contested in the case of real-time video or audio services, where an errored frame can lead to a loss of synchronisation and thus possibly a severe degradation of video quality, or a merely trivial one. The *FER* would thus seem insufficient to describe fully user's QoS requirements. This is, however, not the case when using layered video coding techniques [17]. Each video signal component can then be transmitted separately with its own QoS requirements [18], which enables fundamental components, such as those containing synchronisation information, to be transmitted with a rigorous QoS without having to demand an equally stringent QoS for the less important components such as a high resolution component.

• Throughput bound : W_m

The throughput requirement and its bound are defined as :

$$W \geq W_m$$

where W is the mean throughput actually provided by the AAL and W_m the higher layers specified lower bound on throughput, that is the minimum mean throughput the AAL must offer to the higher layers during the call.

• Delay bound : D_M

$Di \leq D_M$ for all i,

where Di is the delay with which the i-th frame sent by the higher layer is delivered to the destination, and D_M the upper bound on delay.

• Jitter bound : J_M

$Ji = | Di - D | \leq J_M$, for all i,

where D is the target delay that is calculated by the AAL, Ji is the jitter of i-th frame sent by the higher layers when delivered to the destination, and J_M the upper bound on jitter.

2.3. QoS and performance parameters in the ATM layer

At the ATM layer we discern the following QoS parameters and corresponding bounds.

• Cell Loss Ratio (*CLR*) bound : CLR_M

It is an ATM specific parameter that results mainly from the buffer overflow in ATM switches due to congestion.

• Throughput bound : W'_m

The throughput requirements at the ATM layer are given in terms of W'_m parameter. That throughput bound has to be differentiated from ATM traffic contract or traffic specification. It does not relate to the traffic characteristics, but describes the requested capacity of information flow (cell flow in ATM layer) for a connection.

• Delay and Jitter bound : D'_M and J'_M

The delay (D'_M) and jitter bound (J'_M) in the ATM layer have the same definition as in the AAL but they are now concerned with ATM cells, not frames.

• Cell Insertion Rate (*CIR*) bound

The *CIR* is due to errors in the ATM cell header. In this paper we choose to neglect the *CIR*, since the ATM switches detect and discard cells having an errored header, and the AAL function efficiently detects misinserted cells. Therefore the *CIR* is negligible compared to the *CLR* [24].

• Cell Discard Ratio (*CDR*) bound : CDR_M

CDR can be defined in case of AAL type 1 using FEC and/or interleaving. Let *discarded_cells* denote the number of cells that are at most allowed to be lost in the ATM network, in one block. The *CDR* bound is not imposed by the higher layer but specified by the AAL itself such that consecutive lost cells remain within the correction range of FEC (Forward Error Correction) and/or interleaving.

$$CDR = \frac{discarded\_cells}{cells\_in\_block}.$$

The network performance parameters are visible only to the network. They are under study in CCITT [23]. In [2][23], the performance parameters which are currently defined are Bit Error Ratio (*BER*), Cell Error Ratio (*CER*), Cell Loss Ratio (*CLR*), Cell Insertion Rate (*CIR*), Delay, Jitter and Cell transfer capacity.

The *BER* is not an ATM specific parameter. The *CER* means that the errors occurred in the cell payload. We have already chosen to neglect the CIR. The cell transfer capacity describes the flow of cells within an ATM connection, which corresponds to the throughput parameter in our study.

3. QoS PARAMETER TRANSLATION

3.1. QoS parameters in any AAL type

In Section 2, the AAL and ATM layer QoS parameters have been defined for all service classes. The selection of the AAL and ATM layer QoS parameters is basically dependent upon the services demanded. Figure 2 presents one possible solution to selecting the AAL and ATM layer QoS parameters and bounds in any AAL type.

The QoS parameters in AAL type 3/4 (i.e., for data services) will be a subset of the parameters in AAL type 1 and 2 since AAL type 3/4 services normally have no strong requirements for real-time communication. In the case that they do have such requirements, they can be optionally represented by D_M and J_M An additional QoS parameter, CDR_M, is defined in AAL type 1, as compared to AAL type 3/4. In AAL type 2, that is VBR services with timing constraints, the mean throughput requirements may not be useful parameters. This needs a further study.

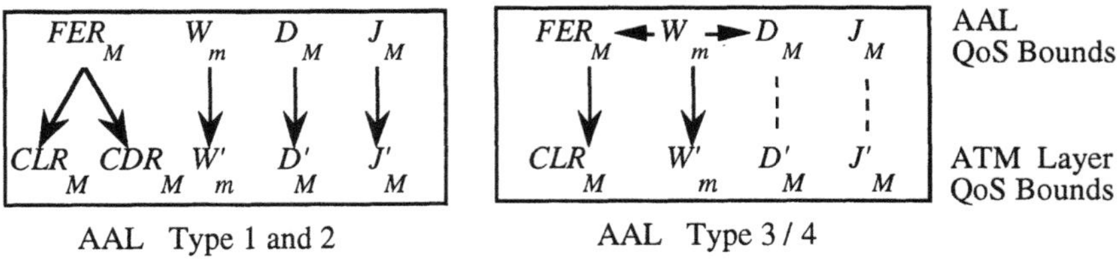

Fig. 2. AAL and ATM layer QoS parameters and bounds

3.2. Translation from SAR Sublayer to ATM Layer

In order to give some general results, independent of the AAL types, it is assumed that AAL does not use CS sublayer and takes the SAR-PDU structure in AAL Type 1, shown in Figure 3. Section 3.3. describes the translation including the CS sublayer with the CBR video service as example, Section 3.4. describes it for the AAL type 3/4. The translation generates the ATM performance requirements (CLR_M, CDR_M, W'_m, D'_M, J'_M) by using the AAL QoS bounds (FER_M, W_m, D_M and J_M) as input.

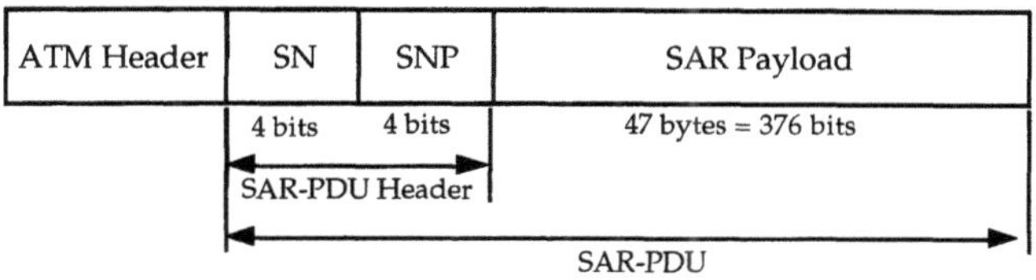

SN : Sequence Number SNP : Sequence Number Protection

Fig. 3. SAR-PDU Structure in AAL Type 1

• *CER* is due to errors in the ATM transmission systems. For the sake of simplicity, let us assume that the bit error occurs independently. Let *CER*(ATM) be the probability that the cell has an errored information field and *CER*(SAR) the probability that the SAR payload is errrored. If $BER << 1$, we have :

$$CER(\text{ATM}) = (1- BER)^{48} (1 - (1- BER)^{384}) \approx 384\ BER.$$

$$CER(\text{SAR}) = (1- BER)^{47} (1 - (1- BER)^{376}) \approx 376\ BER.$$

• *CLR* at the ATM layer is due to three events : (1) cell header error; (2) protection switching (burst error); (3) cell loss by buffer overflow in network queues. However, it is well known that *CLR* is mainly due to buffer overflow in ATM switches.

The SAR-PDU loss ratio at the SAR sublayer is due to *CLR* and incorrigible errors in the SN and SNP field. The second term being negligible so we find : *CLR*(SAR) = *CLR*.

Let us calculate the resulting *FER* at the SAR sublayer.

$$\begin{aligned} FER &= \text{Probability of errored frame} \\ &= \sum_M \text{P(errored frame} \mid M\text{=m) P(}M\text{=m)} \\ &= \sum_M (1 - (1- x)^m)\ \text{P(}M\text{=m)} \\ &= \sum_M mx\ \text{P(}M\text{=m)} \quad \text{(assuming that } x<<1\text{)}, \end{aligned}$$

where x is the probability of SAR-PDU error at the SAR sublayer and M is a random variable which is the number of SAR-PDUs in a frame.

The probability of SAR-PDU error does not follow directly from the *CLR*, since we must take into account any substitution mechanism. For example, if cell losses are detected, dummy cells may be substituted for the lost cells in order to maintain bit count integrity. This also, however, introduces user information bit errors. Thus we have : $x \approx CER(\text{SAR}) + \text{a}\ CLR$, where a (1 < a < 376) is a compensation factor depending on substitution mechanism.

From the previous equation, x is calculated and we can derivethe value of CLR_M that must be guaranteed by ATM switches.

$$CLR_M = \frac{x - CER(\text{SAR})}{\text{a}} = \frac{x}{\text{a}} - \frac{376}{\text{a}} BER.$$

- Throughput bound : W'_m

The throughput bound at the ATM layer, W'_m, must be greater than the throughput bound at the AAL, W_m, in order to account for the SAR and ATM protocol overhead. When using SDH transmission, the throughput bound must be further increased due to SDH transmission overhead.

$$W'_m = \frac{53}{47} W_m .$$

- Delay and Jitter bound : D'_M and J'_M

The D'_M and J'_M bounds must be guaranteed by the ATM layer. On the other hand, they also depend on the playout buffer size of the receiver entity of AAL. So, the bounds that were negotiated by the AAL entities at connection establishment may later be changed, that is into a more strict delay bound and a less strict jitter bound.

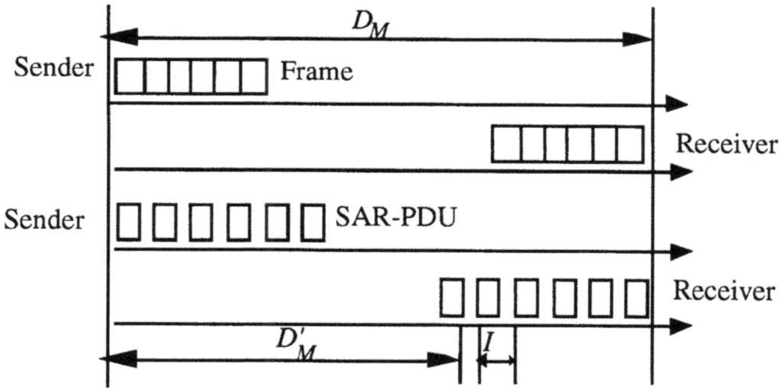

Fig. 4. Relationship between D_M and D'_M

The delay bound D'_M at ATM layer is affected by the SAR sublayer segmentation/ reassembly process, described in Figure 4.

$$D'_M = D_M - I(M\text{-}1) - d_M$$

where I is the interval between SAR-PDUs when a frame is segmented into SAR-PDUs and M is the number of SAR-PDUs in a frame. The d_M is the maximum of playout delay. The delay requirement for the ATM queueing delay is given by $D'_M - D_P$, where D_P is the propagation delay.

The jitter bound J'_M at ATM layer is not affected by the SAR sublayer segmentation/reassembly.

$J'_M = J_M + j_m,$

where j_m is the minimum of jitter compensation from using the receiver playout buffer. The jitter requirement for the ATM queueing jitter is given by $J'_M - j_M$(UNI), where j_M(UNI) is the maximum jitter for ATM cells at the User Network Interface access.

3.3. Translation from CS Sublayer to ATM Layer

We will adjust the translation results from the previous section assuming that the FEC/Interleaving (Figure 5) is supported by the CS sublayer itself according to higher layer requirements in order to provide more secure protection against errors. FEC/Interleaving does not use any fields of CS-PDU.

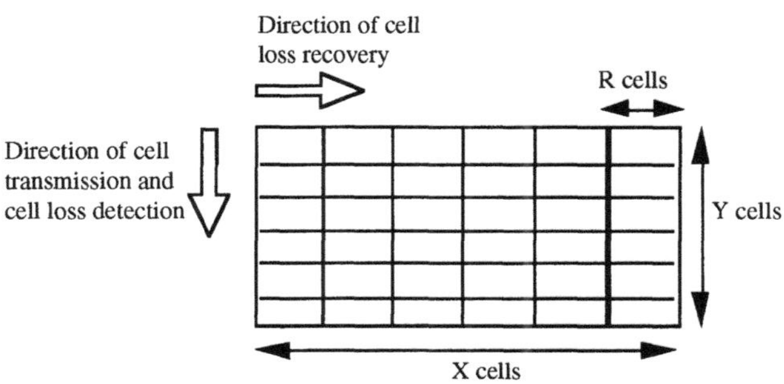

R : the number of cell loss detection cell in a row
K : the number of cells in a block before FEC/Interleaving (XY-RY)
N : the number of cells in a block after FEC/Interleaving (XY)
Fig. 5. FEC/Interleaving

The translation results in the performance requirements (CLR_M, W'_m, D'_M, J'_M) which is the same as the above translation.

- Throughput bound : W'_m

The throughput bound W'_m at the ATM layer must now also take into account the FEC redundancy. We have :

$$W'_m = \frac{53}{47} W'_m(\mathrm{CS}) = \frac{\mathrm{N}}{\mathrm{K}} \frac{53}{47} W_m.$$

- Cell Loss Ratio (CLR)

Note that the overhead by FEC/Interleaving increases the traffic rate. If we let l traffic rate before FEC coding, l(1+N/K) is the traffic rate after coding. However, the impact of traffic rate increase on CLR is negligible. To analytically model cell CS-PDU loss compensation, it is reasonable that the number of lost cells in a block is a random variable with binomial distribution. Knowing that the number of cells in a block is N and the probability of cell loss is CLR (represented by p), we have :

$$\text{Prob}(i \text{ lost cells in a block}) = \text{b}(N,i,p) = C_i^N\, p^i\, (1-p)^{N-i}.$$

If FEC/interleaving can correct up to S cells per block, the mean number of lost packets in a block, L, after decoding is given by (assuming that p<<1) :

$$L = \sum_{i=S+1}^{N} i\, \text{b}(N,i,p) = Np - \sum_{i=1}^{S} i\, \text{b}(N,i,p).$$

The cell loss rate after decoding, CLR(CS), is thus given by :

$$CLR(\text{CS}) = L \,/\, N = \sum_{i=S+1}^{N} i\, \text{b}(N,i,p) \,/\, N.$$

In the case of CBR video, P(M=m) = 1 and M is a constant. So, the CLR(CS) can again be expressed in terms of the FER as :

$$CLR(\text{CS}) \approx \frac{FER_M}{aM}.$$

We finally obtain the bound on CLR_M that must be guaranteed by the ATM network.

• Delay bound : D'_M

The delay bound D'_M(SAR) at the SAR sublayer is affected by the CS sublayer segmentation/reassembly, and Interleaving. That is,

$$D'_M(\text{SAR}) = D_M - d_M(\text{FEC}) - d_M(\text{interleaving}),$$

where d_M(FEC) = $I'(M'-1)$. I' is the interval between blocks when a frame are segmented by FEC blocks and M' the number of blocks in a frame. d_M(interleaving) is a value which can be determined from FEC block size. Also, D'_M can be calculated the same approach.

• Jitter bound : J'_M

The jitter bound J'_M(SAR) at the SAR sublayer is not affected by block segmentation/reassembly, but can be affected by interleaving. Also J'_M is defined the same approach.

$$J'_M(\text{SAR}) = J_M - j_M(\text{interleaving}).$$

3.4. Translation from SAR sublayer to ATM layer for AAL type 3/4

In this Section, we take the SAR-PDU structure in AAL type 3/4, shown in Figure 6 and quickly give an overview on a possible parameters tranlation.

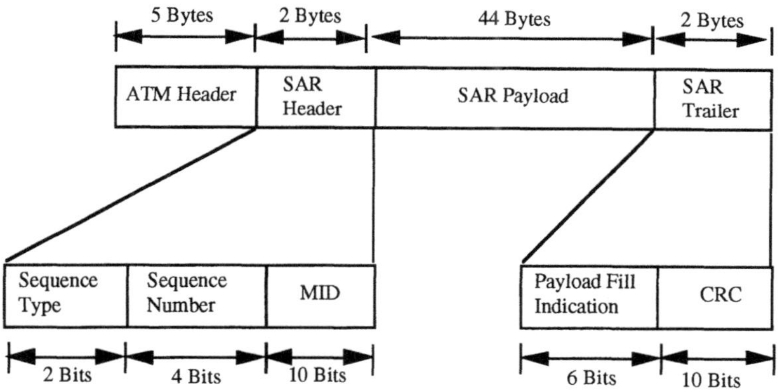

MID : Message ID CRC : Cyclic Redundancy Code

Fig. 6. SAR-PDU Structure in AAL Type 3/4

• There are several Error Protection alternatives : no error protection (protection in higher layers), error detection over MID and segment-type fields, error detection over SAR header, SAR trailer, SAR payload, single bit error correction on MID and segment-type fields, single bit error correction on MID and segment-type fields and error detection on SAR payload.

• The translation of throughput bound can be affected by the flow control mechanisms as well as the error handling mechanisms such as retransmission.

• FER_M is the errored frame ratio bound which means a tolerance of error requirement. As the data service is strict in error requirement, FER_M must be very close to 0. However, in the case that the higher layers protect the user information or tolerate some errors, FER_M can be given to AAL.

• W_m is the throughput bound which means a minimum throughput requirement. We can imagine many cases upon user's protection requirements.

• $W'_m = \frac{53}{47} \frac{W_m}{1\text{-}FER_M}$.

• Cell Loss Ratio (CLR) bound : CLR_M

The resulting FER at the SAR sublayer is given by the same expression than in Section 3.2. The probability of SAR-PDU error does follow directly from the CLR(SAR) and CER(SAR). Thus we have :

$$x = FER/M = CER(\text{SAR}) + \text{b}\ CLR(\text{SAR}),$$

where b (= 376) is a factor depending on the size of SAR-PDU. So, b can be dependent upon SAR-PDU Structure in AAL.

- $CLR_M = \frac{FER_M}{bM} - \frac{376}{b} BER$

4. CONCLUSION

In this paper we have addressed the definition of QoS parameters and their translation from the AAL to the ATM layer. This translation concept came from the QoS framework in which the influence of each protocol layer processing on the QoS should be considered. The QoS concept using parameters has been addressed by research in the area of multimedia [6]. The powerful mechanism, such as QoS parameters, should be supported in order to provide efficient control between distributed multimedia systems and communication subsystems. So, it is expected that the QoS translation will be helpful in the multimedia communication area.

In the above translations we have dealt with the AAL type 1 for a CBR video service and AAL type 3/4 for a data service. Even though our case studies do not cover all possibilities, we have shown that the QoS processing between layers can be analyzed and that we can arrive at the ATM performance requirements.

Our case studies demonstrate not only that the ATM performance requirements resulting from the translation can be very different to the QoS parameters, but furthermore that even for a single service, the ATM QoS parameters are highly dependent on the AAL (or higher layer's) functions chosen. On the one hand, the ATM network must have to be flexible to meet a variety of ATM requirements. On the other hand, these QoS differences (ex. throughput value, semantics in error requirement, and delay and jitter values) should be considered in a performance contract between user and ATM network provider.

In AAL type 3/4, that is connection-oriented and connectionless data services respectively, the translation of throughput bound can be affected by the flow control mechanisms as well as the error handling mechanisms such as retransmission.

In our study, the QoS parameters are not exhaustive in the sense that they are related only to the data transfer phase in call connection. The QoS parameters related to the call connection and disconnection have to be defined in the AAL. Moreover, QoS parameter related to the burst level access control (ex. Burst Blocking Probability) should be eventually defined for applications using the reservation principle [25].

Finally, the translation can be extended to the QoS verification issue. The user can verify if a bearer QoS provided by ATM network is satisfying within the range of performance contract. The reverse translation, that is from the performance indicators to the user QoS parameters, is used. The time when the client accepts to trust service qualities provided by network provider without complaint is in fact gone by, especially in professional communication area. It seems that client QoS verification tools will be more powerful than provider's. B-ISDN and ATM network architects must design a sophisticate QoS framework [20][21][22] to guarantee user's QoS requirements and provide user with QoS values actually supported by the network.

REFERENCES

[1]M. De Prycker, "Asynchronous Transfer Mode : Solution for Broadband ISDN", Ellis Horwood, 1991
[2]R. Handel, M.N. Huber, "Integrated Broadband Networks : An Introduction to ATM-Based Networks", Addison-Wesley, 1991
[3]M.E. Anagnostou, et al., "Quality of service requirements in ATM-based B-ISDNs", Computer Communication, vol 14, no 4, may 1991
[4]D.B. Hehmann, M.G. Salmony, H.J. Stuttgen, "High-Speed Transport Systems for Multi-Media Applications", Protocols for High-Speed Networks, IFIP, 1989, North-Holland
[5]J.I. Jung, D. Seret, R. Picard, "Quality of Service in B-ISDN and Relation with Network Management", IEEE ICC '92, Chicago, June 1992
[6]C. Nicolaou, "An Architecture for Real-Time Multimedia Communication Systems", IEEE JSAC, April 1990
[7]E.W. Biersack, et al., "Gigabit Networking Research at Bellcore", IEEE Network Magazine, March 1992
[8]A.A. Lazar, G. Pacifici, "Control of Resources in Broadband Networks with Quality of Service Guarantees, IEEE Communication Magazine, October 1991
[9]D.C. Verma, Ph. D. Thesis, "Guaranteed Performance Communication in High Speed Network", University of California, Berkeley, December 1991
[10]D. Ferrari, "Client Requirements for Real-Time Communication Services", IEEE Communications Magazine, November 1990
[11] J.I. Jung, D. Seret, "Translation of QoS parameters into ATM performance parameters in B-ISDN", INFOCOM'93, San Fransisco, March 1993
[12] D. Dromard, D. Seret, J.I. Jung, "User's vs Provider's viewpoints of Quality of Service in B-ISDN", submitted to IDNC 94, Portugal, April 1994
[13]A.J. Casaca, M.S. Nunes, "Some Aspects of the Adaptation of Services to an ATM Network", INDC'92, Finland, March 1992
[14]"B-ISDN ATM Adaptation Layer (AAL) Specification", CCITT I.362, Geneva, 1990
[15]"B-ISDN ATM Adaptation Layer (AAL) Functional Description", CCITT I.363, Geneva, 1990
[16]P. Coatanéa, Y. Rouaud, "Installations Privées et Interfaces", L'écho des recherches, 3ème trimestre 1991, spécial ATM, CNET-ENST
[17]W. Verbiest, L. Pinnoo, B. Voeten, "The Impact of the ATM Concept on Video Coding", IEEE JSAC, December 1988
[18]S. D'Agostino, J. David, R. Heiss, W. Verbist, "Architecture Universelle de Codage Vidéo pour Reseau ATM", pp. 186-193, Revue des Télécommunications, Volume 64, N° 2/3, ALCATEL, 1990
[19]P. Adam, J-P. Coudreuse, "ATM et Réseaux : domaines d'application", L'écho des recherches, 3ème trimestre 1991, spécial ATM, CNET-ENST
[20]RACE 1082 QOSMIC, "QOSMIC Exercise : The HDTV Case Study", QOSMIC Deliverable D1.6B, February 1992
[21]RACE 1082 QOSMIC, "QOSMIC Exercise 1", QOSMIC Deliverable D1.6A, February 1991
[22]P.E. Stern, B. Golfier, J.I. Jung, P. Potin, "On the Management of Quality of Service for Telecommunication Based Software", 3rd European Conference on Software Quality, Madrid, November 1992

[23]"B-ISDN ATM Layer Cell Transfer Performance", CCITT Draft Recommendation I.35B, 1992

[24]M.E. Anagnostou, M.E. Theologou, E.N. Protonotarios, "Cell Insertion Ratio Analysis in ATM Networks", Computer Networks and ISDN Systems, Vol. 24, No. 4, 1992

[25]P.E. Boyer, D.P. Tranchier, "A Reservation Principle with Applications to the ATM traffic Control", Computer Networks and ISDN Systems, Vol. 24, No. 4, 1992

Broadband Communications, II (C-24)
S. Tohmé and A. Casaca (Editors)
Elsevier Science B.V. (North-Holland)

High Time-Resolution Cell Loss Measurements of Mixed CBR and VBR Traffic

Experiments carried out on RACE BIPED and RACE PARASOL demonstrators

N. Björkman[a], A. Latour-Henner[b]

[a]Telia Research, 136 80 Haninge, Sweden, Fax: +46 8 707 54 80,
E-mail: nils.bjorkman@haninge.trab.se
[b]Telia Instrument, 123 86 Farsta, Sweden, Fax: +46 8 713 20 24,
E-mail:alex@birgit.p.tvt.se

Abstract

High time-resolution cell loss measurements have been carried out for a conceivable ATM network scenario, realised in the RACE BIPED demonstrator. In this scenario, a high quality CBR (Constant Bit Rate) audio/video traffic stream is multiplexed with two links carrying traffic from a large number of residential VBR (Variable Bit Rate) videophones. The cell loss process of the total traffic multiplex has been monitored and recorded by a novel ATM test tool developed in RACE PARASOL.

An analytical model, with its inherent approximations, provided precise guidance for the experiments. Adequate performance levels, in terms of e.g. long term CLR (Cell Loss Ratio), were readily obtained for the above scenario. The long term CLR does not however provide enough information about the temporal distribution of the cell loss process, which has a significant impact on the actual QoS (Quality of Service) outcome. The results comprise a number of other measures for the cell loss process. In view of the fact that cell loss occurred in bursts we conclude that short term CLR for a cell loss burst may be more applicable. Being a short term cell loss measure though, it has to be complemented by a frequency of occurrence value.

The bursty structure of the cell loss process has a strong impact on the performance of FEC (Forward Error Correction) employed in the experimental codec. Our results show that the FEC code presently standardised for distributive video AAL 1 in ITU-T offers very little protective effect. It seems that the dimensioning of the FEC algorithm limits its ability to cope with the effects of overload situations to those that are already resolved through moderately sized buffers in the network. As the current FEC definition allows correction of only approximately 3% lost cells in a protected block there is a need for employing other measures to cover for the high level of short term CLR (5-10%) as indicated in our experiment. Instead of increasing the degree of protection and introducing more overhead we conclude that careful network dimensioning can efficiently reduce the time between uncorrectable cell loss events to desired periodicity.

1. INTRODUCTION

The traffic handling properties of ATM networks have been debated ever since inception. A number of theoretical models have been applied to predict network performance. This paper gives one of the few available reports comparing theoretical results to practical reality.

An experimental investigation of cell loss performance has been performed for a conceivable network scenario. Traffic streams with different statistical properties have been multiplexed at the output port of an ATM cross connect. The resulting cell loss process has been monitored and recorded. Thus acquired data have been used to assess the necessary capability for FEC (Forward Error Correction) to be applied in ATM networks.

2. NETWORK SCENARIO

The network topology is depicted in Figure 1. The experiment was directed towards quantifying cell loss impact on a high quality CBR (Constant Bit Rate) audio/video traffic stream from an OB (Outside Broadcast) facility. OB is used for transmission of program material from occasional events occurring in geographically remote areas. The network scenario in this case implies access for high quality audio/video transmission in a residential area with large deployment of simple VBR (Variable Bit Rate) video telephones.

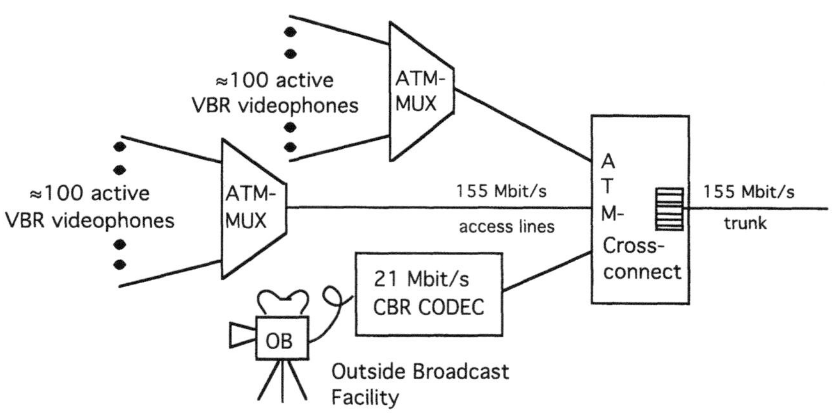

Figure 1 Network scenario for high time-resolution cell loss measurements

The local ATM cross connect is of a non-sophisticated type suitable for residential customer applications. No facilities are provided for differentiated traffic handling, e.g. in terms of cell loss priorities. At the same time it is assumed that spare bandwidth available is restricted due to high utilisation goal in a price sensitive application.

From network planning point of view the situation can be described as pronounced application of statistical multiplexing which is perceived as an alternative for increased efficiency, especially compared to peak bandwidth allocation policy.

The total traffic multiplex comprising a large number of residential VBR audio/video telephones and the high quality CBR audio/video traffic stream is conveyed towards a trunk network interface where the cell loss for the CBR stream is measured and recorded.

3. EXPERIMENTAL NETWORK ENVIRONMENT

The above scenario was emulated in an experimental set-up using equipment from RACE BIPED and RACE PARASOL demonstrators.

3.1 RACE BIPED broadband demonstrator

The BIPED demonstrator (Migotti et al, 1992) consisted of two business CPN (Customer Premises Network), two customer access connections and a switching node, i.e. a complete end-to-end broadband system from the terminal interface (S_B) to NNI (Network Node Interface). It was a hybrid broadband system which could handle pure ATM as well as N-ISDN traffic rates. Figure 2 shows a schematic drawing of the BIPED demonstrator for one access connection.

The ATM cross connect had an internal bus structure with small buffers at the input ports (≈10 cells) and the main buffers at the output ports (≈100 cells). Due to internal cell handling the link rate at the output port was limited to 152.64 Mbit/s.

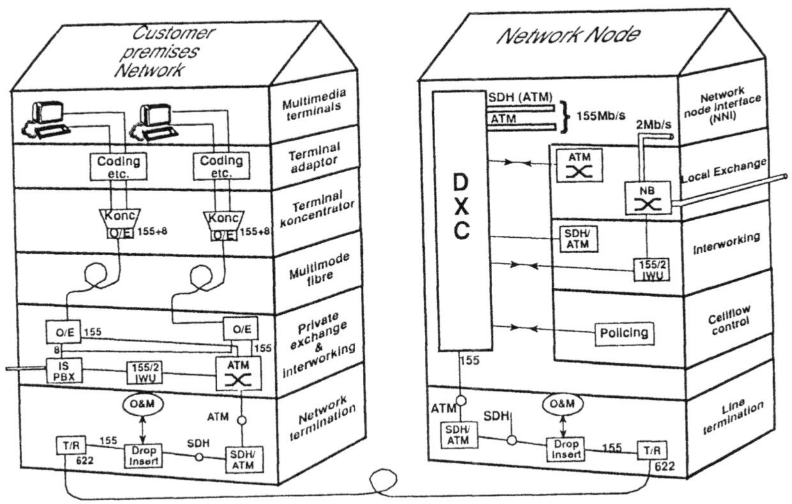

Figure 2 BIPED Demonstrator

Multimedia terminals with audio and video were connected at the S_B-interface via the GTA (Generic Terminal Adapter). The audio signal was digitised by linear PCM to about 0.5 Mbit/s per channel. The digitised video signal was compressed down to a constant data rate of 17 Mbit/s by a video codec using a hybrid coding scheme (specified in RACE HIVITS) intended for TV distribution. After addition of a FEC (Forward Error Correction) code the total bitrate for the combined audio and video was 21.1 Mbit/s.

3.2 ATM test tool

Test equipment used to perform experiments described in this paper is the result of a co-operative development by sixteen (16) partners from ten (10) countries around Europe. The project, RACE PARASOL, culminated in the application of a large system configuration for the purpose of gathering Network Performance data from a realistically loaded ATM network. Another configuration was subsequently applied to prepare and run a number of experiments with perceived QoS. The latter set of experiments will be described in a later paper.

The core functionality of PARASOL configuration used for gathering Network Performance data was provided by two modules:

- Synthetic Traffic Generator (STG)
- Impairment Recorder Module (IRM)

The STG, developed by Norwegian Telecom Research in co-operation with DELAB-Runit of Norway and KTAS of Denmark, is a model based real-time ATM traffic generator capable of producing a traffic stream corresponding to a multiplex of thousands independent sources belonging to at most ca 60 different types. The underlying principle and some implementational details for the STG are provided in (Helvik et.al, 1993).

IRM evaluates in real-time a received stream of test cells and stores separate records of the processes of cell loss and delay variation. The memory buffer coupled with an efficient storage format is sufficient for recording during several days, thus enabling gathering of statistically significant data sets even at low event intensities. IRM has been developed by Telia Instrument.

Recordings from IRM are analysed with the aid of Impairment Data Manager (IDM) which is a software program also developed by Telia Instrument. The program enables visual presentation of recorded data in different time scales ranging from days to µs. Analysis facilities allow evaluation of time domain statistics for the recorded process instances as well as mapping of the effect of impairments into the higher layer protocol structures (e.g. blocks of a coded video cell stream).

For the purpose of subjective evaluation Telia Instrument developed Impairment Generator Module (IGM). The module is capable of imposing controlled impairments, cell loss and cell delay, onto a live cell stream containing user information. The generation of impairments is controlled by a stored program which provides full repeatability without restrictions of temporal distribution of impairment events. The program can be either generated by an off line simulation or be taken from a part or whole of an IRM recording. Facilities for editing of IGM programs are provided through the IDM. Separate programs are control generation of cell loss and cell delay. This enables investigation of the effects of each of these impairment types without interference from the other. In reality though, cell loss and cell delay variation occur simultaneously. IGM has the ability to synchronise the execution of the two programs so that adequate degree of realism is achieved.

3.3 Experimental set-up

The Network Scenario depicted in Figure 1 was created using a subset of the demonstrator equipment: the ATM test tool and the ATM cross connect.

Traffic on three input ports of the ATM cross connect was directed to one of its output ports. The first input port conveys CBR traffic from the foreground generator. The background generators with VBR traffic were connected at the two remaining input ports. The multiplexed cell streams coming from the output ports were analysed by the Impairment Recorder Module in the PARASOL test tool; cell loss events were recorded with respect to their magnitude and time of occurrence.

3.4 Traffic load

The foreground traffic had a bitrate of 21.1 Mbit/s, CBR, thus emulating high quality audio/video traffic from the codec described above.

The background traffic had the following characteristics:

* peak bit rate	2 Mbit/s
* average bit rate	0.4 Mbit/s
* burst duration	0.2 s average, exponentially distributed
* number of sources:	200 - 251

This background traffic was chosen because:

* a good load efficiency can be achieved by statistical multiplexing for ATM-traffic which consists of many small bursty sources, see table 1.
* many ON/OFF sources simulate a mixture of continuos VBR sources.
* if VBR codecs will be used for videophone services, a realistic VBR source of that type could generate 2 Mbit/s peak and 0.4 Mbit/s mean. This bit rate is chosen in order to achieve a high statistical gain on the present 155 Mbit/s links. Bearing in mind that the next generation of switches will be equipped with 622 Mbit/s ports, the result will be representative also for codecs which require 4 times the given peak and mean bit rates.
* the bursts from videophone codecs are estimated to last for a period of less than one second. A burst duration of 0.2 s was chosen.

4. HIGH TIME-RESOLUTION CELL LOSS RECORDINGS

4.1 Dimensioning for a statistically multiplexed network

The measurement task was to investigate how cell losses are distributed over time for a statistically multiplexed network. A peak allocated network, with no load induced cell loss, will result in a low achievable load, as indicated by the first two columns of table 1.

In the experiment the cell loss process was recorded for various load situations. The measured average CLR (Cell Loss Ratio) was in the range from 10^{-4} to 10^{-7}. For CLR below 10^{-7} the number of cell loss events was too low to be statistically reliable, since our total measurement period was maximised to one week.

Table 1
Dimensioning for a statistically multiplexed network

Calculated average CLR	Total load (%)	Number of background sources	Number of lost cells recorded	Number of cells transmitted
$3{,}8 \times 10^{-4}$	80	251	13 381	$3{,}5 \times 10^{+7}$
$3{,}3 \times 10^{-5}$	74	231	12 124	$3{,}0 \times 10^{+8}$
$2{,}7 \times 10^{-6}$	70	214	7 511	$2{,}1 \times 10^{+9}$
$2{,}2 \times 10^{-7}$	66	200	2 700	$8{,}1 \times 10^{+9}$
$1{,}9 \times 10^{-8}$	63	188	n.a.	n.a.
$1{,}8 \times 10^{-9}$	60	178	n.a.	n.a.
peak allocation	31	66	n.a.	n.a.

4.2 Time scale from celltimes to days

As already mentioned in chapter 3.2 our recordings provide a resolution down to μs while spanning over several days. Figure 3 shows the temporal distribution of cell loss events during 2 days recording. Each bar contains a number of cell loss events. Figure 4 shows the temporal distribution of one cell loss event in the recording of Figure 3. This is the maximum resolution, which means that all bars comprise consecutive cell losses. It can be seen that the majority of cell loss occurs as single lost cells. Two cell losses of 3 cells in a row can also be clearly identified.

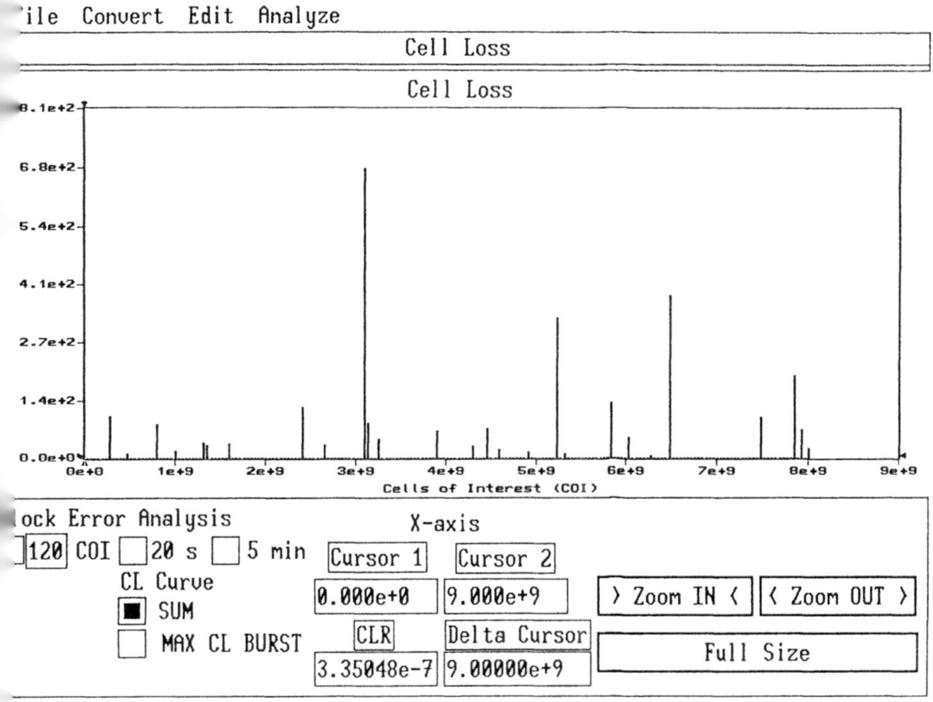

igure 3: Temporal distribution of cell loss events during 2 days recording Long term CLR≈10^{-7}

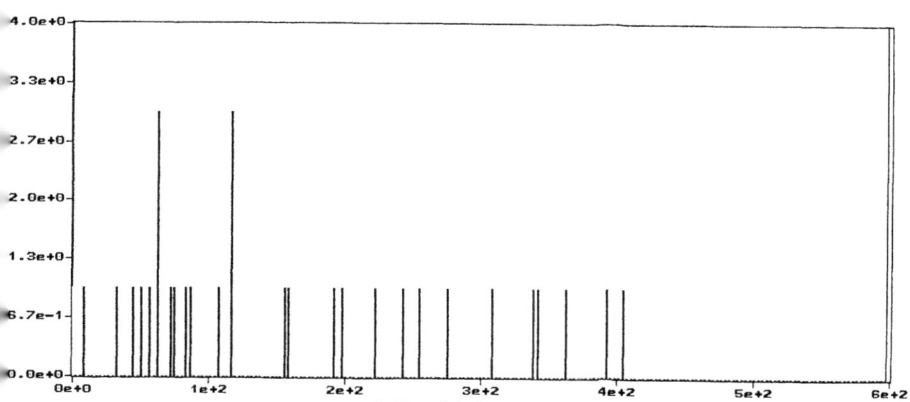

igure 4: Temporal distribution of one cell loss event in the recording of Figure3

5. DEFINITION OF BUSY PERIOD IN AN EXPERIMENTAL CONTEXT

In the theoretical context below the busy period is defined as the period during which a buffer contains cells (belonging to the analysed burst scale problem). During these measurements we had no information about the state of the buffer. Instead we defined that a new busy period commenced when 1 000 COI[1] had been received without a single cell loss.

The value of 1 000 COI was chosen after evaluation of busy period intervals in the recording for a range of busy period definition values. Figure 5 shows the relationship of busy period definition to average length of a busy period.

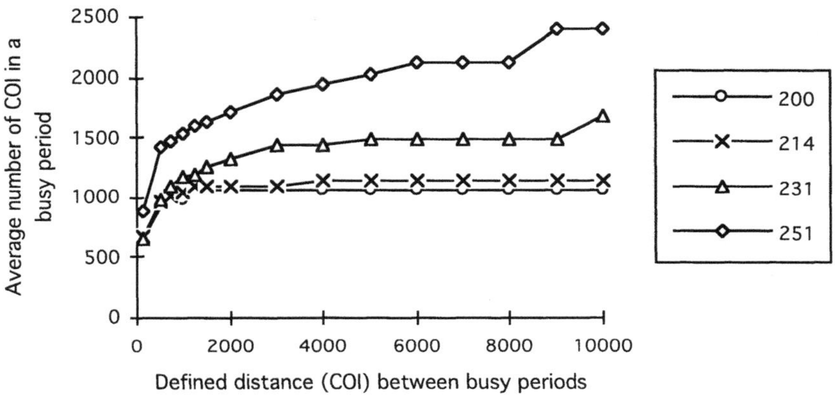

Figure 5 Impact of Busy Period definition on average length of Busy Period

6. COMPARISON TO TRAFFIC THEORY AT VARIOUS TRAFFIC LOADS

A main objective for the evaluation was to verify analytical methods for dimensioning of ATM networks. The whole series of steps in mathematical treatment of statistically multiplexed traffic on ATM-links is given in (Lindberger, 1991). Calculations specifically carried out to guide the measurements described here have been presented in (Björkman et al, 1992) and further elaborated in (Tidblom, 1993). These calculations have for this paper been modified by taking into account that the CBR source will lose relatively fewer cells than the VBR sources. Also, the link capacity has been corrected to the implemented value of 152.64 Mbit/s.

The expected values were calculated while considering the overload situation on a *burst scale* (2). For the implemented buffer sizes (≈100 cells) at the output ports of the ATM cross connect, the size of the buffers is not decisive for the ability

[1] COI- Cells of Interest. Refers to the foreground traffic stream which here is 21.1Mbit/s CBR.

to handle a burst. However, even this size of buffer will prevent many small cell loss events(≈10 lost cells on the monitored CBR source) to occur. Another scale, the *cell scale*, was assumed not to cause any cell losses considering that traffic arrives to the ATM cross connect on only three links.

The ON/OFF background sources provided us with a simple background load by which we could control the average CLR. Figure 6 shows the calculated and measured CLR for various traffic loads. The theoretical model, with its inherent approximations, provided precise guidance for the dimensioning of the experimental setting.

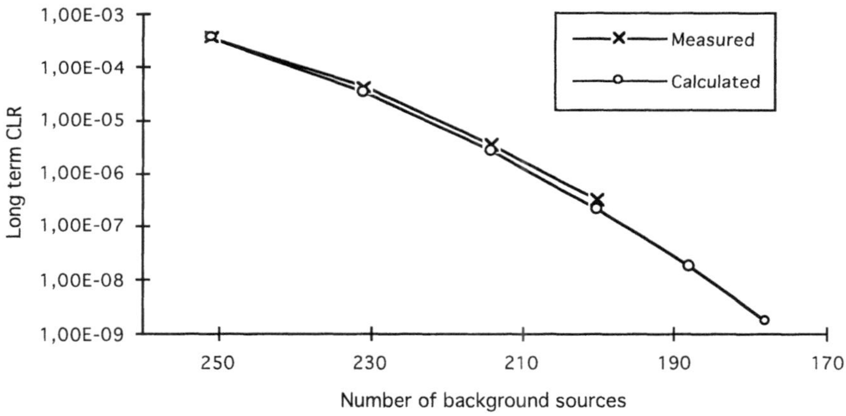

Figure 6 Long term CLR, calculated and measured values

Even though the analytical model proved to be an accurate tool to calculate average CLR for an ATM link, one needs to be cautious to draw general conclusions about the characteristics of a particular source. In this case we have focused our interest on the effect on a high quality CBR source with a bitrate of 21.1 Mbit/s. If we on the other hand choose to study the characteristics of one of the small VBR sources, or the total link behaviour, the absolute values will change. E.g. the number of cells in a busy period and the number of cells lost in a busy period will change proportionally to the difference in numbers of cells monitored. In opposition to the above, the relative values, i.e. long and short term CLR values will not be affected by the change of scale.

7 .HIGH TIME-RESOLUTION MEASUREMENT OF CELL LOSSES

The distribution of cell losses within a busy period will be decisive for the subjectively perceived QoS of the high quality video service. We therefore analysed the temporal distribution of cell losses for various parameters.

Figure 7 shows the distribution of the short term CLR within a Busy Period for a long term measurement at CLR = $3x10^{-7}$. From the graph we can see that 5-10% of the cells were generally lost when the state of the buffers became saturated and cell losses started to appear. (Single lost cells, having short term CLR=1, have been excluded) As the output link had a capacity of approximately 150 Mbit/s, 10 % of the cells lost means that the link was overloaded by 15 Mbit/s or equivalent to 7.5 of our background sources (peak bitrate 2 Mbit/s). Our interpretation of this result was that an overload situation of just one or two background sources will either rapidly be increased by more sources or vanish before the buffer is filled.

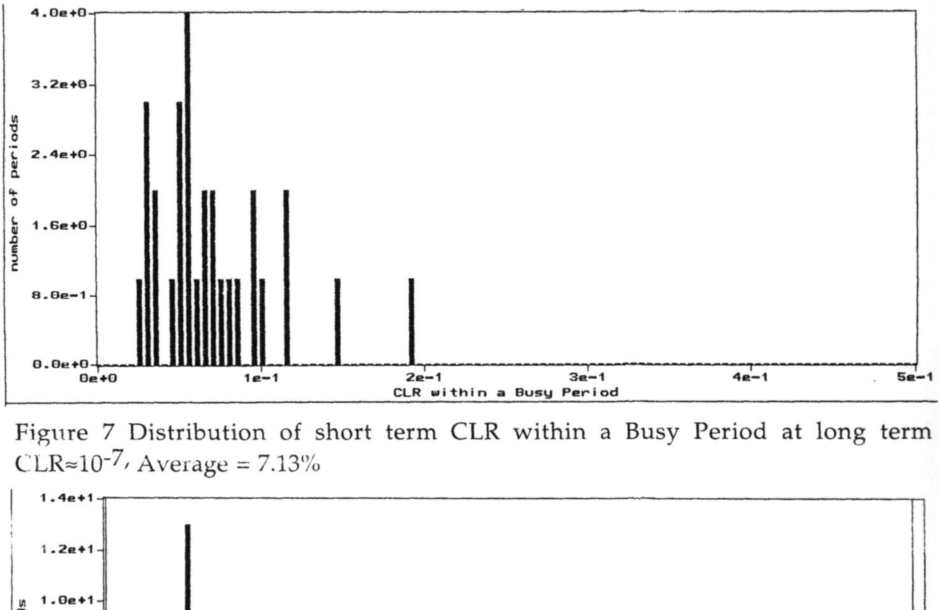

Figure 7 Distribution of short term CLR within a Busy Period at long term CLR≈10^{-7}, Average = 7.13%

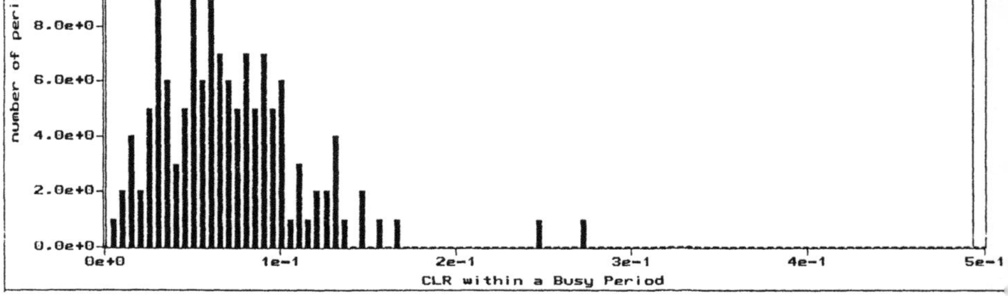

Figure 8 Distribution of short term CLR within a Busy Period at long term CLR≈10^{-5}, Average = 7.11%

The intensity of cell losses within a busy period proved to be very similar at all our measured long term CLR levels. For comparison the short term CLR distribution for long term CLR = $3x10^{-5}$ is shown in Figure 8.

As mentioned, the length of a Busy Period will vary with the proportion of the link we are monitoring. In this case, we monitored a rather large part of the link, i.e. 21.1 Mbit/s on a 152.64 Mbit/s link. In Figure 9 we can see how the length of the Busy Periods varied from 1 to 3500 COI for a measurement at long term CLR= $3x10^{-7}$. For comparison the similar distribution for CLR = $3x10^{-5}$ is shown in Figure 10.

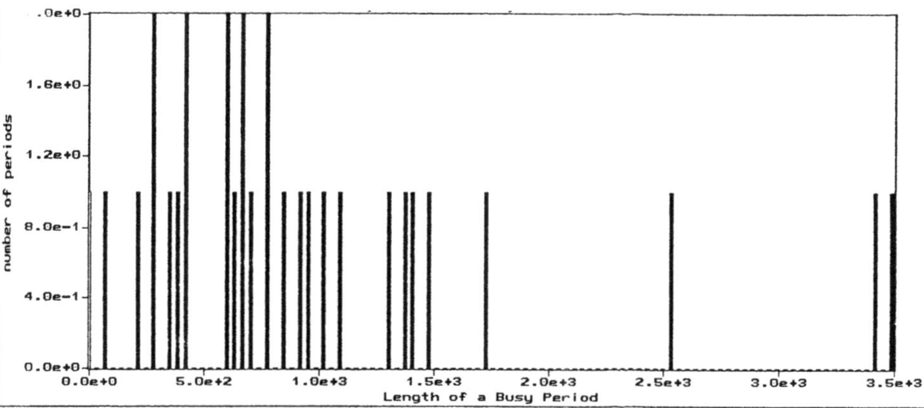

Figure 9 Distribution of length of a Busy Period at long term CLR≈10^{-7}
Average = 983 COI, standard deviation = 859 COI

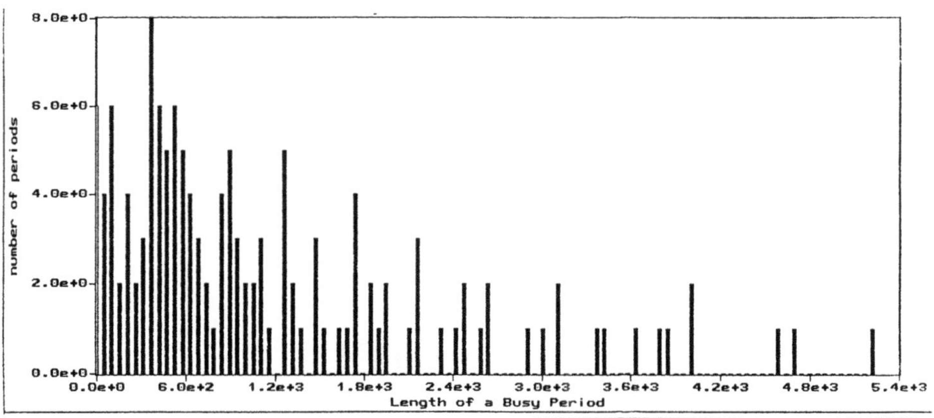

Figure 10 Distribution of length of a Busy Period at long term CLR≈10^{-5}
Average = 1180 COI, standard deviation = 1130 COI

In Figure 11 we can see how the number of lost cells per busy period varied from 1 to 680 COI for a measurement at long term CLR= $3x10^{-7}$. Figure 12 shows a similar distribution for CLR = $3x10^{-5}$.

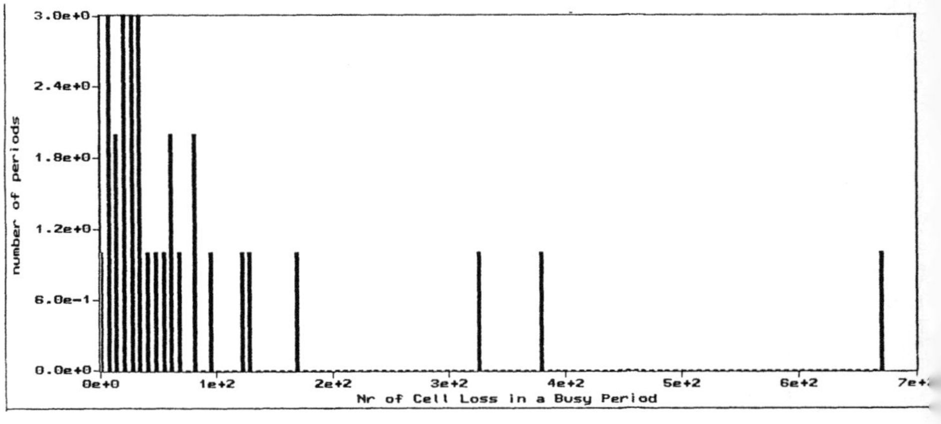

Figure 11 Distribution of number of lost cells at long term CLR≈10^{-7}
Average = 90 lost cells, standard deviation = 140 lost cells

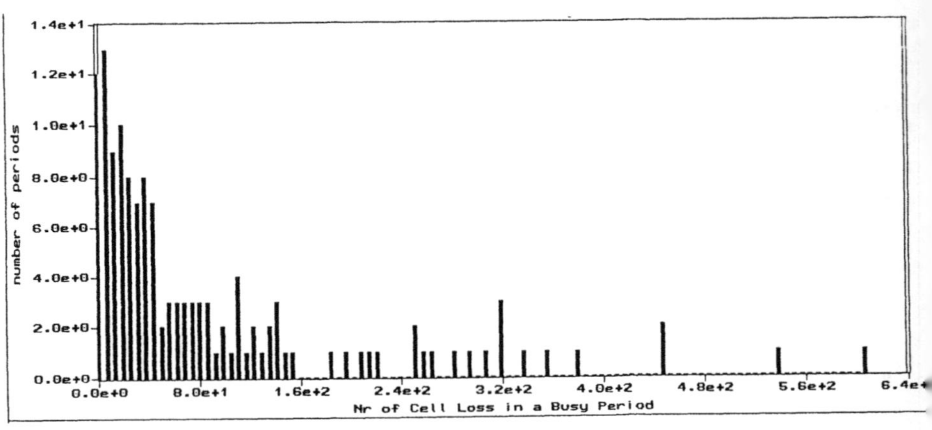

Figure 12 Distribution of number of lost cells at long term CLR≈10^{-5}
Average = 90 lost cells, standard deviation = 116 lost cells

8. IMPACT ON FEC FOR AUDIO/VIDEO TRAFFIC

The decisive criteria for an acceptable cell loss level can only be given by a service user. In a complementary experiment observers assessed the subjectively perceived quality of service for various cell loss induced impairments. In the following sections we look at the objectively measurable consequences on FEC (Forward Error Correction) on audio/video traffic and discuss how stimuli have been selected for subjective evaluation.

8.1 FEC (Forward Error Correction)

FEC Reed-Solomon (227,235) code is applied to the digital audio and video. The correction method combines FEC and octet interleaving in such a way that up to 4 cell losses in a frame of 120 cells can be corrected. The method is similar to the FEC Reed-Solomon (124,128) specified for CS AAL 1 (CCITT I.363). The used FEC code has an overhead of 4 cells per 116 cells (opposed to 4 cells per 124 cells in the present AAL 1).

8.2 Impact of cell losses on FEC

The short term CLR in Figure 7 indicated that for this particular traffic mix most of the busy periods had a cell loss intensity higher than 4 cell losses within a frame of 120 cells. In other words, for most of the busy periods the cell loss intensity was too high to be corrected by the implemented FEC. It seems that the buffer at the output port performed the function FEC was intended for, i.e. protection from short busy periods of moderate cell loss intensity.

Furthermore, the short term CLR within one busy period was not governed by the long term CLR value (corresponding to the total number of sources), as shown in Figure 13. This is explained by the fact that it is the number of active sources which determines the short term CLR - the number of inactive sources idling in the background does not influence the buffer state.

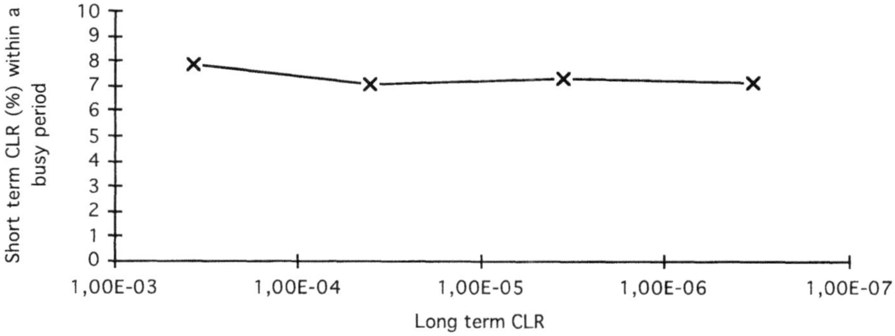

Figure 13 Short term CLR (%) within busy periods for various long term CLR

On the other hand the long term CLR, regulated by the total number of sources determined the frequency by which the FEC code was violated. Correct network dimensioning proved to be of most importance for control of the frequency of uncorrectable cell loss events on the high quality CBR audio/video channel. From Figure 14 we can predict the average time between violated FEC frames to be in the order of 100 hours for CLR=10^{-9} (≈175 sources). If we on the other hand dimension the link for CLR=10^{-5} (≈220 sources) we will have to accept in the order of 100 violated FEC frames every hour.

However, not every violated FEC frame is detectable by an observer. In fact, the severe picture impairments, caused by loss of cells carrying picture synchronisation, showed to be comparatively rare. This suggests that correct network dimensioning should be complemented by giving a high degree of protection to the cells carrying the most crucial information. This can be achieved either by prioritised treatment in the network or careful codec design, e.g. a layered coding scheme..

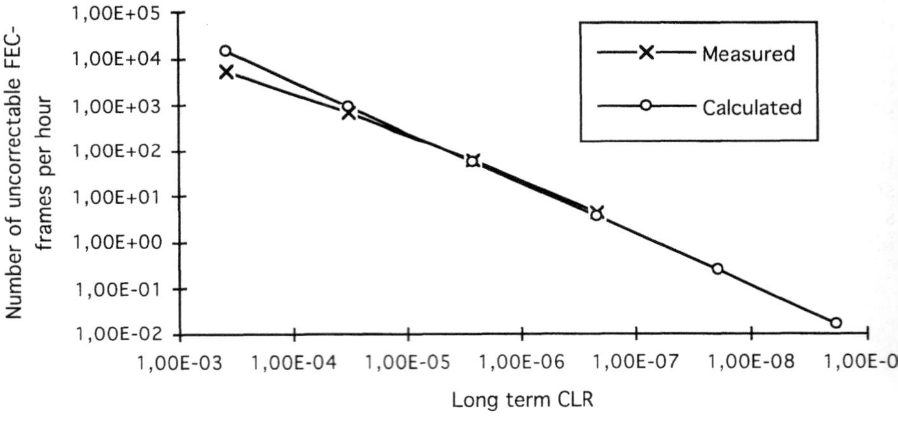

Figure14 Uncorrectable FEC-frames per hour

Finally, we also looked at the relevance of the Sequence Number indication in AAL 1. The SN is conveyed by 3 bits, thus allowing no more than 8 cell losses in a row. This was found to be an sufficient value for this case. Never during all four measurement did more than 8 consecutive cell losses occur. The average number of consecutively lost cells in a busy period was approximately 3 for all presented load situations.

8.3 Subjective evaluation of cell loss induced impairments

Since periods of cell losses have been proved to be rare occasions, with hours between cell loss events at CLR=$3x10^{-7}$, it is impractical to let observers assess audio, video or service quality in a real time process. The approach taken in a complementary experiment was instead to impose recorded cell loss events into a stream of audio and video traffic. Three busy periods were selected from the recordings described in this paper; Busy Period Length=155 Cells Of Interest, BPL=393 COI and BPL=871 COI. With this approach we could deterministicly impose a cell loss distribution, coming from a real loaded network.

9. DISCUSSION

It must be realised that the outcome of an evaluation like this will have inherent limitations. The verification of traffic theory and characterisation of network performance using a laboratory demonstrator provides a level of realism somewhere between a computer simulation and real full scale network; the demonstrator offers a complete network but with a minimum set of equipment.

As in many cases, our early investigations of new phenomena based on new technology was plagued by imperfections of the prototype equipment employed. Our original intentions included measurements spanning several days in a row. This proved impossible due to lack of stability in some of the demonstrator components - the longest measurements lasted one night. Fortunately, the dignity of the instabilities was so large that we could easily discriminate between periods of valid recording and those that were unusable. Apart from the easily discovered anomalies there might also be others that we did not find. Because these failures appeared relatively seldom and the average performance values measured and theoretically predicted correspond well, we conclude that the recordings used for analysis describe the queue overload phenomenon we originally wanted to investigate.

The traffic theory as well as the experiment are based on a specific statistical behaviour model for the background sources. The conclusions available are thus limited to a network environment where this assumption remains correct. In general one can state that investigations of network performance in a laboratory set-up are inseparably independent of the extendibility for the chosen background traffic models into the real world.

In our case the background sources being ON/OFF could be perceived as unrealistic. This shortcoming is though to a large extent remedied by their large number.

The experiment is set from the point of view of the 21.1 Mbit/s CBR traffic stream. This is conditioned by our intention to evaluate, in a subsequent experiment, perceived QoS for a video codec with the same traffic parameters. If we instead choose to measure cell loss for a source with another bitrate, may it be VBR or CBR, the absolute values of some of the evaluated cell loss parameters will obviously change. To conclude about the resulting QoS for these other traffic streams we still need to evaluate them at the service level.

10. CONCLUSION

This paper presents results of an early experiment investigating traffic handling properties of ATM networks. Even though restricted in scope, it provides opportunity to draw a number of interesting conclusions.

Future efficiency of ATM based networks strongly depends of resource allocation strategies adopted. Our experiment indicates that adequate performance levels, in terms of e.g. long term CLR (Cell Loss Ratio), can readily be obtained when statistically multiplexing VBR (Variable Bit Rate) sources. While maintaining control of performance it is possible to mix a CBR (Constant Bit Rate) stream, that occupies a significant part of the link, with a large number of VBR sources.

Even though frequently used, the long term CLR does not provide enough information about the phenomenon in question. The temporal distribution of events constituting the cell loss process has a significant impact on the actual performance outcome. Our results comprise a number of other measures for the cell loss process. In view of the fact that cell loss occurred in bursts we conclude that short term CLR for a cell loss burst may be more applicable. Being a short term cell loss measure though, it has to be complemented by a frequency of occurrence value. It has to be remembered that any measure of cell loss, however elaborate, still does not provide information about the user perception of the cell loss impact on a service.

The bursty structure of the cell loss process has a strong impact on the performance of FEC (Forward Error Correction) code employed in the experimental codec. The FEC code employed in this codec strongly resembles the one currently standardised in ITU-TS for distributive video AAL 1. Our results show that this FEC code had very little protective effect. It seems that the dimensioning of the FEC code limits its ability to cope with the effects of overload situations to those that are already resolved through moderate sized buffers in the network.

As this FEC code allows correction of only approximately 3% lost cells in a protected block there is a need for employing other measures to cover for the high level of short term CLR (5-10%) as indicated in our experiment. Reduced link utlisation did not improve the short term CLR significantly - a buffer overflow event showed a similar loss distribution at all measured load levels. This is explained by the fact that it is the number of active sources which determines the short term CLR - the number of inactive sources idling in the background does not influence the buffer state. However, the time between these severe cell loss events, caused by statistical multiplexing, was readily controlled by careful dimensioning of the network utilisation.

ACKNOWLEDGEMENTS

This work is dedicated to the memory of S. Brevik, Norwegian Telecom Research, who greatly contributed to the achievements accounted for in this paper, both in development of the STG as well as during the experimental sojourn in Madrid.

We also wish to thank P. Tholin, Telia Research, for major contributions to the evaluation of the BIPED demonstrator; K. Lindberger and S-E Tidblom, Telia Research, for their precise theoretical guidance; S. Ösund, Telia Instrument for the analysis software IDM and together with his colleagues C Gisgård and J. Elldin for providing the Impairment Recorder Module with every possible functionality. The evaluation was made possible thanks to T. Migotti and his staff at Ericsson I+D in Madrid, who supported the BIPED demonstrator well beyond their obligations. Special thanks goes to our friends in Norway; Leif Morland, SINTEF DELAB for generating the background traffic and Ole Melteig, Norwegian Telecom, not only for his "helado sin cerveza".

REFERENCES

1 T. Migotti, V. Teacher, N.Björkman "Performance evaluation of broadband connections and services under varying traffic loads" Electronic&Communication Eng Journal 4, p.243-251, 1992.
2. B. Helvik, O. Melteig, L. Morland, "The Synthezised Traffic Generator; Objectives, design and capabilities", IBCN&S, Copenhagen, 1993.
3 K. Lindberger, "Analytical Methods for the Traffical Problems with Statistical Multiplexing in ATM-network". 13th International Teletraffic Congress, Copenhagen, p.807-813, 1991.
4. Björkman, Goldstein, Hedman, Latour-Henner, Tholin and Gil, "Network Performance and its relationship with Quality of Service in an experimental broadband network", Proc. 1st IFIP TC6 Workshop on Broadband Communications, Estoril, p.157-168, 1992.
5. S-E Tidblom, "Hur totala cellförlustsannolikheten på en ATM-länk påverkar ett enskilt bredbandssamtal under en begränsad mätperiod", Proc 11th Nordic Traffic Seminar, Stockholm, 1993.

Broadband Communications, II (C-24)
S. Tohmé and A. Casaca (Editors)
Elsevier Science B.V. (North-Holland)

On Issues Related to Real-Time Applications Support in the ORBIT Gbps Ring

Jeane S.-C. Chen and Roch Guérin

IBM T.J. Watson Research Center
P.O. Box 704
Yorktown Heights, NY 10598, USA

ABSTRACT

This paper considers the influence of network packet format, e.g., fixed or variable packet sizes, on the delay jitter experienced by user messages in the context of ORBIT [1], a Gbps buffer insertion ring developed in IBM Research. ORBIT is to be used to interconnect access agents and workstations involved in video conference applications as well as regular data transmissions. For these applications, real-time traffic such as voice and video are carried on the ring, and the key performance measure to be optimized is the delay jitters experienced by these real-time messages. Messages are the native data units generated by applications, and they represent the basic entities for which performance guarantees should be provided. The focus of this work was on the impact of network packet format on delay jitter as seen by the applications. We examined the use of fixed size ATM format and variable size packet format. Both these formats can be supported by ORBIT. The study was conducted by means of simulations and various traffic scenarios were investigated. The results provide a frame work in understanding how to provide real-time applications with certain quality of service in the broadband networking environments.

Keywords:Performance Evaluation, Local and Metropolitan Area Networks, Multimedia Services and Systems, ATM

1 Introduction

The many significant technological progresses that have taken place in recent years have drastically altered most of the traditional assumptions on which communications systems were built. For example, the availability of fibers and optical transmitters/receivers increased transmission speeds by several orders of magnitude and required a rethinking of many communication protocols designed with bandwidth limitations in mind. Similarly,

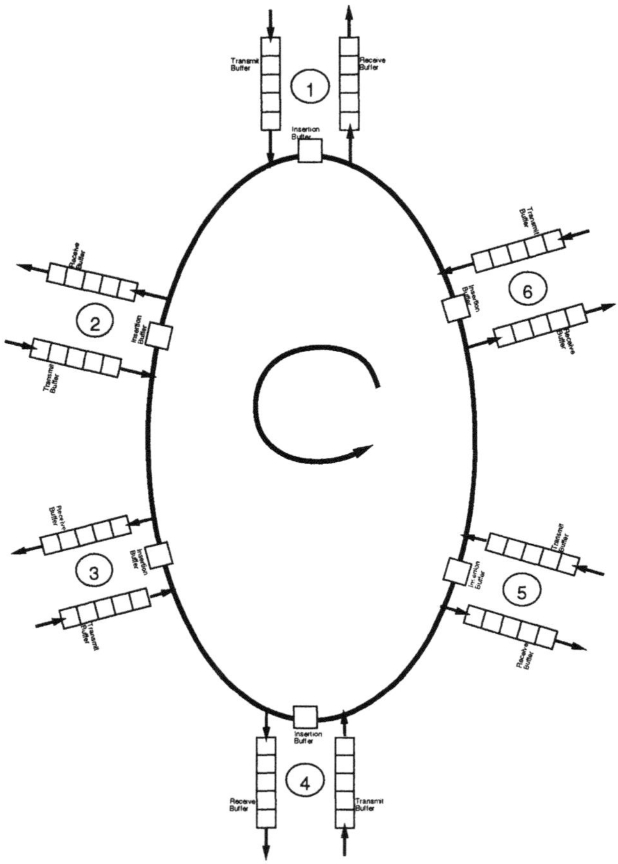

Figure 1: Overview of buffer insertion ring structure

fast-packet switches led to the introduction of the integrated network concept. Integrated networks have the ability to carry a wide range of traffic, e.g., voice, data, video, etc., over a single platform. This not only has the advantage of consolidating many of the existing communication facilities, but also enables the development of many new applications, e.g., multimedia, distributed computing, etc.

These new developments have been reflected in the emergence of several standard and architecture proposals for both wide area [2, 3, 1], and local area [4, 5, 6, 7] networks. In the context of local area networks (LANs), the availability of higher communication speeds significantly changes many of the assumptions traditionally made when comparing different approaches [8, 9]. In particular, transmission and buffering delays now only represent a small fraction of the overall delay, with propagation delay being often the dominant factor. This shift in the relative importance of the different delay components has led to a renewed interest in buffer insertion rings [10, 6, 11, 7, 12] (see Figure 1), which previously had been disadvantaged by the potential impact of multiple insertion buffers on the end-to-end delay. With this problem essentially eliminated by the higher transmission speeds, the many other desirable properties of buffer insertion rings make them attractive candidates for high speed LANs. In particular, they provide for a simple implementation of *spatial reuse*[1] [6, 7, 13], which can significantly improve throughput by allowing multiple simultaneous transmissions.

There has been many previous studies devoted to buffer insertion rings and their throughput characteristics, e.g., [8, 6, 7]. This paper differs from these earlier works as its focus is on the impact of the ring packet format on several key performance parameters. In particular, while fast packet-switched networks provide tremendous flexibility and efficiency in integrating different types of traffics, they also introduce new problems. For example, the statistical sharing of network resources requires new approaches to bandwidth and buffer allocation [14] to guarantee performance to individual connections. Of special importance to many *real-time* connections such as voice or video[2], is the *delay jitter* introduced by the many buffering points inside the network. The greater the jitter, the more difficult it is to meet the delay requirements of real-time traffic, e.g., playback buffers must be dimensioned accordingly. While the use of transmission priorities and/or different service policies can help in controlling the delay jitter of real-time connections [16, 17], other factors such as packet format also influence jitter performance.

Network packet formats can typically be divided in two major categories that reflect the origins and motivations of the network. Traditional packet-switched networks have supported a variety of packet formats and sizes, which were motivated by the heterogeneous needs of the many different *data* users they served. A natural extension is then to adopt this variable size packet format in the new world of fast packet-switched networks. This has indeed been the approach taken in a number of standards and architecture proposals [4, 2, 1]. On the other hand, the use of a unique fixed-size packet format has also been proposed and is the solution of choice in a number of standard proposals [3, 5].

It is beyond the scope of this paper to argue the respective merits of each approach, even for the limited case of buffer insertion rings (see [18] for a related study in the WAN environment), and it is furthermore likely that both will coexist at least in the foreseeable future. Rather, we focus on the influence of packet format on both system throughput and delay jitter in a buffer insertion LAN, ORBIT [1], which has the ability to support both types of packet formats. ORBIT link speed is about 1 Gigabit/sec and is to be used to interconnect workstations either directly or through access agent (routers) attaching more traditional data networks (Ethernet, Token Ring, FDDI, etc.), and therefore provide high bandwidth connectivity between remote users. In addition to allowing traditional

[1] Also termed destination release.

[2] See [15] for examples and discussions on how such connections are supported across fast packet-switched networks.

data communications, it is also expected that ORBIT will carry a significant volume of video conferencing and image transfer applications. Video conferences and image transfers will involve workstations and servers either attached directly to ORBIT [19], or connected through lower speed LANs, e.g., E/N, T/R, FDDI, [20, 21, 22, 23] which will be attached to ORBIT through access agents using it as a high-speed LAN-to-LAN interconnect.

The goal is then to better understand the impact of the ORBIT packet format from the point of view of these potential users. Specifically, system throughput and jitter should be investigated for different types of connections and system configurations, and the influence of packet format on these performance measures should be identified. This study is carried out by means of simulations and the results provide a better intuitive understanding of the performance of buffer insertion based high speed LANS (ORBIT) from a application point-of-view.

In particular, while the lower transmission overhead often gives the variable packet size format an expected throughput advantage, the impact of this variability on delay jitter needs to be better understood, especially in the case of applications for which it might be critical, e.g., voice and/or video. The simulation results provide a better insight into the impact of various system parameters, and reveal some initially surprising behaviors. It is found that contrary to initial expectations, a fixed-size packet format need not result in better application level jitter characteristics. Although individual network packets do experience a lower jitter, this does not necessarily translate into a reduced jitter for the reconstructed application data unit, i.e., the jitter of multiple network packets needs to be accounted for. This effect is observed even in the absence of any difference in overhead between the two network packet formats. Simulations are carried out for a number of scenarios that provide a better understanding of the scope and magnitude of this phenomenon.

The organization of the paper is as follows: Section 2 describes the buffer insertion ring model used in this study and introduces relevant notations. The traffic scenarios and assumptions used in the simulations are also presented and motivated. Section 3 represents the bulk of the paper and is devoted to numerical examples that illustrate the findings of the paper. The influence of packet format on throughput and/or access delay is first briefly discussed, and the rest of the section is then devoted to the study of jitter characteristics. This study is carried out essentially for periodic sources which are most sensitive to the effect of jitter, and the impact of various system parameters such as ring load, number of nodes, etc., is investigated. Source characteristics are also varied and, in particular, the impact of the size of application data units (vs. network packet size) is examined. Section 4 briefly summarizes the findings of the paper.

2 Model Definition

The operation of buffer insertion rings is a well documented topic in the literature, e.g., [8, 10, 9, 7], and we therefore only summarize its main features as needed to establish a common understanding of the system under study. The typical configuration of a buffer insertion ring is shown in Figure 1. As indicated by its name, the network has a ring topology, with stations or nodes connected in series by transmission facilities. Each node consists of a ring or insertion buffer, a receive buffer, a transmit buffer, and some logic to control packet transmissions and receptions.

Transmission of a packet between two nodes requires that the source node gains access to the ring and only involves the links located between the two nodes. The source node puts the packet on the ring just after its insertion buffer and, conversely, the destination node removes the packet just before it enters its own insertion buffer. For example, a packet to be transmitted between nodes 1 and 3 in Figure 1, is put on the ring just after the insertion buffer of node 1 and removed from it just before entering the insertion buffer of node 3. Note that limiting transmission to links between the two nodes ensures that other links remain free for the transmission of additional packets. In general, several simultaneous packet transmissions can take place, provided they do not interfere on the ring. This is the spatial reuse feature mentioned earlier.

Once on the ring, packets cut through the insertion buffers of intermediate nodes , i.e., packet transmission around the ring is not store-and-forward, unless they encounter an ongoing local packet transmission. In that case, the local packet transmission is allowed to proceed and the ring packet is buffered in the insertion buffer until its own transmission can resume. The resulting delay lasts for at most the transmission time of one maximum size packet, which limits the size of the insertion buffer to one such packet. In other words, ring traffic is given non-preemptive priority over local traffic, and local packets can only be put on the ring when their insertion buffer is empty or idle.

The performance measures of interest are the access delay experienced by user messages upon their arrival to the transmit buffer of a node and the delay jitter[3] that the network introduces in the transmissions between nodes. Note that this study is carried out in the context of a *stable* system. In particular, it is assumed that the load on all ring links is such that each node is always provided with sufficient ring access, i.e., the total offered traffic never exceeds capacity on any link. Before proceeding any further with the details of the models used in the study, we introduce some terminology used throughout the rest of the paper. First, the terms *messages* and *packets* are used to distinguish between user and network data units, respectively. As mentioned above, our focus is on the performance experienced by end-users, i.e., network performance quantities such as packet delay and jitter still need to be translated into corresponding message level measures.

Depending on the message size and the ring network packet format the necessary conversion or adaptation can take either one of two forms. In the case of a variable size packet format, which is one of the two modes of operation of the ORBIT ring, messages which do not exceed the maximum allowable packet size[4] are transmitted as is after adding the necessary network routing information. This typically allows a one-to-one correspondence between user messages and network packets. Messages that exceed the maximum packet size must, however, still be segmented into multiple packets, each of maximum size ex-

[3]By *jitter*, we mean the standard deviation of the associated delay.

[4]Most network impose such a limitation because of practical buffer sizing considerations.

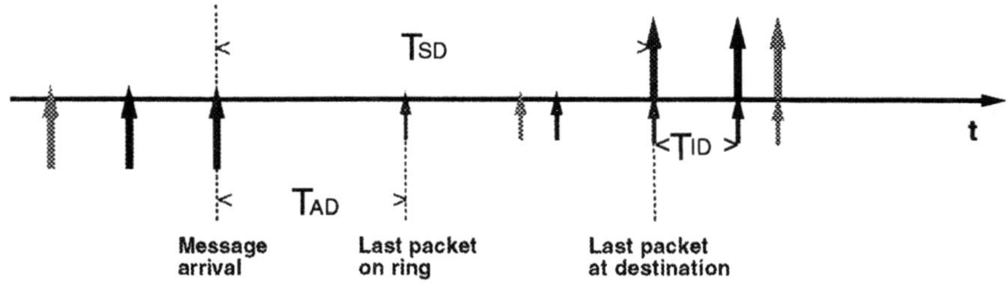

T_{AD} : Total access delay (from time of first bit arrival to last bit on ring)

T_{SD} : System delay (from time of first bit arrival to last bit at destination)

T_{ID} : Interdeparture time (time between arrivals of last bit of consecutive packets)

Figure 2: Message performance measures in buffer insertion rings

cept for the last one which only carries the remainder of the message, i.e., no "padding" is present. In the rest of this section, we use the common notation of PTM (Packet Transfer Mode) to refer to networks with a variable size packet format.

In networks that rely on a fixed size packet format, e.g., ATM, which is the other possible mode of operation of the ORBIT ring, segmentation and padding are typically always required. In the specific case of the ATM formats considered in this paper, user messages must be sent as integer multiples of 53-byte cells, where in each cell at least 5 bytes are devoted to network related information, i.e., the usable data payload is at most 48 bytes. In this paper we typically assume the minimum overhead of 5 bytes per cell, i.e., an AAL5 adaptation layer. In some instances, we also consider adaptation layers with a greater network overhead, e.g., AAL4, and for comparison purposes provide results for a system with zero bytes of overhead. Note that fixed size network packets also imply some amount of padding for most messages. For the purpose of this paper, the more important aspect is, however, segmentation as the performance seen by individual segments may differ from that provided to messages.

A goal of this paper is to understand the impact of segmentation (and padding) on the performance offered to applications. The first aspect that is investigated is the impact of segmentation on the access delay, i.e., the time it takes from the arrival of a message at the transmit buffer of a node until it has been fully transmitted on the ring (see Figure 2.) Full transmission implies that the last "bit" of the message has been put on the ring. In the case of an ATM format, this requires transmission of the last cell generated when segmenting the message.

Differences in access delay due to the packet format are two-fold. The first one comes from the higher ring load that the fixed size packet format induces for a given data load, i.e., the per-message overhead is typically greater. The second reflects the fact that multiple ring accesses are required to transmit all the cells from a message, while only a single access is necessary if the message is transmitted as one packet. A single *packet* access delay is likely to be greater with PTM than ATM formats because of the larger packet sizes[5], but it is unclear if this result also holds at the message level. The results of this paper help shed some light on this issue.

After investigating access performance, we concentrate on performance measures that account for the impact of packet format on the complete transmission path from source to destination. In particular, we focus on the variability in end-to-end transmission delays as a function of the network packet format. As discussed earlier, this delay variability or jitter is an important criteria for real-time traffic such as voice or video. One interesting finding of this study is that short (fixed) packet sizes need not translate into lower message jitter, when messages are likely to fit into more than one packet. While this result is obtained and illustrated for a buffer insertion ring, ORBIT, it provides some insight into this issue and point to the fact that it may often be difficult to accurately map network performance to application performance.

In order to better reflect the nature of real-time connections which are most sensitive to jitter, periodic arrivals and constant size messages were assumed. Messages may, however, still translate into several packets, depending on both their size and that of network packets. Performance measures were gathered for a *tagged* connection with the above characteristics. The rest of the network traffic remained unchanged with exponentially distributed message lengths and Poisson message arrivals. This allows us to concentrate on the impact of packet format on the transmission delay experienced by a single "real-time" connection when the rest of the background traffic remains the same. In particular, we can identify how the presence of potentially large and variable size packets, influences the jitter experienced by a constant rate connection in a PTM network.

Jitter in buffer insertion rings has two major components. The first one is the access jitter, i.e., the variability in ring access. This aspect is likely to favor the PTM format as it results in fewer ring accesses per message (a single ring access for the scenarios assumed in this paper.) Note that with an ATM format, the variation in access delay experienced by each packet or cell from a message contributes to variations in the overall message delay as it influences the time at which subsequent packets can access the ring. The significance of this factor clearly depends on the size of the message itself, i.e., if the message fits in a single ATM cell, a PTM format clearly offers limited advantage. It is, therefore, of interest to investigate the influence of message size on this factor.

The second component is the insertion jitter, i.e., the delay variations introduced when a packet is required to wait in an insertion buffer because of the ongoing transmission of a local packet. This aspect is likely to favor an ATM format as the smaller size of the insertion buffer limits the range of possible variations. Furthermore, the difference will depend on

[5]The lower network load may, however, compensate for some of this.

the length of the path between the source and destination nodes, i.e., the more insertion buffers are present the greater the difference will be, and we investigate this aspect.

In order to better understand the characteristics of jitter for ATM and PTM formats, we focus on two different measures. The first one, *system delay,* provides a network oriented perspective, while the second, *inter-delivery jitter,* reflects the point-of-view of a tagged connection. These two quantities are again illustrated in Figure 2.

System delay gives the total end-to-end message delay from the time of arrival at the source transmission buffer until the last bit is delivered to the receive buffer of the destination node. Both the mean and the standard deviation of the system delay are obtained for various network loads and distance (hop count) between source and destination nodes. The standard deviation is representative of the variability in end-to-end delay that the network introduces in the transmission of messages from the same connection.

Inter-delivery jitter represents the variability in how messages are delivered at their destination. In the case of periodic arrivals, the average time between successive arrivals at the destination equals the inter-arrival time at the source. Messages, however, are not delivered as regularly as they are generated because of the randomness introduced by the network. This jitter in the delivery of successive messages at the destination is particularly significant for real-time connections, where it determines the sizing of the playback buffer.

The next section describes the many simulation scenarios used to investigate the issues discussed above and provides many illustrative examples.

3 Performance Results

In this section, we compare various delay components as seen by application messages in buffer insertion rings using either variable-size or fixed-size packet formats. In particular, we obtain the delay-throughput curves achieved by these two formats and compare the standard deviations of the system or end-to-end delay, the access delay, the ring delay, and the inter-delivery time. The difference in network load induced by the two formats are illustrated through their respective delay-throughput curves, while the levels of jitter incurred by messages at different stages of transmission are demonstrated by the standard deviations of the various delay components. Jitter characteristics are further investigated by varying hop count, ring size, ring load, message length, and overhead size. The simulation was written using the RESQ simulation language [24], which provides a flexible simulation environment and supports many standard features such as confidence interval guarantees, setting of initial discard portion, number of replications, etc. Due to lack of space, no further details will be provided on the simulator.

3.1 Load Utilization

In this subsection, we give simulation results that illustrate the relation between message delay and ring throughput for the variable-size and fixed-size packet formats. Here, throughput refers to the net payload throughput, excluding any header or padding overhead. As explained in the previous section, each packet is prepended with a network header so that it can be routed to its destination. In the variable-size format, a packet consists of a whole message, subject to the maximum length constraint. In the fixed-size format it is a fixed size cell. As mentioned earlier, this typically results in a greater overhead when the message size is either smaller than the cell size (because of padding), or larger than the cell size (multiple network headers.) This in turn, often yields a lower throughput for the fixed-size format.

We studied a uniformly loaded ring with 10 nodes with a distance of about 1km between nodes (propagation delay of about 5 μsec). The link bandwidth of the ring is assumed to be about 1 Gbps. (Note that a lower bandwidth, e.g., 100 Mbps, would result in higher delay and jitter). Each node in the ring transmits to all other nodes, excluding itself, with equal probability. The message length is exponentially distributed with a mean of 1 kbytes. In the variable-size format, a header of 10 byte is added to each message, to account for the overhead due to control and routing information [1]. In the fixed-size format, we assumed a 53 byte fixed-size cell transport as in the ATM environment, and we investigated three header sizes of 0, 5, and 9 bytes, respectively. A 5 byte header is needed for ATM AAL5 adaptation layer, while a 9 byte header is needed when ATM AAL4 adaptation layer is used. Results for ATM without header, where the load impact due to header overhead is eliminated, is included for comparison purposes.

3.2 System Jitter

System jitter is a phenomenon caused by the random delays experienced by a message as it travels from its source to its destination. As mentioned in the previous section, system jitter in a buffer insertion ring has two major components, reflecting the contributions of the two steps involved in the transfer of a message. These two components are the access jitter and the ring jitter. Access jitter accounts for the delay variations in accessing the ring at the source node. The range of these variations is clearly a function of the number of ring accesses needed to transmit the full message. It is, therefore, expected that this component will be sensitive to message length in networks using fixed-size packets. Ring jitter accounts for the delay variations experienced by packets on the ring. These variations are created by the possible buffering of ring packets in a node insertion buffer when transmission of a local packet is in progress upon the arrival of the ring packet. The magnitude of these variations clearly increases with the length (hop-count) of the path between source and destination nodes.

In addition to message and path lengths, system jitter is also affected by the size of the packet overhead, the size of the ring, and the loading of the ring. We investigate the impact of these parameters on system jitter via simulations. In the simulations, we tag a certain connection and let other nodes transmit to each other, creating a uniform background load. The ring loading is controlled by adjusting the level of the background load. Since the issue of jitter is more important in the context of real time traffic, the tagged connection is assumed to be a 50 Mbps video connection, periodically generating fixed size messages of 1 kbytes, at a rate of 0.00625 messages per μsec. Note that we use fixed size messages for the tagged connection in order to filter out contributions to the jitter from variations in message size. The background traffic, however, is still assumed to consist of exponentially distributed messages with a mean of 1kbytes.

3.2.1 Impact of Hop Count

For the comparisons shown in this section, a 10-byte packet header is assumed for the variable-size format and 53 bytes ATM cells with a 5-byte header are chosen for the fixed-size format. In Figures 3, 4, and 5, the system, access, and ring jitters are represented by the standard deviation of the respective delays and plotted with respect to the hop count of the tagged connection for both the fixed-size and variable-size formats. The data is for a 10-node ring with a background loading of 1 Gbps. It is seen that the fixed-size format results in larger system and access jitters and smaller ring jitter than the variable-size format. For the fixed-size format, the ring jitter is almost two orders of magnitude smaller than the access jitter, and therefore the system jitter is seen to stay constant with respect to the hop count as it is dominated by the access jitter which is independent of the hop count. On the other hand, for the variable-size format, the ring and access jitters have the same order of magnitude, and the resultant system jitter is seen to increase with the hop count of the connection.

Figures 3 shows the advantage of the variable format over the fixed-size format for up to 10 hops. However, they also show that the curves for the variable-size format are approaching those of the fixed-size format as the hop count increases. The obvious question is then: Are they, and if yes when, going to intersect and reverse the advantage? To answer this question, we simulated a larger ring with 50 nodes. The results for a background load of 1 Gbps are shown in Figure 6. It is seen that the advantage of the variable-size format is still preserved even for up to 50 hops. An intuitive explanation for this behavior,is that the greater number of nodes increases the variability in packet destinations. This means that for the same load a given node will receive fewer packets and, therefore, have longer times between successive ring accesses.

3.2.2 Equal Overhead Comparison

The advantage of the variable-size format shown in the last subsection is partly due to the larger header overhead for the fixed-size format. In order to eliminate this contribution, we simulated both systems without header. Figure 7 shows the mean and standard deviation of the system delay for both the fixed-size and variable-size formats, again for messages with fixed length of 1kbytes, but no header overhead. It is seen that at the beginning, i.e., small hop-count, the variable-size format still outperforms the fixed-size format for both the mean and standard deviation of the delay. As the hop-count of the connection increases, the two mean delay curves eventually intersect and the advantage in mean delay is reversed. This reversal, however, does not take place for the standard deviation even for the maximum hop-count of 20 (recall that we assume a 20 nodes ring).

As stated at the beginning of this section, the fixed-size format is penalized in case of large messages, by the larger number of ring accesses required to transmit the message. In order to investigate the sensitivity to message size, we reproduced the above study with smaller message sizes. The results are shown in Figure 8, where the size of the messages generated by the tagged connection was chosen to be 125 bytes, i.e., 1/8th the length used in Figure 7. The background traffic was, however, still assumed to generate messages of exponentially distributed length with mean 1 kbytes. We see that the performance of the fixed-size format is improved by the smaller message size. For the mean system delay, the fixed-size format now outperforms the variable-size format for connections with more than 6 hops. In addition, a similar reversal is also observed for the standard deviation, i.e., system jitter, where the fixed-size format shows better performance for connections with more than 13 hops.

3.2.3 Impact of Message Size

The last subsection illustrated the fact that message size is a major factor in the performance comparison of the two formats. To further investigate this aspect, we simulated a fixed configuration, i.e., fixed hop count, ring size, and load, and varied the message size of the tagged connection. We again assumed no header overhead for both formats, to eliminate any contribution from that factor. We considered a ring with 20 nodes generating a background load of 1 Gbps. The path length of the tagged connection was taken to be 10 hops. The standard deviation of the system delay for both formats is plotted in Figure 9 as a function of message size, over a range from 60 bytes to 1 kbytes. It is seen that for this configuration, the variable-size format outperforms the fixed-size format over the whole range of message sizes. It is further seen that the jitter in the fixed-size format grows linearly with the message size, while it is nearly insensitive to message size in the variable-size format. This is again due to the fact, that with the fixed-size format the system jitter is dominated by the access jitter which increases with the message size, while for the variable-size format the dominant factor is the ring jitter which is relatively constant for this configuration. It can be argued, that the ability to keep the jitter of messages from the same connection independent of their size is a desirable, if not important, feature.

3.3 Inter-Delivery Jitter

Besides system jitter, another important measure that needs to be investigated is the jitter of successive packet deliveries, i.e., the inter-delivery jitter. This jitter is again caused by the random transmission delays incurred in the connection's path, and it alters the periodic nature of the original message stream at the source node. This is an important factor in the context of both rate control, i.e., it can change the "burstiness" of a connection, and the sizing of receive or playback buffers for real-time connections. This jitter is defined as the standard deviation of the inter-arrival time of packets from the same connection at the destination node, where message arrival time has been defined as the arrival time of the last packet of a message. The inter-delivery jitter was gathered for all the simulation cases previously described, and the results are shown in Figures 10 to 14. It is seen that the conclusions that can be drawn when comparing the inter-delivery jitter for the fixed-size and variable-size formats are essentially similar to those obtained for the system jitter.

4 Conclusion

In this paper we have investigated the impact of packet format on various performance measures in a buffer insertion ring. This was achieved by means of simulations, which allowed us to accurately model the operation of a buffer insertion ring, assuming either fixed-size (ATM) or variable size (PTM) packet formats. The goal of the paper has been to study application-related measures such as message access delay and end-to-end jitter, rather than simply evaluate "raw" network performance. This focus on message rather than packet level performance measures was motivated by the need to better understand how to operate the ORBIT network, so as to provide applications such as video conferencing with the best possible quality of service.

The study resulted in a number of potentially interesting observations. In particular, it was found that significant differences could be observed between message and packet jitters in scenarios where packet and message sizes differed. In such cases, it was found that the ability to map a message into a single packet helped reduce end-to-end jitter. This conclusion must clearly be qualified, and as mentioned earlier depends on both the assumption on where the adaptation layer function is performed and on the access mechanism of the buffer insertion ring. However, in addition to illustrating possible advantages of a variable size packet format for some scenarios, the results also point to the difficulty in precisely mapping network performance into application performance (see [18] for a related study focusing on message vs packet losses). This problem will have to be faced by providers of broadband networks when attempting to define quality of service parameters for both applications and existing networks. For example, this is likely to be raised when defining how to support Frame Relay [2] connections over ATM networks. In general, much work remains to be done to understand this complex issue, and it is hoped that the preliminary results of this paper will trigger further work in this important area.

5 Simulation Results

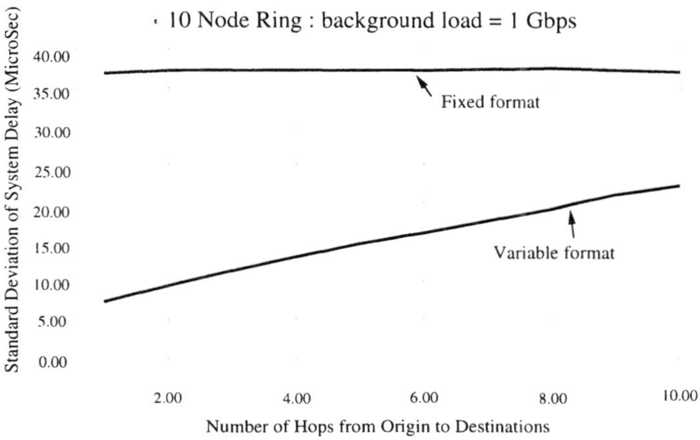

Fixed–size messages: 1 Kbyte
Variable format header: 10 byte/message
Fixed format header: 5 byte/cell

Figure 3: System jitter: background load = 1 Gbps

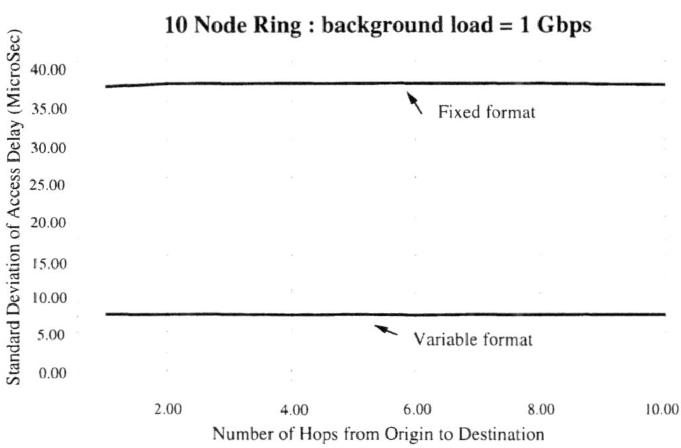

Fixed–size messages: 1 Kbyte
Variable format header: 10 byte/message
Fixed format header: 5 byte/cell

Figure 4: Access jitter: background load = 1 Gbps

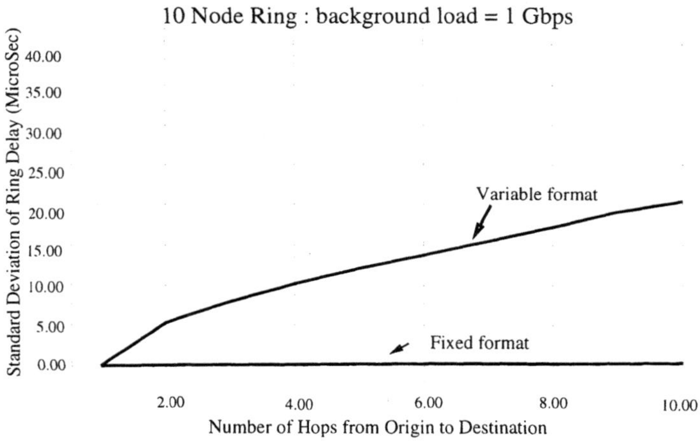

Fixed–size messages: 1 Kbyte
Variable format header: 10 byte/message
Fixed format header: 5 byte/cell

Figure 5: Ring jitter: background load = 1 Gbps

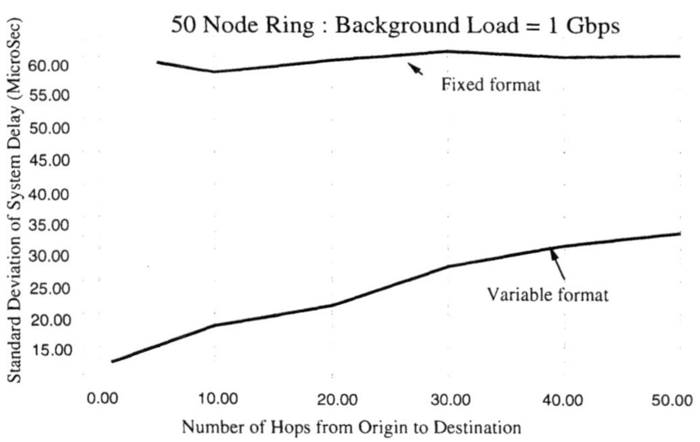

Fixed–size messages: 1 Kbyte
Variable format header: 10 byte/message
Fixed format header: 5 byte/cell

Figure 6: System jitter: impact of hop count

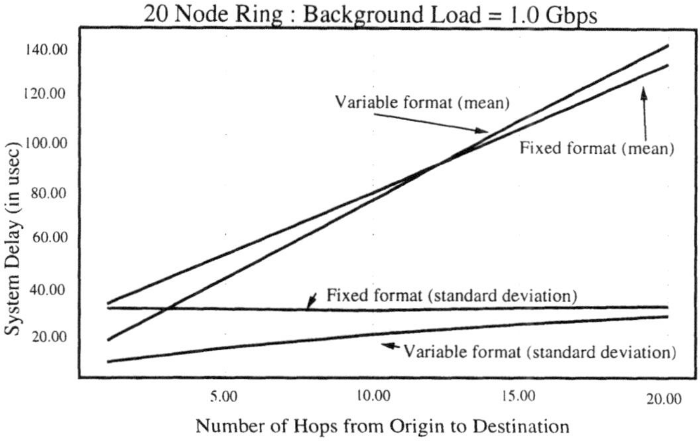

Tagged connection: fixed size messages (1 Kbyte)
Background traffic: exponential messages (mean: 1 Kbyte)
No header overhead for both formats

Figure 7: Equal overhead comparison: large messages

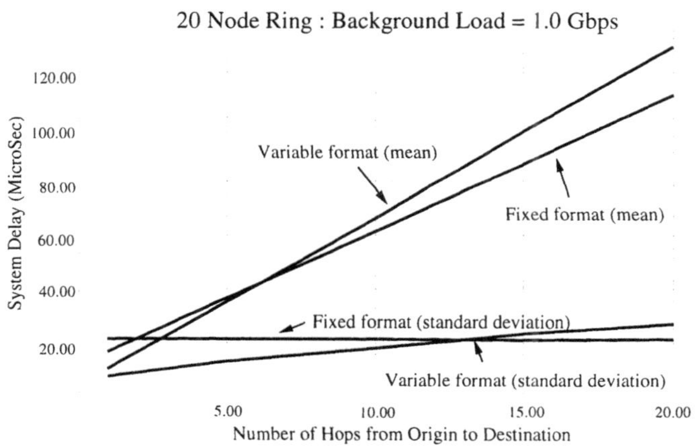

Tagged connection: fixed size messages (125 bytes)
Background traffic: exponential messages (mean: 1 Kbyte)
No header overhead for both formats

Figure 8: Equal overhead comparison: small messages

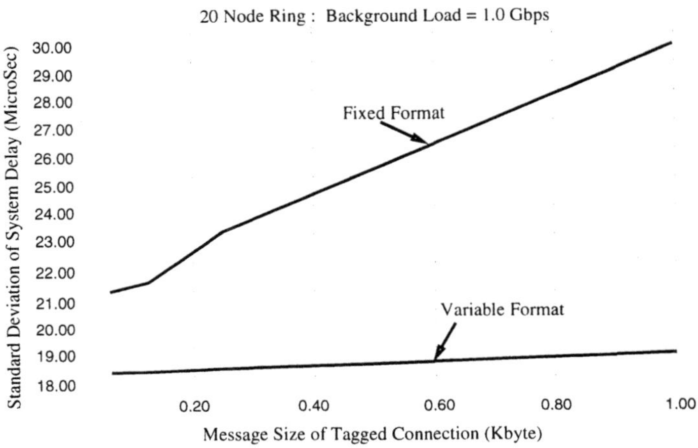

Tagged connection has 10 hops from origin to destination
Background traffic : exponential messages (mean: 1 kbyte)
No header overhead for both formats

Figure 9: System jitter: impact of message size

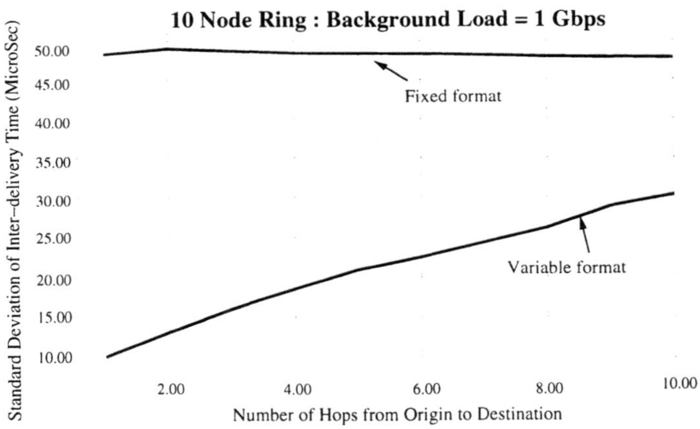

Fixed-size message: 1 Kbyte
Variable format header: 10 byte/message
Fixed format header: 5 byte/cell

Figure 10: Inter-delivery jitter: background load= 1 Gbps

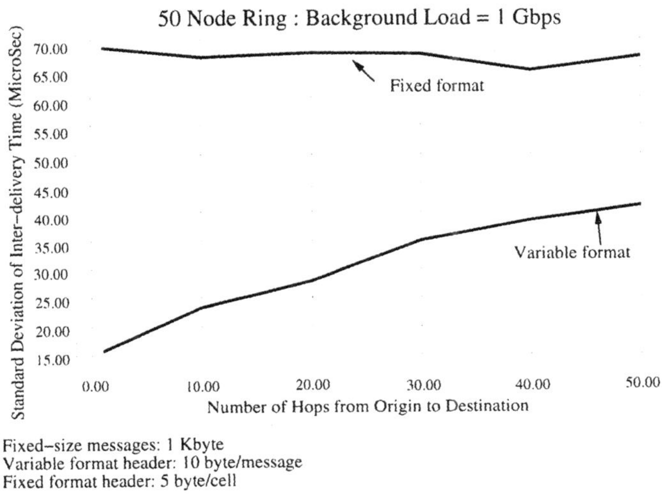

Figure 11: Inter-delivery jitter: impact of hop count

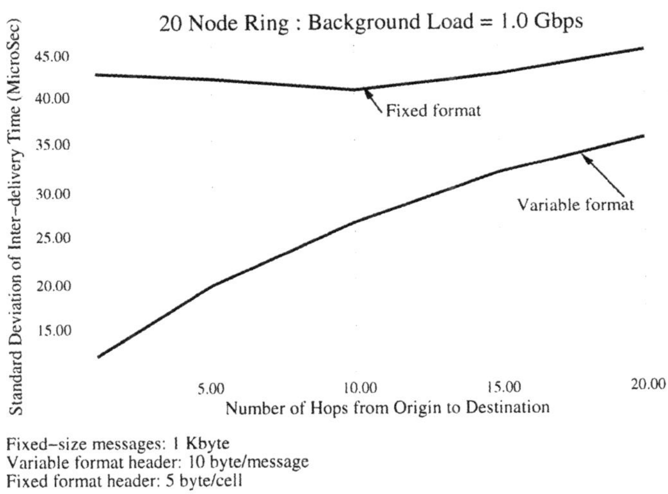

Figure 12: Inter-delivery jitter: no overhead, large message

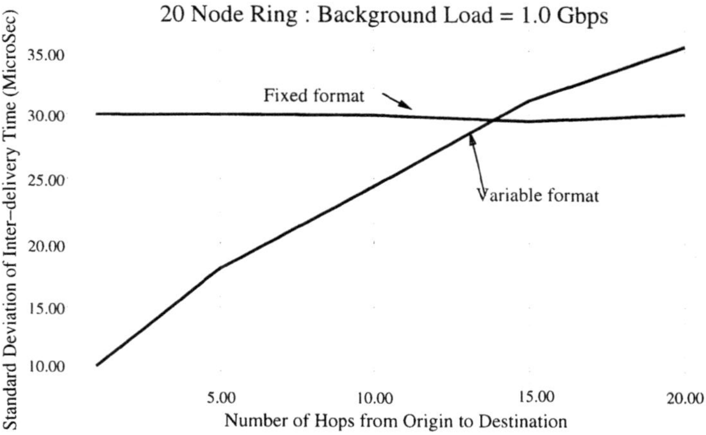

Fixed–size messages: 125 byte
Variable format header: 10 byte/messsage
Fixed format header: 5 byte/cell

Figure 13: Inter-delivery jitter: no overhead, small message

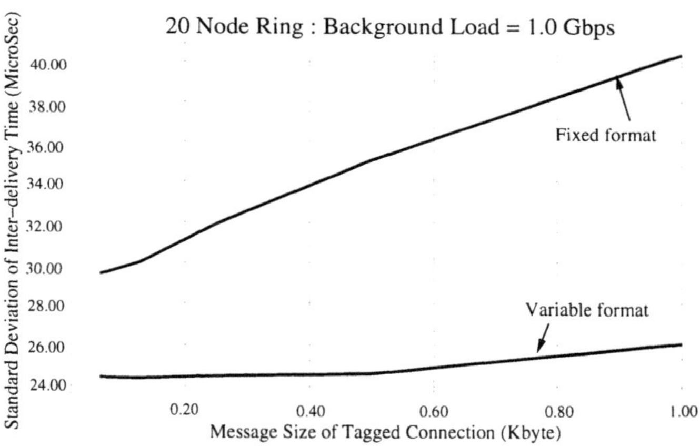

Tagged connection has 10 hops from origin to destination
Background traffic: exponential messages (mean: 1Kbyte)
No header overhead for both formats

Figure 14: Inter-delivery jitter: impact of message size

References

[1] I. Cidon, I. Gopal, P. M. Gopal, R. Guérin, J. Janniello, and M. Kaplan. The plaNET/ORBIT high speed network. Technical Report RC 18270, High Performance Computing and Communications, IBM T. J. Watson Research Center, P.O. Box 704, Yorktown Heights, NY 10598, August 1992. Available from the authors.

[2] Framework for providing additional packet mode bearer services. CCITT recommendation I.122, CCITT Subworking Party XVIII/1-2, 1988.

[3] Draft - general B-ISDN aspects. Report R 34, CCITT Study Group XVIII, June 1990.

[4] FDDI token ring - media access control. ANSI X3.139, American national Standard, November 1986.

[5] Distributed queue dual bus (DQDB) subnetwork of a metropolitan area network (MAN). Standard document P802.6/d14, IEEE Project 802.6, July 1990.

[6] R. M. Falconer and J. L. Adams. Orwell: A protocol for an integrated service local network. *Br. Telecom Technol. J.*, 3(4):21–35, April 1985.

[7] I. Cidon and Y. Ofek. Metaring – a full duplex ring with fairness and spatial reuse. In *Proc. Infocom'90*, pages 969–981, 1990.

[8] W. Bux. Local area subnetworks: A performance comparison. *IEEE Trans. Commun.*, COM-29(10):1465–1473, October 1981.

[9] W. Bux. Performance issues in local-area networks. *IBM Systems Journal*, 23(4), 1984.

[10] D. E. Huber, W. Steinlin, and P. J. Wild. SILK: An implementation of a buffer insertion ring. *IEEE J. Select. Areas Commun.*, SAC-1(5):766–774, November 1983.

[11] T. Takeda and K. Tanno. Register insertion/self-token protocol for high speed ring LANs. In *Proc. 15th Conference on Local Computer Networks*, pages 60–64, 1990.

[12] P. Heinzmann, H. R. Müller, D. A. Pitt, and H. R. van As. Buffer-insertion cell-synchronized multiple access (BCMA) on a slotted ring. In *Proc. Second Intl. Conf. Local Commun. Sys.: LAN and PABX*, Palma, Balearic Islands, Spain, June 26–28 1991.

[13] M. J. Karol and R. D. Gitlin. High-performance optical local and metropolitan area networks: Enhancements of FDDI and IEEE 802.6 DQDB. *IEEE J. Select. Areas Commun*, 8(8):1439–1448, October 1990.

[14] R. Guérin, H. Ahmadi, and M. Naghshineh. Equivalent capacity and its application to bandwidth allocation in high-speed networks. *IEEE J. Select. Areas Commun.*, SAC-9(7):968–981, September 1991.

[15] Special issue on packet speech and video. IEEE J. Select. Areas Commun., June 1989. SAC-7,(5).

[16] I. Cidon, I. Gopal, and R. Guérin. Bandwidth management and congestion control in plaNET. *IEEE Commun. Mag.*, 29(10):54–63, October 1991.

[17] A. K. J. Parekh. *A Generalized Processor Sharing Approach to Flow Control in Integrated Services Networks.* PhD thesis, Laboratory for Information and Decision Systems, Massachusetts Institute of Technology, Cambridge, MA 02139, February 1992. No. LIDS-TH-2089.

[18] M. Naghshineh and R. Guérin. Fixed versus variable packet sizes in fast packet-switched networks. In *Proc. INFOCOM'93*, San Francisco, CA, May 1993. To appear.

[19] M.-S. Chen, Z.-Y. Shae, D. D. Kandlur, T. P. Barzilai, and H. M. Vin. A multimedia desktop collaboration system. Technical Report RC 17951, IBM Research Report, May 1992.

[20] J. Zdepski, K. Josheph, and D. Raychaudhuri. Packet transport of VBR interframe DCT compressed digital video on a CSMA/CD LAN. In *Proc. of GLOBECOM'89*, volume 2, pages 886–892, Dallas, TX, November 1989.

[21] R. Morin, G. Forbes, D. Ghering, J. Salutz, and W. Pavlicek. Present status and future directions of the Mayo/IBM PACS project. In *Proc. of SPIE'91*, volume 1446, pages 436–441, Bellingham, WA, February 1991. Int. Soc. for Optical Eng.

[22] R. Ahuja, S. Gaitonde, and D. Jacobson. Design of an integrated LAN architecture. In *Proc. of ICC'91*, volume 2, pages 830–835, Denver, CO, June 1991.

[23] L. B. Sung, C. B. Bee, and L. A. Beng. Image transfer services for PC LAN. In *Proc. of ICIE'91*, volume 2, pages 859–867, Singapore, December 1991.

[24] C. H. Sauer, E. A. MacNair, and J. F. Kurose. The research queueing package version 2. Technical Report RC 138 and 139, IBM Research Report, 1986.

PROTOCOLS AND TESTING

Broadband Communications, II (C-24)
S. Tohmé and A. Casaca (Editors)
Elsevier Science B.V. (North-Holland)

Traffic Control for Best Effort Data Communications in ATM Networks: Simulation Comparison of Alternative Schemes

G. Bagnoli [2], M. Listanti [1], R. Winkler [2]

[1]: Università "La Sapienza" - INFOCOM Dept, Via Eudossiana, 18 - 00184 Roma (Italy)
Tel. +39 - 6 - 44585458 Fax. +39 - 6 - 4873300

[2]: Fondazione Ugo Bordoni - Via Baldassarre Castiglione, 59 - 00142 Roma (Italy)
Tel. +39 - 6 - 54803414 Fax. +39 - 6 - 54804404 e-mail. fubpos4@itcaspur.caspur.it

Abstract

The identification of an efficient access control mechanism is a major issue for the support of broadband data communications in ATM networks. In this paper we compare the performance of open loop and closed loop mechanisms, as applied to best effort data communications. The loss performance are analyzed both at the cell level and at the burst level, for homogeneous as well as heterogeneous traffic mixes. The results show that closed loop control schemes, able to co-ordinate the activity of the access node to that of the sources, obtain better loss and delay performance than the open loop traffic control schemes. Moreover, it is shown that the performance of closed loop controls are sensitive to the source parameters and heterogeneous sources are not always fairly handled.

Keyword codes: 1
Keywords: Asynchronous Transfer Mode Networks, Performance Evaluation

F.U.B. contribution was carried out in the framework of the agreement between the Italian PT Administration and the Fondazione Ugo Bordoni and under the partial financial support of the National Research Council (CNR) in the frame of the Telecommunication Project.

1. INTRODUCTION

The ATM Forum is introducing the Best Effort Service (specifically called *Available Bit Rate Service*) to support data communications in ATM networks. This is a MAC-like type of service, that is made available to serve the existing connectionless service providing networks (i.e. IEEE LAN's and MAN's and IP-based WAN's) and to support the future gigabit data applications.

The communications that made use of the best effort service share a common capacity, made up of the bandwidth unused by the connections handled with guarantee of resources and, eventually, of the capacity allocated to the best effort traffic by means of management procedures. The best effort communications do not reserve any buffer or bandwidth resource. Consequently, they receive a quality of service highly depending on the status of the network.

A number of open issues should be solved for a complete definition of the best effort service. Among them we have:

- identification of the mechanism(s) required to handle, at the ATM layer, the coexistence between communications handled with call admission and resource reservation and the best effort ones, so that the requirements of first ones are never jeopardized;
- the specification of the Usage Parameter Control action: is it still required and, if so, on what parameters is it performed ?
- the selection of a traffic control scheme suitable to handle the best effort traffic.

This paper deals with the last issue. We are concerned by the fact that the users of the best effort service are not able to foresee their bandwidth requirements or to describe in advance the traffic characteristics of their data and that makes unsuitable the methods based on the preventive resource allocation. The solution is to design a reactive method, able to dynamically adjust the rate of the ongoing best effort connections as a function of the offered load and the available capacity.

For this purpose the performance of two traffic control schemes, already known from the literature, are compared and discussed in this paper: the *Open Loop Shaper (OLS)* [1] and the *BackPressure Scheme (BPS)* [2]. The OLS is an open loop traffic control scheme that uses a shaping device to guarantee loss-less operation at the source. All the control functionalities are concentrated in the source equipment. The BPS is a closed loop scheme, in which a network access node stops all the attached sources upon congestion. It guarantees loss-less network operation, given the knowledge of the peak bit rate of the best effort connections.

Further on, the performance of the two mechanisms are compared to those attained by the *uncontrolled* policy, here considered as the reference method. According to it, the source equipment does not perform any control action and the network nodes drop the excess cells in the case of buffer overflow.

Some performance results are already available for both the BPS and the OLS, but they have been obtained from different scenarios and are not suitable for a thorough comparison of these mechanisms when best effort communications share the resources left available by resource guaranteed communications. Instead, we have identified a simulation scenario, suitable to stress the pros and cons of these control schemes when they are applied to best effort communications. The following assumptions are peculiar of the considered scenario:

- at the ATM connection set-up the sources describe themselves with their peak bit rate and their maximum burst length (the former depends on the physical arrangement of the user equipment, the latter depends on the adopted high level protocols);
- a traffic handled with performance and resource guarantees is superposed to the best effort traffic and introduces alternating periods of link capacity availability and unavailability as far as the service of the best effort traffic is concerned; separated buffering is applied to protect the performance requirements of the guaranteed class of service;
- in the case of the BPS, the upper layer protocols in the user equipment do not interpret the congestion signals generated by the network access node; consequently cells can be lost during sustained overload periods;
- in the BPS, the cells belonging to corrupted AAL-PDU's are dropped before entering in the network; cell dropping has not been introduced in the OLS because of the resulting increased complexity and of the constraints imposed on the internal switch architecture.

Both delay and loss performance have been measured by means of simulation. The loss performance have been evaluated both at the ATM and at the AAL levels, but the latter is here considered as the most meaningful QoS parameter. In fact the AAL Protocol Data Unit (AAL-PDU) is the basic information unit in data applications in which the user information is arranged in frames and the cell loss probability does not provide enough information on the actual network behaviour and on the performance appreciated by the customer.

In the section 2 the main features of the OLS and of the BPS are summarized. The system model and the results of the performance comparison are presented in the section 3. The conclusions are given in the section 4.

2. OUTLINE OF THE CONTROL SCHEMES

The performance of the control schemes introduced in the section 1 have been evaluated at the access of an ATM network by means of simulation, according to the model of figure 1.

The MUX multiplexes a number of ATM workstations or external connectionless networks onto the same ATM switch port, to attain a reasonable access cost. It is supposed to be the bottleneck of our system. The user equipments are equipped with a rate shaper, operating at the ATM level, that consists of a buffer and a rate adapter. The former stores the cells generated by the source and waiting to enter in the MUX, the latter is driven by the reception of congestion notification signals emitted by the MUX (in the BPS) or by the local buffer (in the OLS).

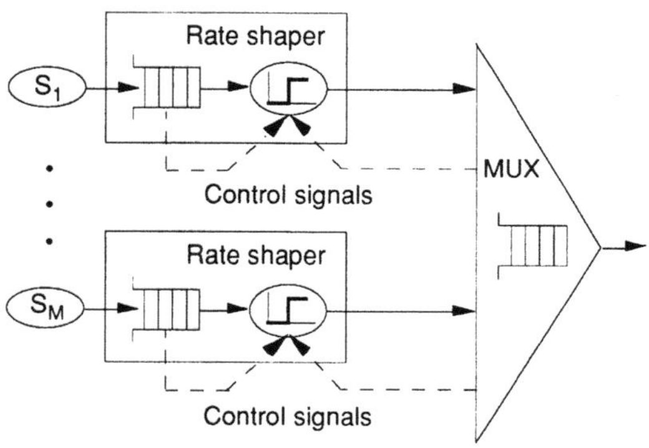

Figure 1. Configuration of the network access interface

In the *uncontrolled scheme*, the MUX simply drops the incoming cells when its buffer overflows. No shaper is provided in the user equipment and no congestion notification is generated by the MUX.

No congestion notification is generated by the MUX in the case of the *OLS approach*, too. The shaper output rate F_s is a function of the level of occupation of the shaper buffer, with the constraint that no cell is lost in the shaper. In [1], F_s is equal to the source peak bit rate F_p when the shaper buffer is full, it is equal to *0* when the buffer is empty and is set to an intermediate value F_i, close to the source mean bit rate F_m, otherwise. In this paper, F_i is equal to the fair share of the mean capacity available to the best effort traffic among the active best effort connections, so as to take into account that F_m is not known.

In the BPS, two thresholds are set in the MUX buffer occupation level to regulate the shaper operation. When T_h cells are in the MUX buffer, the MUX emits a *congestion onset signal* to stop all the shapers. T_h is the worst case buffer drift during the longest round trip propagation delay and is:

$$T_h = B_{mux} - \sum_{i=1}^{m} M_i * \left\lceil \frac{2 * \tau_p * F_p^i}{L_c} \right\rceil$$

wherein B_{mux} is the MUX buffer dimension (cells), τ_p is the propagation delay (seconds) between the MUX and the farthest source, m is the number of different source peak rates, F_p^i is the i-th peak rate (bit/s), M_i is the number of sources with peak rate F_p^i, L_c is the length (bits) of an ATM cell.

The shapers are again enabled to transmit when they receive the *congestion expired signal.* This signal is emitted by the MUX as soon as the number of cells in its buffer goes below the low water mark T_l, given that the MUX buffer depletion is begun (see the figure 2). To avoid the starvation of the MUX buffer, it results:

$$T_l \geq \left\lfloor \frac{2 * \tau_p * C}{L_c} \right\rfloor$$

where C is the gross capacity available for the NR traffic, expressed in bit/s.

This threshold setting guarantees fair operation also if the sources are not equidistant to the MUX. In fact, that entails only a displacement of the emission and blocking periods of the shapers, but all the submitted data are always accommodated in the MUX buffer. Conversely, if the threshold design does not exclude cell losses in the MUX [e.g. 3], the quality of service perceived by a source depends on its distance from the MUX since the sources closest to the MUX have better chance of finding available buffer places upon congestion.

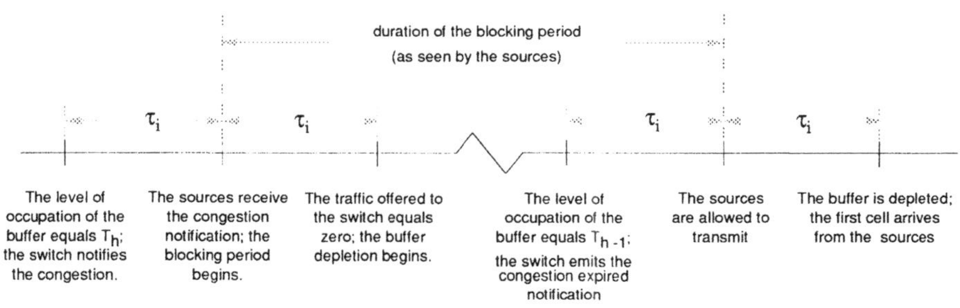

Figure 2. Time evolution of the backpressure scheme

While it could be argued that some level of hystheresys could help in lowering the MUX buffer congestion rate, thus improving the loss performance, a number of simulation experiments, performed with different values of T_l and T_h, have led to: *i)* longer block periods forced, on the average, by the MUX and *ii)* a lower level of utilization of the MUX buffer resource, since the mean MUX buffer occupation level is reduced. Consequently the performance figures of the BPS presented in this paper have been obtained for $T_l = T_h - 1$, that is, a blocking period ends as soon as the level of occupation of the buffer begins to decrease.

T_h can be evaluated "once in a life" for each access node and changes only if an user equipment moves away or attaches to it. This is a very conservative approach, that takes into account the maximum possible load generated by all the attached sources. More efficiently, the congestion threshold may be dynamically adapted to the number and the characteristics of the established connections. In this case we can take into account the present value of the threshold and the dimension of the MUX buffer available for the best effort traffic to introduce also a very simple connection admission control criterion for best effort communications. This control does not aim at checking the availability of resources to be assigned to a data source; its targets are to avoid the over commitment of the buffer resources and the inefficient MUX buffer utilization. Assuming that x is the present level of occupation of the MUX buffer, T is the current value of the congestion onset threshold and T' is the new value of the congestion onset threshold, this connection admission control states that:

1) $x < T'$: the new call may be set-up
2) $x > T'$: the new call is set-up as soon as the condition 1) is met;
3) $T' < \left\lfloor \frac{2 * \tau_p * C_p}{L_c} \right\rfloor$: the new call is rejected (C_p is the MUX peak bit rate).

The condition 2) ensures that the congestion onset threshold never becomes lower than the present level of occupation of the MUX buffer, since that could cause losses in the MUX. The condition 3) guarantees that the cells emitted by an enabled source arrive at the MUX before the MUX buffer becomes empty.

Finally, as already introduced in the section 1, the sources do not interpret the congestion signals emitted by the MUX and may feed the shapers also if these have been stopped. This means that the BPS is supposed unable to prevent losses in the user equipment. To limit the number of AAL-PDU's affected by cell losses, we have enhanced the BPS with an AAL-PDU handling discipline, called *corrupted AAL-PDU dropping*. For this purpose, the AAL type 5 protocol is utilized to guarantee that no interleaved AAL-PDU's appear at each MUX input. Moreover, as soon as an AAL-PDU is corrupted, the AAL-PDU end indication in the ATM cell header is used to reject all the incoming cells of this AAL-PDU that follow the first overflowing cell and to drop all the cells of the same AAL-PDU that are already stored in the shaper buffer. To effectively implement this facility the shaper buffer shall be long enough to contain an entire AAL-PDU. The same facility has not been considered for the OLS for the resulting MUX complexity and of the constraints imposed on the internal MUX architecture.

3. SYSTEM MODEL AND PERFORMANCE EVALUATION

A simulation study has been carried out to evaluate the following performance parameters:

- the *AAL-PDU loss probability*, defined as the ratio of the number of corrupted AAL-PDU's to the number of AAL-PDU's submitted to the network (an AAL-PDU is corrupted if at least one cell obtained from its segmentation is lost because of buffer overflow);
- the *cell loss probability*, defined as the ratio of the number of ATM cells lost for buffer overflow to the number of ATM cells submitted to the network;
- the *normalized AAL-PDU mean queuing delay*, defined as the time elapsing from the instant the first cell of an AAL-PDU enters in the shaper buffer to the instant the last cell of the same AAL-PDU is emitted to the MUX, normalized to the mean AAL-PDU transmission time at the source peak bit rate.

The data sources generate data according to an on-off process, described by the peak bit rate F_p (bit/s), the mean bit rate F_m (bit/s) and the mean burst length L_b (cells). The burst length follows a truncated negative exponential distribution with maximum $L_{b,max}$ (cells). The set-up and release phases of a call are not simulated in our experiments; the best effort connections are active for the whole duration of a simulation.

The parameters of the considered sources are summarized in the table I below.

	F_p (Mbit/s)	F_m (Mbit/s)	L_b (cells)	$L_{b,max}$ (cells)
S1	20	4	100	200
S2	20	4	1000	2000
S3	20	4	50	100

Table 1 Source parameters for the performance evaluation

The source shaper is modelled as a single server queuing system, with buffer size B_s (cells) and variable transmission rate F_s (bit/s). B_s is always assumed to be equal to $L_{b,max}$. The propagation delay between a source and the shaper is equal to zero.

The MUX service is modelled as an on-off process, the "on" periods representing the intervals in which no cell generated by the resource guaranteed connections is waiting for transmission. The burstiness of this on-off process is equal to the ratio between the link capacity C_p and the mean capacity available for the best effort traffic C_m; these are supposed equal to *150 Mbit/s* and *65 Mbit/s*, respectively. Two values of the mean duration of the MUX unavailability $T_{off,\ mux}$ have been considered: *10* and *100* cell times. A buffer in the MUX is dedicated to (and shared by) the best effort connections. Two values of the dimension B_{mux} of this buffer have been considered: *25* and *200* cells.

The sources are equidistant from the MUX. All the simulations have been run assuming a source to MUX distance of *2 km* ($\tau_p = 10\mu s$). This is a conservative position when a customer premises network is considered and a reasonable configuration for the User to Network interface of a wide area network.

3.1. Performance evaluation for homogeneous load conditions

A first set of simulations has been carried in the case of homogeneous sources.

The AAL-PDU and the cell loss probability are plotted in the figure 3, assuming that the corrupted AAL-PDU drop facility is not applied and $T_{off,\ mux} = 100$ cells.

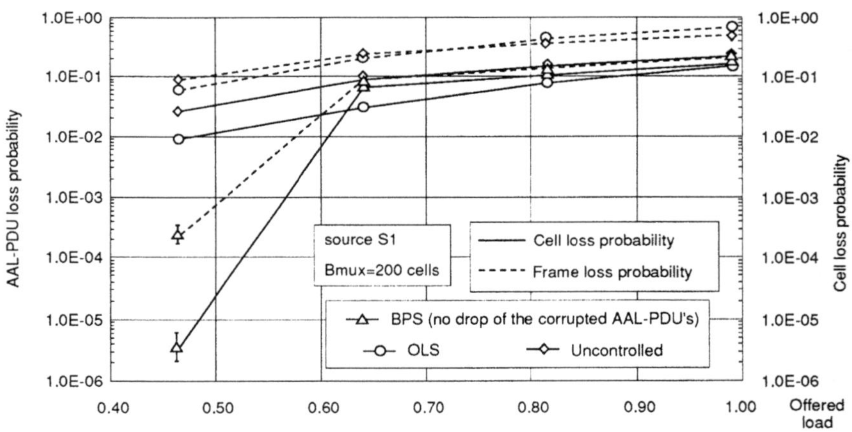

Figure 3. AAL-PDU and cell loss probability vs. the offered load

The figure 3 shows that:

- the BPS shows the lowest AAL-PDU loss probability, whatever the offered load is;

- there is almost no difference between the AAL-PDU loss probability of the uncontrolled approach and that of the OLS; from that derives the ineffectiveness of mechanisms, like the OLS, in which the MUX does not control the sources when best effort and guaranteed service traffics coexist and the latter is given low priority;
- as for the cell loss probability, the BPS behaves better under medium offered loads, while the OLS behaves better under increasing loads; this is because the BPS blocks the sources as soon as T_h is exceeded, but that does not necessarily means that the MUX is congested and causes a probability of overflow of the shaper buffer higher than that of the OLS, in which cells are lost only during actual congestion periods.

The performance attained when the corrupted AAL-PDU's drop facility is introduced are shown in the figures 4 and 5 for the source S1.

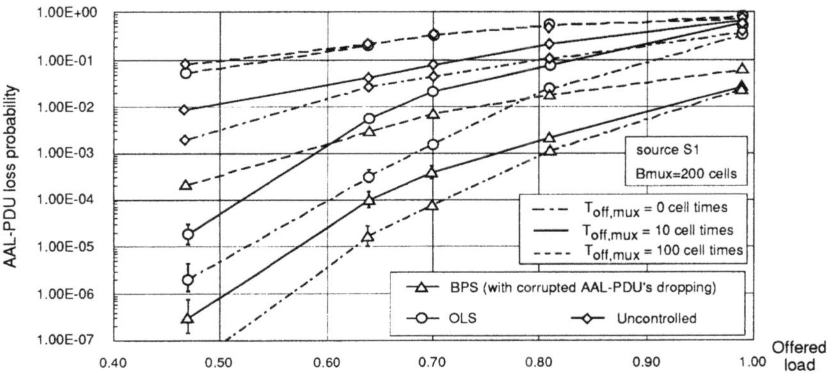

Figure 4. AAL-PDU loss probability applying the corrupted AAL-PDU drop facility

These curves stress the usefulness of the corrupted AAL-PDU's drop facility in decreasing the AAL-PDU loss probability (compare the dashed curves in the figures 3 and 4). Moreover, these plots highlight the impact of the value of $T_{off,mux}$ on the performance: the AAL-PDU loss probability values measured for $T_{off,mux}$=10 cell times are higher than those obtained when a part of the MUX capacity is dedicated to the best effort traffic ($T_{off,mux}$=0) and the increment is even dramatic when $T_{off,mux}$=100. This worsening is particularly evident when the OLS is applied and the offered load is higher than *0.7*, since the loss performance become even worse that those relevant to the uncontrolled scheme. Conversely, the BPS outperforms the uncontrolled scheme in any load condition and also when the corrupted AAL-PDU drop facility is not applied.

The AAL-PDU loss probability increases dramatically if the MUX buffer is very short (see the figure 5).

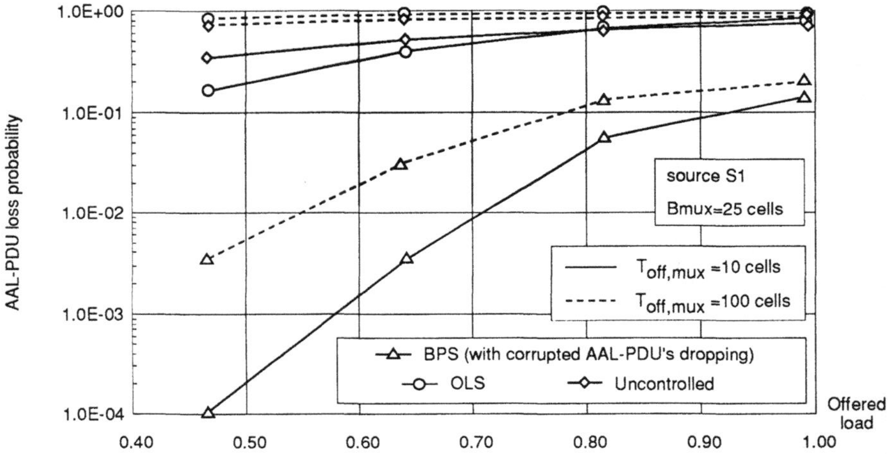

Figure 5. AAL-PDU loss probability vs. the offered load for long burst sources

The impact of the burst length on the system performance is shown in the figure 6, where a comparison between the loss performance of sources S1 and S2 is performed.

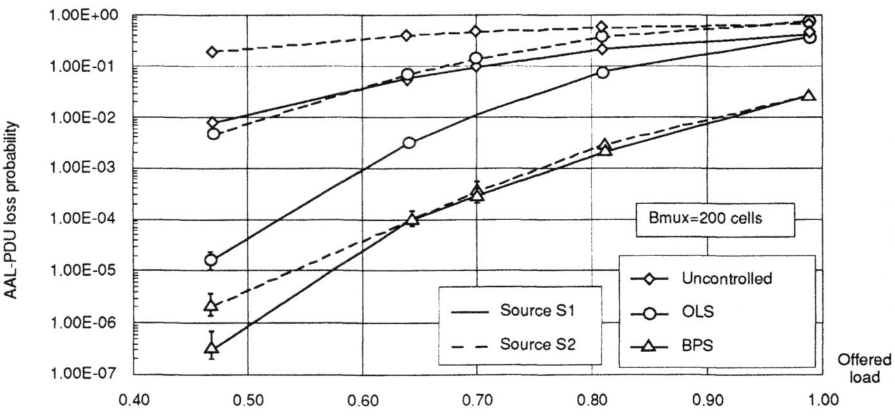

Figure 6. Impact of the burst length on the AAL-PDU loss probability

The plots in the figure 6 can be explained by noting that long bursts are quite likely to superpose at the MUX input, thus increasing the risk of buffer overflow. That causes the increase of the number of corrupted AAL-PDU's in the case of the OLS and an higher number of blocking periods in the case of the BPS.

The normalized AAL-PDU mean queuing delay is plotted in the figure 7 for the source S1 and B_{mux} = 200 cells, assuming that the corrupted AAL-PDU drop facility is applied.

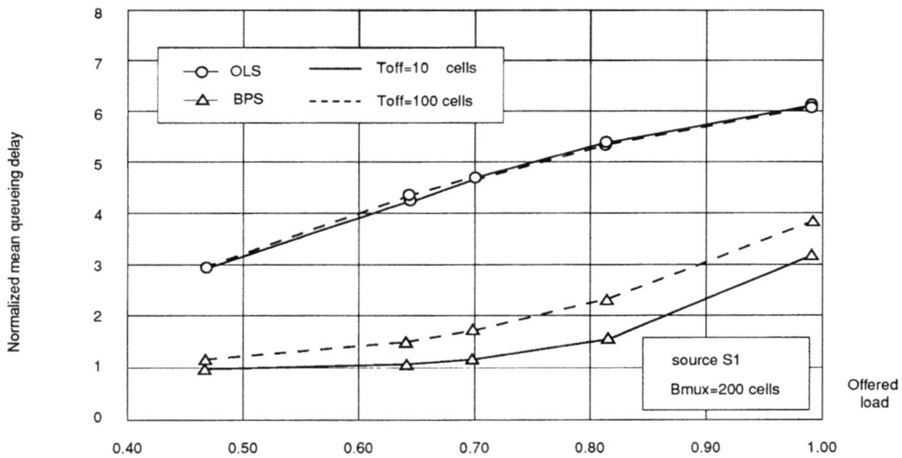

Figure 7. Normalized mean queuing delay for the corrupted AAL-PDU drop facility (B_{mux}=200)

Once again this figure highlights the superiority of the BPS, whose main cause for delay is the occurrence of blocking periods imposed by the MUX. These periods are longer and more frequent when short buffers are introduced at the MUX and when a long MUX unavailability is caused by the resource guaranteed traffic. The OLS delay performance do not depend on the MUX availability (the shaper is not slaved to the MUX) but it suffers for generally higher delays because the rate smoothing reduces the offered traffic also under medium load conditions.

3.2. Performance evaluation for heterogeneous load conditions

The performance of the BPS and the OLS have been evaluated also under heterogeneous load conditions, to analyze the sensitivity of these traffic control methods to the system and the source parameters.

Taking into account the prominent role played by the burst length in influencing the loss performance (refer to the figures 4 and 6), we have started by measuring the loss performance in the case the contending sources differ only for the value of the mean burst length L_b. For this purpose, a number of simulations have been carried out in which S1 sources are contending the MUX with as many sources with different mean and maximum burst length. The mean burst lengths considered for this second group of sources are 10, 25, 50, 75, 100, 300 and 500 cells. As usual $L_{b,max}$ is twice the value of L_b and the shaper buffer dimension is equal to $L_{b,max}$.

The results are shown in the figure 8 where $T_{off,mux}$=10 cells and the offered load is equal to 0.8, that correspond to fourteen sources as a whole (similar plots, down-shifted of about 2 decades, have been obtained for a total offered load equal to 0.6, i.e. 10 sources as a whole).

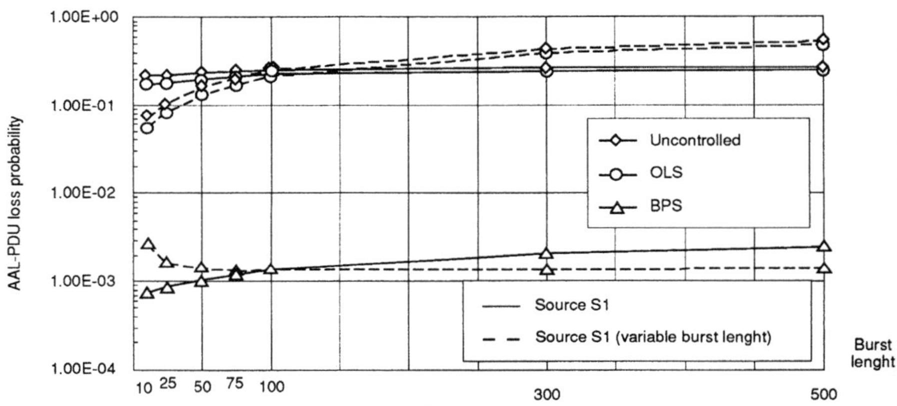

Figure 8. AAL-PDU loss probability for heterogeneous load conditions vs. the burst length

This figure shows that:

- the AAL-PDU loss probability relevant to the OLS and the uncontrolled method are higher than those of the BPS; that confirms the results previously discussed for the homogeneous load conditions;
- the AAL-PDU loss probability is higher for longer burst sources than that of short burst sources when the OLS and the uncontrolled methods are considered, the opposite holds when the BPS is applied;
- in the OLS and the uncontrolled method, the loss probability of the S1 sources is almost independent from the mean burst length of the other group of sources.

This can be explained by considering that, in the case of the OLS and the uncontrolled method (where all the cells are lost at the MUX), short bursts have better chance than longer ones to find some room available in the MUX buffer and segmentation can help the sources in reducing their AAL-PDU loss probability.

Viceversa, as far as the BPS is concerned, a source generating a number of bursts during a single blocking period (resulting from the superposition of the MUX unavailability and of its congestion status) experiences a higher loss probability than a source featuring a longer burst duration, provided that these sources have the same burstiness. This unfairness could be recovered by dimensioning the shaper buffer to a number *K* of maximum burst lengths, being *K* equal to the ratio between the mean blocking period imposed by the MUX and the source mean duty cycle (i.e. the sum of the mean off and the mean on periods). Unfortunately, the value of K is not available *a priori.*

Another set of simulations has been performed considering the sources S1 and S3. In these experiments, given a target load of 0.8, we have evaluated the loss probability of the two groups of sources varying the traffic mix (i.e. the relative number of S1 and S3 sources).

The results (see the figure 9) confirm all the considerations already performed.

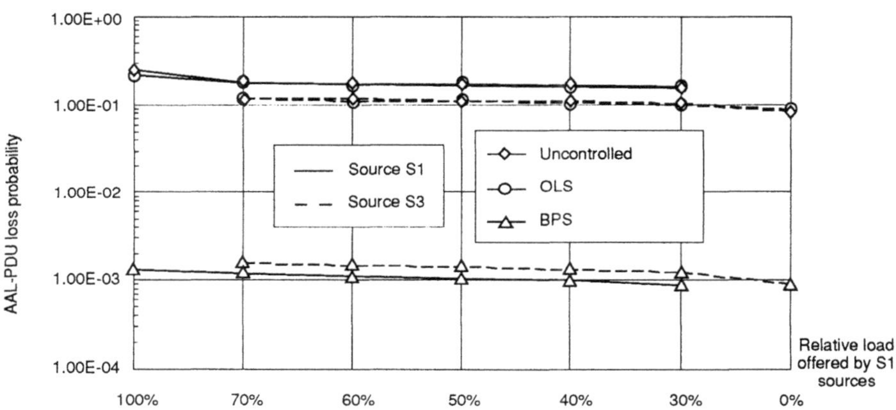

Figure 9. AAL-PDU loss probability for heterogeneous load conditions vs. the traffic mix

First, the BPS outperforms the other mechanisms. Second, short burst sources are handled better by the uncontrolled method and the OLS than the BPS. Last, in the BPS, the AAL-PDU loss probability measured when no S3 source is active is higher than that measured when the traffic contribution of S1 sources is null.

4. CONCLUSIONS

We have presented the results of a performance evaluation study of two traffic control mechanisms, applied to control best effort communications at the access of an ATM network. This service is able to use efficiently the leftover bandwidth after communications handled with resource guarantees have been served. Moreover, it solves in many cases the problem of identifying a set of traffic descriptors, thus simplifying the connection admission control and the usage parameter control procedures.

In this study we have focused our attention on the AAL-PDU loss performance, whose importance derives from the relaxed delay requirements of this traffic class and the observation that the AAL-PDU is the basic information unit in a scenario in which user data are arranged in frames.

The results show the superiority of backpressure schemes, that coordinate the MUX and the source operation to obtain lossless MUX operation. Conversely it has been shown that open loop schemes (e.g. the uncontrolled and the shaper based control schemes) do not provide any performance advantage in real operating scenarios, in which the best effort communications receive only the residual bandwidth not used by the QoS guaranteed traffic. Further benefits are obtained if the corrupted AAL-PDU's are directly dropped by the source equipment, so as to avoid the injection of useless traffic in the network.

The role played by the source burst length in determining the loss performance has been stressed, under homogeneous as well as in heterogeneous traffic conditions. In particular, it has been shown that short burst sources are slightly unfairly handled by the backpressure scheme, if limited buffering capabilities are provided in the source equipment. This is an interesting topic in a domain in which heterogeneous traffic sources coexist and will be furtherly considered in future papers.

REFERENCES

[1] F.Bernabei, L.Gratta, M.Listanti, M.Testa: "Analysis of two level shaping for multiplxing of on-off ATM sources"; *Proceedings of ICC'93*, Geneva, pp. 1380-1385.

[2] J.Cherbonnier, J.Y.Le Boudec: "A GFC protocol for congestion avoidance in the ATM connectionless service"; *Proceedings of EFOC/LAN '92*, Paris, pp. 305-309.

[3] Y.T.Wang, B.Sengupta: "Performance analysis of a feedback congestion control policy under non-negligible propagation delay"; *Proceedings of SIGCOMM '91*, Zurich, pp. 149-157.

Broadband Communications, II (C-24)
S. Tohmé and A. Casaca (Editors)
Elsevier Science B.V. (North-Holland)

A New Selective Repeat ARQ Scheme for Multicast Communication

M. Aghadavoodi Jolfaei, U. Quernheim

Department of Computer Science (Info IV), Aachen University of Technology
Ahornstraße 55, D-52074 Aachen, Germany
Tel.: +49 241 80 21413, Fax: +49 241 8888220
E-mail: masoud@informatik.rwth-aachen.de

Abstract

Many of the communication protocols require an error free transmission of data which cannot be guaranteed by using only a forward error correction technique. Therefore, automatic repeat request schemes must be applied. This paper presents a new selective repeat ARQ scheme, called XOR selective repeat strategy, suitable for multicast communication over broadcast links. The idea of the XOR selective repeat strategy is to physically combine several blocks negatively acknowledged by different receivers by XORing (i.e. modulo 2 addition) them to minimize the number of retransmissions and to increase throughput. The analytical and simulative results show that in terms of throughput XOR strategy yields remarkably better results than the pure selective repeat.

Keyword Codes: C.2.1; C.2.2
Keywords: Network Architectures and Design, Packet Switching Networks, Wireless Communication, Performance Evaluation; Network Protocol, Protocol Architecture

1. INTRODUCTION

A broadcast link is characterized by the property that a single transmission from a source station may be received simultaneously by several destination stations. Examples of broadcast links include various forms of local area and high speed networks, satellite channels, and packet radio networks. As data communication over a link suffers either from bit errors or loss of packets due to congestion in the gateways and bridges (or even buffer overflows at the receiver stations), protocols for achieving efficient, reliable communication between a transmitter and the corresponding receiver(s) are very important. Many of communication protocols such as file transfer require an error free transmission of data which cannot be guaranteed by using only a forward error correction technique.

One widely used technique for handling transmission errors at the data link layer in data communication systems is error detection with automatic-repeat-request (ARQ). This scheme is simple and provides high reliability. The protocol is usually classified in the following three basic categories: 1) Stop-and-Wait , 2) Go-back-N and 3) Selective repeat. Much work has been done to analyze these basic protocols [BeF 64], [BrM 86], [Bus 72], [Sast 75a]. Variants to the three basic protocols have also been thoroughly studied [Mor 78], [Sast 75b], [SOSS 89], [ToM 87], [YuL 81], [YuL 82], [Wel 82]. One principal idea of those variants is to send multiple copies of a block instead of just sending a single copy.

All of the work mentioned above has concentrated on single destination type communication. For multi-destination type communication, Calo and Easton in [CaE 86] have suggested a broadcast protocol for large file transfers in a satellite network. Gopal and Jaffe in [GoJ 84] have presented three different go-back-N ARQ schemes suitable for point-to-multipoint communication. Wang and Silvester [WaS 88] also have studied a series of optimal adaptive go-back-N ARQ protocols suitable for point-to-multipoint communication. The protocols are based on those discussed by Gopal and Jaffe, but instead of sending a single copy of the data block, the sender transmits multiple copies to the receivers in order to maximize throughput.

In section 2 we discuss the system scenario and the assumptions made. A description of the protocols is found in section 3. Some XOR strategies are presented in section 4. The pure selective repeat (pure SR) and XOR selective repeat (XOR SR) protocols for point-to-multipoint communications are analyzed in section 5. Finally, in section 6 we present the analytical and simulative results and conclude the paper.

2. SYSTEM SCENARIO

As a scenario we consider a heterogenous network consisting of high speed and satellite links (see figure 1). While on the terrestrial links cell losses may occur, the satellite link is dominated by bit errors. We assume that in case of an ATM network a block consists of k cells. A block is said to be erroneous, if it has at least one cell missing or contains at least one erroneous bit.

The scenario we consider consists of one sender and K receivers, where the communication between the sender and the receivers takes place over a multicast/broadcast channel. The sender transmits data blocks which can be received simultaneously by all K receivers. Each data block includes a cyclic redundancy check (CRC) to enable the receivers to detect errors. A receiver that receives the data block first checks the CRC code and sends an acknowledgment block

(ACK) back to the sender (on a feedback channel) if the block is successfully decoded or a negative acknowledgment block (NACK) if decoding of the data block fails. The timeout mechanism is responsible for the signaling the loss of packets in the network. Each timeout packet will be considered as a negative acknowledged packet. The feedback channel is assumed to be error free. We also assume a saturated sender. We state that the nodal processing time is negligible and the time-out period is set to exactly the round trip propagation delay (RTD). During one round trip delay, S data blocks can be sent. After a round trip delay, an ACK or NACK is always received. Each data block has a sequence number which uniquely identifies the data blocks and its position in the transmission sequence.

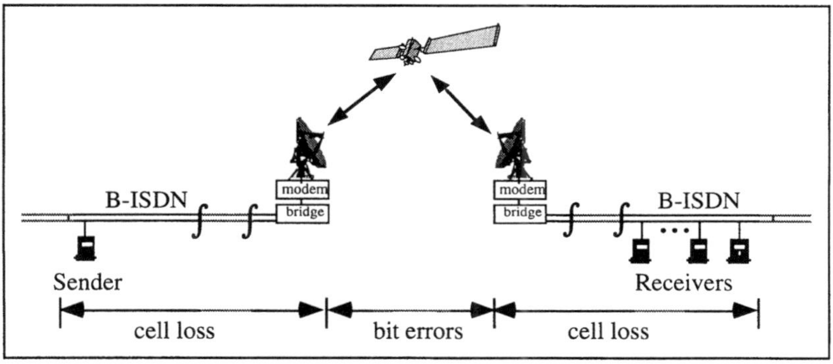

Figure 1: Connecting B-ISDN networks via satellite

For error processes on the link, we make the simple assumptions that bit errors are independent (burst errors are not considered) [AgKM 92] and that two kinds of errors can occur: 1) errors on the uplink channel, with probability p_u, which cause a data block to be received in error by all the receivers, 2) errors on the down link channel, with probability p_i, which cause an erroneous data block being received by one particular receiver i ($1 \leq i \leq K$). We further assume that the down link error processes are independent for all receivers. In this paper we consider the special point-to-multipoint communication case where the same blocks should be transmitted to K receivers (multicast), but it can also be adapted for the general case.

3. PROTOCOLS

With pure SR, the sender transmits data blocks to the receivers continuously and keeps a copy of the block in its retransmission buffer until a positive acknowledgment (ACK) for the block is received.

The receivers (see figure 2) check the validity of each received block, and each sends either an ACK or a NACK to the transmitter. In figure 2 an ACK represents the positive acknowledgment of all receivers while a NACK means that at least one of the receivers transmits a negative acknowledgment. After one round trip delay, the sender can check whether all the receivers have successfully received the data block. If not, the sender retransmits the data block.

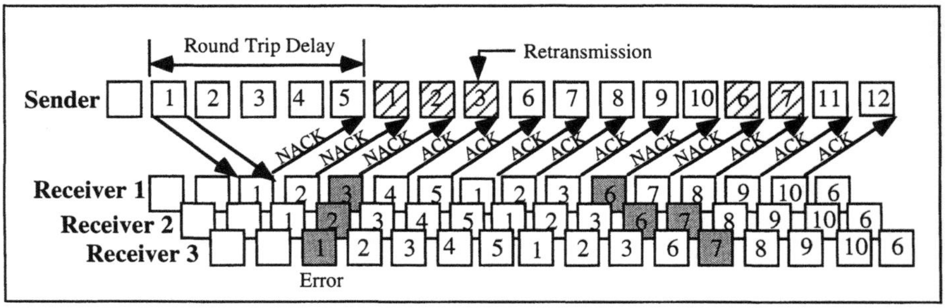

Figure 2: Point-to-Multipoint selective repeat strategy

With XOR-SR, the sender does not retransmit NACKed blocks immediately upon reception of a NACK, but it gathers a number of ACK/NACK acknowledgments and then decides on retransmitting not acknowledged blocks. The number of acknowledgments gathered is determined by the size of the so called XWindow (for XOR window). The idea of the XOR-SR strategy is to physically combine several NACKed blocks by XORing (i.e. modulo 2 adding) them in order to minimize the number of retransmitted blocks and hence to increase throughput. The resulting block is called an XOR block. Of course, a receiver expecting a particular block must be able to reconstruct it from the XOR block, i.e. it must know all other blocks participating the XOR block. Applying the XOR operation again to the XOR block with all the participating and correctly received blocks the receiver obtains the expected block. Thus, the basic difficulty of the strategy is to combine as many NACKed blocks in an XOR block as possible assuring that each intended receiver will know all but one of the participating blocks. The algorithm works as follows:

Sender:

- Search in XWindow for negatively acknowledged blocks and remove the positively acknowledged blocks.
- Decide which of these NACKed blocks can be XORed (combining by modulo-2-addition) so that each intended receiver of the XOR block will be able to reconstruct the block it is waiting for.

- Transmit the XOR blocks.
- Keep a copy of the original data blocks participating the XOR block in the retransmission buffer.

Receiver:

- keep a copy of correctly received data blocks in the local buffer for a certain time.

Figure 3 presents an example of XOR-SR. The size of the XWindow is 3 data blocks. The sender sums the NACKed blocks 1,2, and 3 by elementary modulo 2 addition (XOR) to an XOR block and transmits this XOR block instead of the single blocks 1, 2, and 3. After errorfree reception of the XOR block, each receiver checks the header of the XOR block in order to determine the sequence numbers of the data blocks participating in the XOR block. With this information it can reconstruct the expected block from the XOR block as described. E.g., for receiver 1 this is block 3, which can be extracted by modulo 2 addition of the XOR block with blocks 1 and 2.

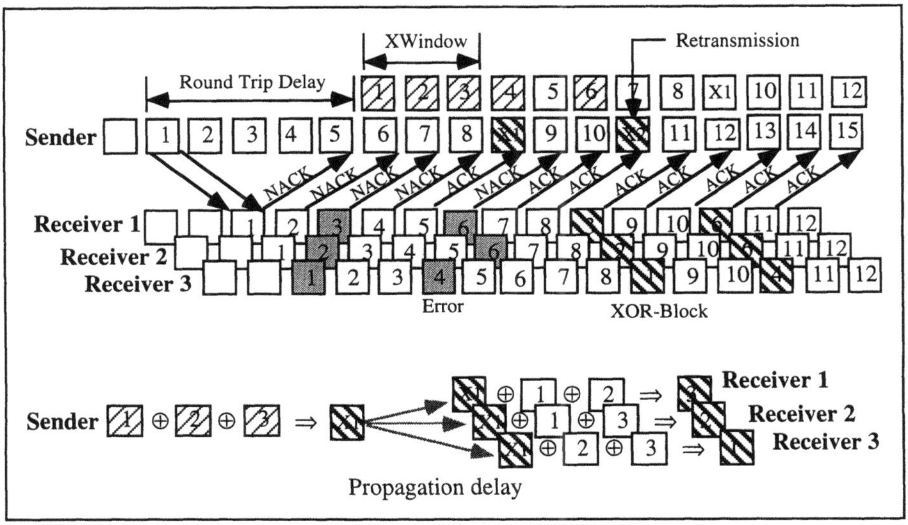

Figure 3: Point-to-Multipoint-XOR selective repeat strategy

4. XOR METHODS

For the analysis of our strategy, we have to describe mathematically which stations need the retransmission of what blocks. For this, we use a so called *error matrix*, generated after completion of an XWindow with receiver numbers as rows and block numbers as columns. If the matrix element at position (i,j) is set to one, then station i has received block j in error, otherwise the element would be set to zero. We will interpret this matrix as the incidence matrix of a hypergraph. By that we link the problem of generating XOR blocks from a given error matrix to the edge coloring problem of graph theory. In this context, we give approaches for solving edge coloring problems for hypergraphs in order to deduce heuristics for generating XOR blocks. We start with some definitions.

Let $V = \{v_1,...,v_K\}$ and $E = \{e_1,...,e_m\}$ be finite sets with $e_i \subseteq V$, $i = 1,...,m$, then the tuple (V, E) is called a hypergraph H [Ber 89]. The elements of V are termed the *vertices* and the elements of E the *edges* of H. H can be represented by an incidence matrix $J = (a_{ij})_{i=1,...,K,j=1,...,m}$ with each element of J from the set $\{0,1\}$, where the columns of J represent the edges and the rows of J the vertices of H, respectively. $a_{ij}=1$ and $a_{ij}=0$ mean either $v_i \in e_j$ or $v_i \notin e_j$, respectively. Two edges e and e' of H are called adjacent if they have a common vertex. Two vertices $v \neq v'$ are called adjacent if they are both elements of a single edge $e \in E$. The degree $\Delta(v)$ of a vertex v is defined as

$$\Delta(v) := |\{v';\ v' \text{ adjacent to } v\}|.$$

The vertex degree of H is defined as $\Delta(H) := \max_{v \in V} \Delta(v)$. Further we set

$$\Delta(e) := \sum_{v \in e} \Delta(v)$$

the edge degree of $e \in E$. The chromatic index $\chi(H)$ of H is the smallest number of colors necessary for coloring the edges of H in such a way that two adjacent edges have different colors. Associating colors with XOR blocks, $\chi(H)$ is the optimum (i.e. smallest) number of XOR blocks for the given error matrix J. The problem of determining the chromatic index of a hypergraph is NP-complete [Ho 81]. For simple graphs, a theorem of Vizing states that $\chi(G) \in \{\Delta(H), \Delta(H) + 1\}$. It is possible to determine an edge coloring with $\Delta(H) + 1$ colors in $O(|V|^4)$ steps. In the general hypergraph case such an efficient algorithm seems to be impossible. In our case the problem of minimizing the number of necessary transmissions using the XOR strategy is equivalent to finding an optimal edge coloring of the hypergraph H indexed by the error matrix. This problem is, of course, NP-hard. Implementing the XOR strategy we are facing the problem of efficiently determining an edge coloring close to the optimum. We have implemented several heuristics for solving this problem. In the following we give an algorithm for assigning colors to the edges of the hypergraph. Colors are identified by integers.

Input: Hypergraph H=(V,E) with $V = \{v_1,...,v_K\}$ and $E = \{e_1,...,e_m\}$.

Output: Edge coloring of H.

Algorithm:

```
begin
        assign color 1 to e1;
        col:=1;
        for i:=2 to m do
        begin
                assign to ei the smallest possible color c with 1 ≤ c ≤ col + 1;
                if c = col+1 then col := col + 1;
        end;
end;
```

Before the above algorithm can be applied, the input sequence of the edges must be determined, i.e. the indexing of the elements of E. This indexing is decisive for the effectiveness of the algorithm, and this is the point which makes the edge coloring problem NP-complete. We have considered different methods to index the edges of H.

- **Random**

Determine the indexing of E randomly.

- **Degree**

Determine an indexing of E fulfilling the condition

$$|e_1| \geq |e_2| \geq ... \geq |e_m| .$$

More heuristics will be considered in [AgMM 93] and their complexity in terms of computing time consumption and effectivity in terms of mean number of blocks participating an XOR block will be investigated simulatively. The simulation results presented in the in this paper were measured using the Degree algorithm. Example 1 demonstrates the functionality of the XOR algorithms. With the exemplary error matrix

$$\begin{array}{c} \\ V1 \\ V2 \\ V3 \\ V4 \\ V5 \end{array} \begin{array}{c} \begin{array}{ccccc} E\,1 & E2 & E3 & E4 & E5 \end{array} \\ \begin{pmatrix} 1 & 0 & 0 & 1 & 0 \\ 0 & 1 & 1 & 0 & 0 \\ 0 & 0 & 1 & 1 & 0 \\ 1 & 0 & 0 & 0 & 1 \\ 1 & 0 & 0 & 0 & 1 \end{pmatrix} \end{array}$$

the corresponding hypergraph is

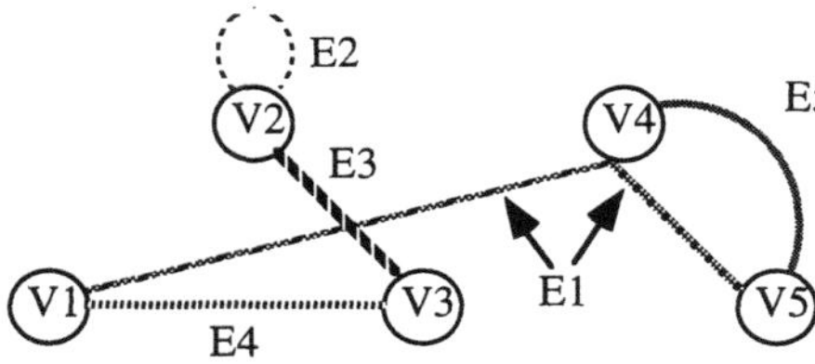

With help of the Random and Degree XOR algorithms we get the following XOR blocks:

1) Sorting the edges/blocks;

Random :	E2, E1, E4, E3, E5
Degree:	E1, E3, E4, E5, E2

2) Coloring/Combining the vertices/blocks in this order;

Random :	x1 = E1⊕E2, x2 = E4⊕E5 and x3 = E3
Degree:	x1 = E1⊕E3 and x2 = E2⊕E4⊕E5

This example shows that the number of retransmissions can be reduced by using Degree heuristics (instead of three transmissions in case of Random method using Degree heuristic only two transmissions are needed).

5. THROUGHPUT ANALYSIS OF XOR SELECTIVE REPEAT

For analyzing the throughput, let the random variables (rv's) X_i describe the number of transmissions for a particular block up to its successful reception by receiver i. In our model, these rv's are independent. A block is successfully received by all stations after $X = \max_{1 \le i \le K} X_i$ transmissions. The throughput T is usually defined as

$$T = \frac{1}{E[X]} = \frac{1}{1 + \sum_{x=2}^{\infty} (x - 1)\, P(X = x)} . \quad (1)$$

This definition must be generalized for an analysis of our XOR strategy. The reason for this can be demonstrated by the following example using K = 2:

Transmissions	1st	2nd	3rd
Block number	1	2	1 XOR 2
Station 1	NACK	ACK	ACK
Station 2	ACK	NACK	ACK

Packets 1 and 2 take two transmissions each, but in total only three transmissions are needed instead of four. To avoid a multiple count of particular packets, a weighted count for transmissions of XORed packets must be introduced.

Therefore, we assign to X the rv's $(Y_{ix})_{1 \leq i \leq x, x \in N}$. Y_{ix} describes the number of XORed packets in the ith retransmission of the considered block under the condition that x is the number of retransmissions needed for the reception of that block by all K stations. Let in the above example X describe the first block. Then $Y_{11}=2$, because in the first retransmission of block 1 two packets have been XORed. Our generalized throughput is defined as

$$T = \frac{1}{1 + \sum_{x=2}^{\infty} \left(\sum_{i=1}^{x-1} \frac{1}{E[Y_{i\,x-1}]} \right) P(X = x)} . \tag{2}$$

Obviously, in the general non-XORing case $E[Y_{ix}] = 1$ holds for all i,x, therefore

$$\sum_{i=1}^{x-1} \frac{1}{E[Y_{i\,x-1}]} = x - 1 . \tag{3}$$

This shows that the usual definition is a special case of our new one.
We are now going to analyze the throughput of the XOR selective repeat strategy compared to that of the selective repeat strategy.

Selective Repeat:
By standard arguments ([GoJ 84], [Dav 70]) it holds that

$$E[X] = \sum_{x=1}^{\infty} x\, P(X = x) = \sum_{x=0}^{\infty} (1 - P(X \leq x)) = \frac{1}{(1 - p_u)} \left(1 + \sum_{x=1}^{\infty} (1 - \prod_{i=1}^{K} (1 - p_i^x)) \right) \tag{4}$$

where, as above, p_u is the uplink block error probability and p_i the downlink block error probability for station i, hence

$$T_{SR} = \frac{1 - p_u}{1 + \sum_{x=1}^{\infty} (1 - \prod_{i=1}^{K} (1 - p_i^x))} . \tag{5}$$

XOR Selective Repeat:

There seems to be no general way how to determine $E[Y_{ix}]$ as this value strongly depends on the employed heuristic of XORing. Therefore, we assume for this first analysis that $E[Y_{ix}] = M \geq 1$ (not depending on i, x).

This results to

$$E[X] = 1 + \sum_{x=2}^{\infty} \frac{x-1}{M} P(X = x) = 1 - \frac{1}{M} + \frac{1}{M} \sum_{x=1}^{\infty} x \, P(X = x)$$

$$= \frac{1}{1-p_u} \left(1 + \frac{1}{M} \sum_{x=1}^{\infty} (1 - \prod_{i=1}^{K} (1 - p_i^x)) \right) \tag{6}$$

therefore

$$T_{XOR} = \frac{1 - p_u}{1 + \frac{1}{M} \sum_{x=1}^{\infty} (1 - \prod_{i=1}^{K} (1 - p_i^x))} . \tag{7}$$

It always holds that $E[Y_{ix}] \leq K$, thus we obtain

$$T_{XOR} \leq \frac{1 - p_u}{1 + \frac{1}{K} \sum_{x=1}^{\infty} (1 - \prod_{i=1}^{K} (1 - p_i^x))} , \tag{8}$$

as an upper bound for the throughput in the XOR case.

It is to mention, that for an exact consideration some additional overhead must be assigned to each XOR block so as to take up the sequence numbers of the basic blocks participating the XOR block. While the block length of the basic blocks is the sum of the data part and the header part of the block in bits, the XOR blocks consist of the data part, header part and additionally M-1 block identifiers (sequence numbers); the header of a basic block already contains one sequence number. Since XOR blocks are longer than basic blocks, the analysis must take this into account. By substitution of the block error probability p_i $1 \leq i \leq K$ with p_{XOR_i} (which is a higher block error probability because of the additional overhead of XOR blocks) in the throughput equation (7), the effect of additional overhead is taken into account. Now the throughput is no longer defined in terms of the number of correctly transmitted blocks per time unit, but in terms of data bits, so that the differing block lengths of the various strategies do not interfere. Throughput efficiency is defined as follows:

$$T_{eff} = \frac{\text{Data (bits)}}{\text{Data (bits) + Overhead (bits)}} \frac{1}{E[X]}$$

6. RESULTS

During an XWindow, the sending station determines how to combine negatively acknowledged blocks of the current XWindow by using one of the XOR heuristics described above. For this purpose, the transmitter keeps a list of the acknowledgements received during the XWindow. When the XWindow is over, it sends the previously assembled blocks, keeping the corresponding data blocks in the retransmission buffer for possible further repetitions. If more than one retransmission is necessary for a particular block, it is sent as a simple data block, i.e. combining the block into an XOR block takes place only at the first retransmission.

The receiver keeps copies of correctly received blocks in its buffer for a certain time in order to be able to decode XOR blocks consisting of these blocks if necessary. To achieve the same throughput as the pure selective repeat strategy, a minimum receive buffer size of S + (1.5 * XWindow size) is required, because after sending the acknowledgment it may take at most a time period of S + (1.5 * XWindow size) until the corresponding retransmission is received (this time results from the round trip delay, plus one XWindow until the transmitter actually repeats the block, plus one half XWindow if every two blocks of the interval are combined to an XOR block).

For our investigations we have assumed the same down link error rates ($0 \leq p_i \leq 1$) for all receivers and no uplink errors ($p_i = 0$). In the following simulation results the XWindow size is set to 20 blocks. In figures 4-10, the ideal case is assumed. That means that the additional overhead for XOR blocks is neglected, and an infinite buffer is assumed. In figures 11 and 12, the additional overhead for the XOR blocks depending on the number of basic blocks included in the XOR block (32 bits block identifier) as well as a finite buffer capacity of $2 \cdot S$ are considered. For the simulation, 99% confidence levels are presented.

Figure 4 shows a comparison between simulative and analytical throughputs. For a direct comparison of the results, the appropriate values of M (average number of blocks participating in an XOR block) corresponding to each block error probability must be known. These values were measured by simulation (see figure 5). For a particular block error probability and the corresponding M value (as taken from figure 5), simulation and analysis achieve the same throughput (in figure 4 the curves of analysis and simulation intersect at these points). This diagram shows that the difference in throughput between the pure SR and XOR-SR strategies increases with M.

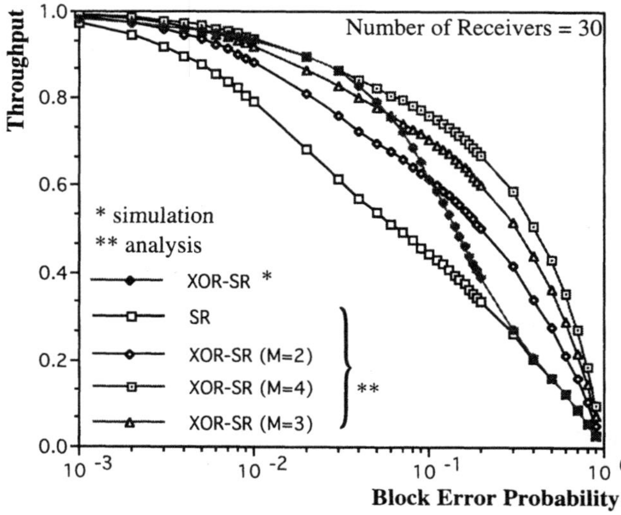

Fig 4: Comparison of the throughput as a function of block error probability

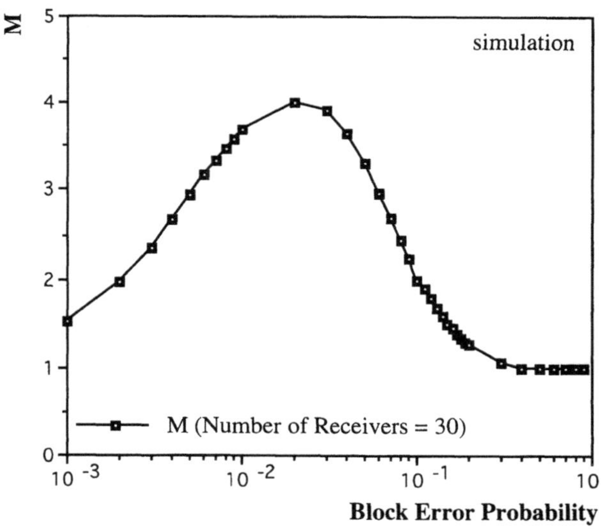

Fig 5: Simulatively measured M values for 30 receivers

Figures 6 and 7 show analytical and simulative results for various numbers of receivers and a constant block error probability in comparison.

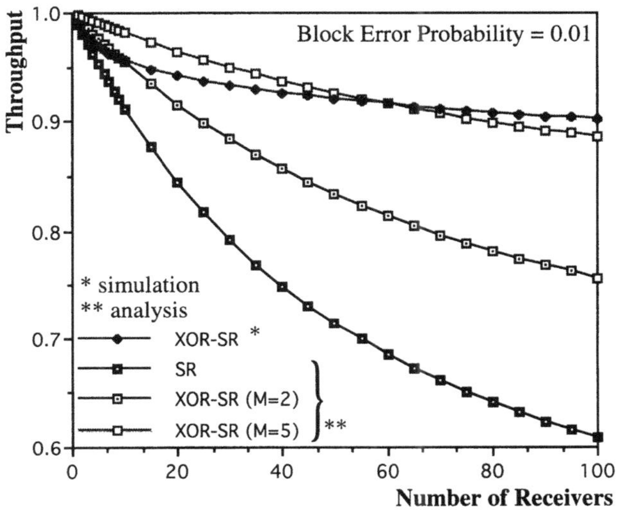

Figure 6: Simulative and analytical throughput as a function of the number of receivers at a fixed block error probability of 0.01

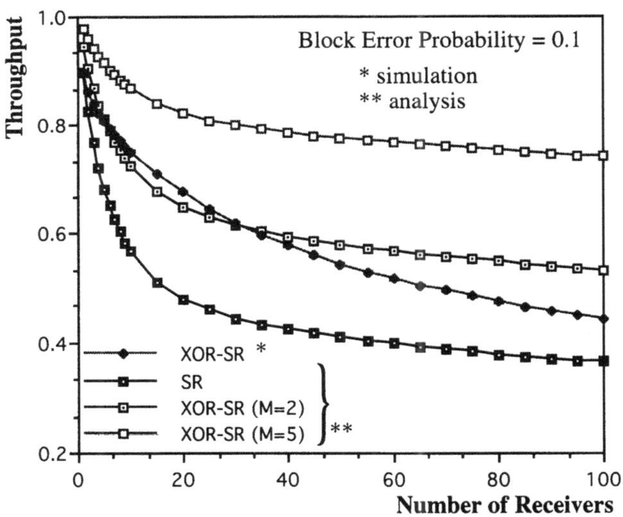

Figure 7: Simulative and analytical throughput as a function of the number of receivers at a fixed block error probability of 0.1

At an average block error probability of 0.01, the gain in throughput of XOR-SR as compared to SR increases drastically with the number of receivers, while at higher block error probabilities the difference decreases with a growing number of receivers. This is a consequence of the fact, that the average number of blocks participating in an X block (M) increases (see figure 8) with a growing number of receivers at medium block error probabilities (e.g. 0.01), but decreases at high block error probabilities (e.g. 0.1), because in the latter case there is a higher probability that several blocks are incorrectly received by different receivers.

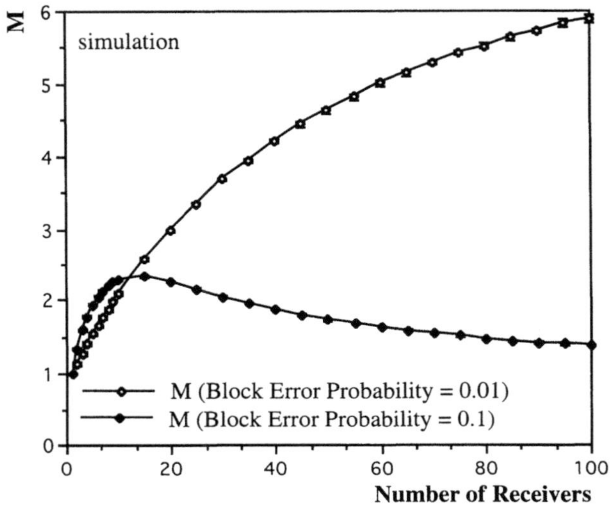

Figure 8: Simulatively measured M values as a function of the number of receivers

Figure 9 shows the throughput of the pure SR and the XOR-SR strategies, respectively, depending on the XWindow size for 30 receiver stations and various block error probabilities. With growing XWindow size, the average number of blocks participating in an XOR block (M) increases (see figure 10). This results in a higher throughput for larger XWindow sizes. Of course the throughput of the pure SR strategy remains constant regardless of XWindow sizes.

In figures 11 and 12 we consider the throughput for a scenario, where the stations have a buffer capacity of 2·S blocks (S = number of data blocks that can be transmitted during one round trip delay). We presume the use of the TCP/IP protocol, with an overhead for 20 bytes for TCP and another 20 bytes for IP, altogether 320 bits. The data part has a size of 4096 bits, and the length of a block identifiers is 32 bits.

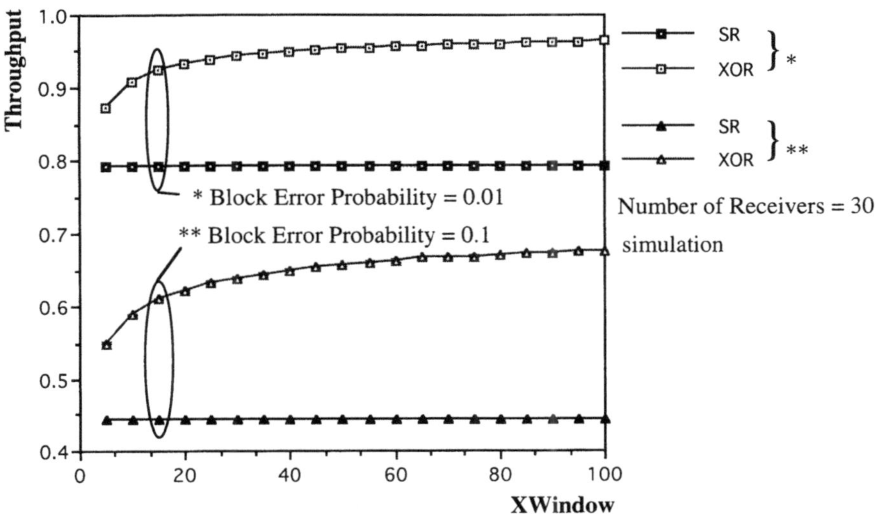

Figure 9: Comparison of the throughput between XOR-SR and pure SR strategies as a function of the XWindow size for various block error probabilities

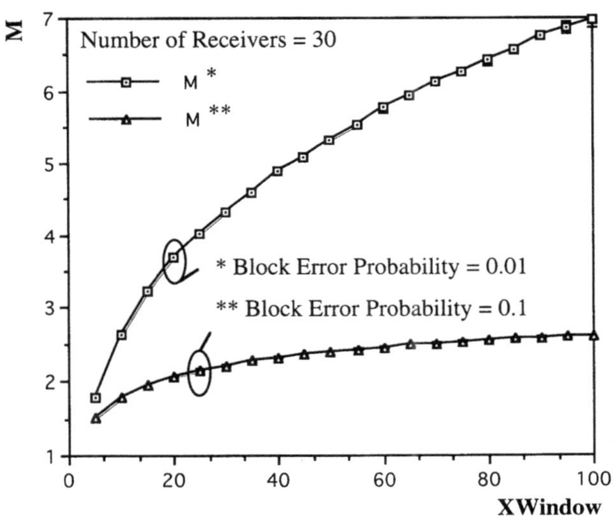

Figure 10: M values measured as a function of the XWindow size

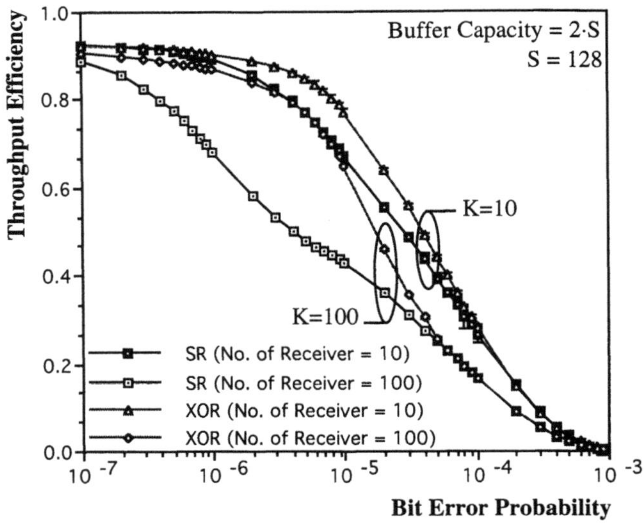

Figure 11: Simulative throughput comparison of SR and XOR-SR with 10 and 100 receivers

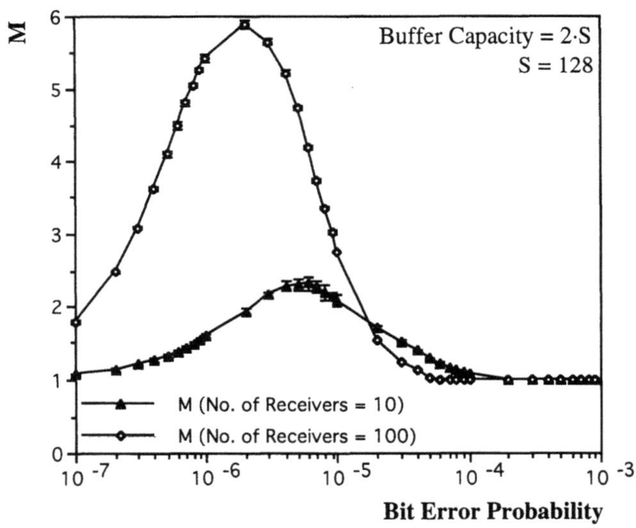

Figure 12: Measured M values for XOR-SR with 10 and 100 receivers

The throughput improvement of the XOR-SR strategy increases significantly with a growing number of receivers. Figure 11 shows that for 100 receivers and up to a certain block error probability the XOR strategy achieves results in a similar range as the SR strategy with 10 receivers. This is a result of the increasing compression factor M (compare figure 11). The M values for 10 and 100 receivers exhibit a large difference (see figure 12).

7. CONCLUSION

In this paper we have discussed a new approach of selective repeat schemes for point-to-multipoint communication, called the XOR selective repeat strategy. In this strategy the sender does not retransmit negatively acknowledged (NACKed) blocks immediately after the reception of a NACK, but it gathers a fix number (XWindow size) ACKs/NACKs and then decides about retransmitting NACKed blocks. The idea of the XOR selective repeat strategy is to physically combine several negatively acknowledged (NACKed) blocks by XORing (i.e. modulo 2 addition) them to minimize the number of retransmissions and to increase throughput. The XOR protocols differ in the way the sender determines which NACKed data blocks are combined. The methods aim to minimize the number of XORed data blocks as far as possible. The analytical and simulative results for selective repeat strategy show the increase in throughput of the XOR strategy with growing XWindow, number of participating stations, and, up to a certain point, block error probabilities. The more receivers are involved the better is the gain. Nevertheless, the number of participating receivers is not unlimited, because the acknowledgements must be coordinated.

REFERENCES

[AgMM 93] M. Aghadavoodi Jolfaei, S. C. Martin, J. Mattfeldt: "A New Efficient Selective Repeat Protocol for Point-To-Multipoint Communication"; Proceedings ICC '93, pp. 1113-1117, Genf, Schweiz, Mai 93.

[AgKM 92] M. Aghadavoodi Jolfaei, D. Kreuer, O. Maly, U. Quernheim: "Two Time Variant Models for Satellite Channels"; Proceeding Supercom/Icc 92, Chicago, USA, pp. 314.31.1- 314.6.3

[BeF 64] R. J. Benice, Jr. A. H. Frey: "An Analysis of Retransmission System"; IEEE Trans. Commun. Technol., 135-145, Dec. 1964

[Ber 89] C. Berge, Hypergraphs, Amsterdam: North-Holland, 1989

[Bol 79] B. Bollobás, Graph Theory, New York: Springer, 1979

[Bre 79] D. Brelaz: "New Method to color the Vertices of a Graph"; Comm. ACM, vol.22, pp. 251-256, 1979

[BrM 86] H. Bruneel, M. Moeneclaey: "On the Throughput Performance of Some Continuos ARQ Strategies with Repeated transmissions"; IEEE Trans, Commun., 244-249, Mar. 1986

[Bus 72] H. O. Burton, D. D. Sullivan: "Errors and Error Control"; Proc. IEEE, pp. 1293-1303, Nov. 72.

[CaE 86[S. R. Chandran, S. Lin: "A Selective-Repeat ARQ Scheme fot Point-To-Multipoint Communications and It´s Throughput Analysis"; Proc. ACM SIGCOM Conference, pp. 292-301, Stowe, VT, Aug. 1984

[Dav 70] M. A. David, Order Statistics, New York: Wiley, 1970

[GoJ 84] I. S. Gopal, J. M. Jaffe: "Point-To-Multipoint Communication Over Broadcast Links"; IEEE Trans. Commun., Com-3, pp. 1034-1044, Sep. 1984

[Ho 89] I. Hoyer: "The NP-Completeness of the Edge-Coloring"; SIAM J. Comput., vol. 10, pp. 718-720, Nov. 1981

[LiCo 83] S. Lin, D. J. Costello: "Error Control Coding: Fundamentals and Applications"; Prentice Hall, Englewood Cliffs, NJ, USA, 1983.

[Mor 78] J. M. Morris: "On Another Go-Back-N ARQ Technique for High Error Rate Condition"; IEEE Trans. Commun., pp. 187-189, Jan. 1978

[SaSc 85] K. Sabnani, M. Schwartz: "Multidestination Protocols for Satellite Broadcast Channels"; IEEE Transactions on Communications, Vol. COM-33, No. 3, März 1985.

[Sast 75a] A. R. K. Sastry: "Improving Automatic Repeat-Request (ARQ) Performance on Satellite Channels under High Error Rate Conditions"; IEEE Transactions on Communication Electronics, pp. 224-231, April 1975.

[Sast 75b] A. R. K. Sastry: "Performance of Hybrid Error Control Schemes on Satellite Channels"; IEEE Transactions on Communications, Vol. COM-23, pp. 689-694, Juli 1975.

[SOSS 89] A. Shiozaki, K. Okuno, S. Satoh, T. Segawa: "Automatic Repeat Request/Adaptive Forward Error Correction System for Satellite Communications"; The Transactions of the IEICE, Vol. E 72, No. 5, pp. 565-566, Mai 1989.

[Tom 87] D. Towsley, S. Mithal: " A Selective Repeat ARQ Protocol for a Point to Multipoint Channel"; Proc. INFOCOM, pp. 521-526, San Franscisco, CA, Mar. 1987

[Tows 79] D. Towsley: "The Stutter Go-Back-N ARQ Protocol"; IEEE Trans. Comm., vol. COM-27, pp. 869-875, Jun. 1979

[WaS 88] J. L. Wang, J. A. Silvester: "Optimal Adaptive ARQ Protocols for Point-To-Multipoint Communication"; IEEE INFOCOM 88, pp. 704-713, LA, 1988

[Wel 82] E. J. Weldon: "An Improved Selective-Repeat ARQ Strategy"; IEEE Trans. Comm., vol. COM-30, pp. 480-486, Mar. 1982

[Wu 84] W. W. Wu: "Elements of Digital Satellite Communication"; Band 1, Computer Science Press, 1984

[Wu 85] W. W. Wu: "Elements of Digital Satellite Communication"; Band 2, Computer Science Press, 1985

[YuL 81] P. S. Yu, S. Lin: "An Efficint Selective Repeat ARQ Scheme for Satellite Channels and Its Throughput Analysis"; IEEE Trans. Commun., pp. 353-363, Mar. 1981

[YuL 82] P. S. Yu, S. Lin: "A Hybrid ARQ Scheme with Parity Retransmission for Error Control of Satellite Channels"; IEEE Transactions on Communications, Vol. COM-30, No. 7, Juli 1982

Broadband Communications, II (C-24)
S. Tohmé and A. Casaca (Editors)
Elsevier Science B.V. (North-Holland)
1994 IFIP

Performance Evaluation of Frame Relay based on the Packet Trains model

M. Gagnaire[a] and T. Uso[b]

[a]Ecole Nationale Supérieure des Télécommunications,
46 rue Barrault, 75634, Paris cedex 13 - FRANCE

[b]Consultant for GEC-Alsthom,
38 Avenue Kléber, 75795, Paris cedex 16, FRANCE

Abstract

In the short term, frame relay is one of the most promising techniques for high speed LAN interconnection. Several performance evaluations of frame relay are available in the literature. In most cases, these evaluations assimilate the traffic issued from a bridge/router to an Interrupted Poisson Process (IPP). The experimental analysis of a LAN traffic pattern reveals the lack of reliability of the Poisson assumption. Raj Jain and Shawn Routhier have proposed an alternative model called the "packet trains" model which describes more faithfully a LAN traffic. In this paper, we adapt the packet trains model to the traffic exchanged between two end-users connected on distant LANs. These two LANs are themselses interconnected by means of a frame relay network. A realistic network configuration has been adopted in considering the characteristics of the TRT CP200 frame relay switch. Simulation results dealing with the influence of various parameters of the packet trains model on network performance are presented.

Keywords: Internetworking, Performance evaluation

1. INTRODUCTION

The packet switched networks based on the X25 protocol stack interface have been used during these last decades to interconnect LANs via the public domain. Limited to a few kilobit/s, the X25 networks are not well suited to satisfy the end-to-end delay constraints required by the new applications implemented on high performance workstations. Such applications are generally characterized by a high burstiness. Leased lines (T1 or E1) allow higher throughput and optimize the end-to-end delay. However, they remain too expensive and too rigid in terms of topology to be considered as the right solution for LAN interconnection. The coming Broadband ISDN based on the ATM technique is designed to satisfy in the long term the constraints of a high performance LAN interconnection. Frame relay which is a simple updating of packet switched networks seems to be a solution to this problem in the short term. For that purpose, frame relay relies on the recent developments in the field of data transmission and of end-user devices. Fiber optics offer a very large bandwidth and a very reliable transmission on long range distances. The high performance of RISC architectures allows the end-user workstations to manage several processing tasks traditionally devoted to the network itself. In frame relay, two major functionalities are thus shifted from the network nodes towards the intelligent end-users : flow control and error recovery. Frame relay networks are connection-oriented. The dialogue between two end-users transits through a permanent virtual connection (PVC). The principle of frame relaying will not be recalled, several papers dealing with this topic being available in the literature ([2], [1], [3], [4], [5]).

Like ATM networks, frame relay networks are vulnerable to congestion phenomenon. Statistical multiplexing as well as the absence of link-by-link flow control considerably increase the risks of overflow in the various buffers of the network. Congestion management in both frame relay and ATM networks strongly depends on the source behaviour. Several performance evaluations of frame relay have been proposed in the literature ([6], [13]). In most of these investigations, a common assumption is that packet arrivals are independant and unpredictable. Thus, the traffic generated by end-users is assimilated to an Interrupted Poisson process (IPP). Traffic measurements on LANs ([7], [8], [12]) reveal that the arrival processes are neither Poisson nor Poisson with batch arrivals. Jain and Rouhtier propose an alternative model called the "packet trains" model which describe more faithfully the behaviour of a real-world traffic sources [8].

This paper presents a performance analysis of LAN interconnection via frame relay by using an adaptation of the packet-trains model. Frame relay standard ([10], [11]) includes several mechanisms to prevent or to control any congestion phenomenon. These mechanisms related to admission control, congestion notification and traffic enforcement are not considered in the followings. Section 2 of this paper describes the source and network modeling. The packet-trains model is briefly recalled before presenting our adaptation of this model to the case of a LAN interconnection. Numerical values corresponding to real traffic characteristics are assigned to the various parameters of the model. We use a realistic network configuration based on bridge/routers, switches and 2 Mbit/s links. The frame relay switches modeling takes into account the main characteristics of the TRT CP200 equipment. Section 3 of this paper presents our results of simulation. Performance of the network is evaluated in terms of average throughput, frame loss rate and response time. The influence of sources' burstiness on this performance is discussed. Finally, one gives our conclusions about this study in section 4 of the paper.

2. LAN INTERCONNECTION MODELING

2.1. Source modeling

The reliability of source modeling is of a major importance in the field of packet switched networks resource management. The more precise is this modeling, the better is the prediction of packets arrivals, and the easier is the resource requirements prevision. A good traffic prediction allows to facilitate the choice of the optimal congestion prevention mechanism for the network.

2.1.1. Principle of the packet trains model

Jain and Routhier classify source models in two sets and call them "car" and "packet trains" models. Car models are characterized by the independance of the frame arrivals. The Poisson arrival model is a well-known car model where the frame inter-arrival times are exponentially distributed. Packet trains models are characterized by the exchange of frame groups (called packet trains by Jain and Routhier) between end-user pairs. In packet trains model, the assumption of independance of frame interarrivals is no longer valid. Jain and Routhier have monitored frame interarrival times on a token ring [8]. The same kind of measurements has been carried out for an Ethernet LAN [9]. For both these experiments, one observes that the logarithmic histogram of frame interarrival times has three linear components (Figures 1 and 2).

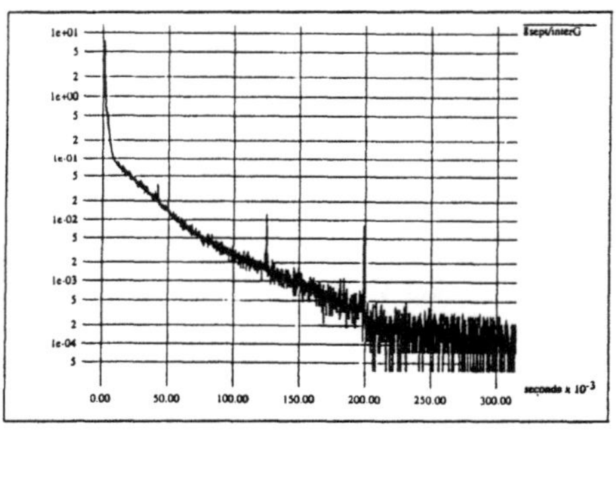

Figure 1. Logarithmic histogram of packet interarrival delays on an Ethernet network

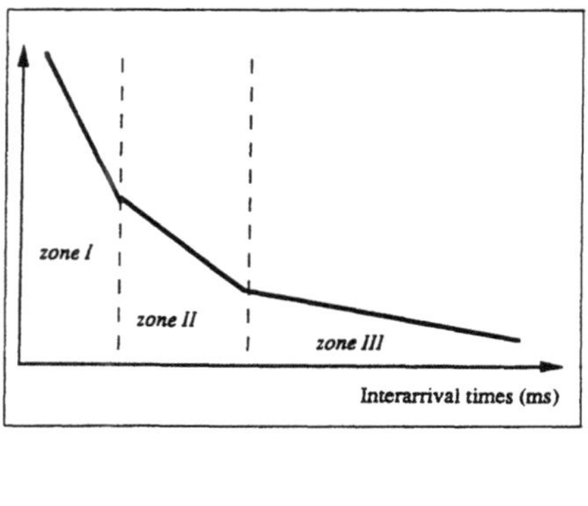

Figure 2. The three working zones associated to the packet trains model

After a statistical analysis of these results, Jain and Routhier conclude that packet trains models are more suited to model a LAN traffic than car models. Let us consider the traffic pattern between two end-users A and B where only frames traveling from A

towards B are represented (Figure 3). This traffic pattern may be interpreted at three different time scales. At the highest level (macroscopic time scale), data traffic is a succession of "bursts" and silences. Inside a burst (second level), one observes that a burst is itself composed of a succession of "mini-bursts" and mini-silences. Finally, at the third level (microscopic time scale), a mini-burst is composed of several isolated frames of variable length.

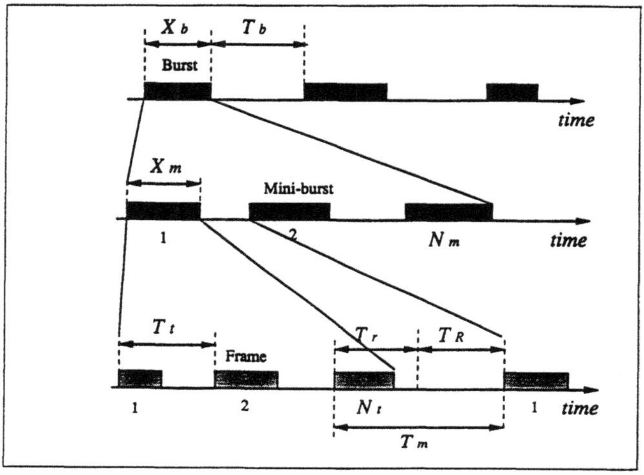

Figure 3. The three time scales of the packet trains model

Figure 3 allows to interpret Figures 1 and 2. Component I (the most frequent interarrival delays) corresponds to the frame interarrival delays in a given mini-burst. Component II corresponds to the interarrival delays between the last frame of the k-th mini-burst and the first frame of the (k+1)-th mini-burst. Finally, component III corresponds to the interarrival delays between the last frame of the k-th burst and the first frame of the (k+1)-th burst , which are of course the less frequent. Although the frames of a same mini-burst are close to each other, they cannot be assimilated to bulk arrivals. The inter-burst time depends on the user's behavior. The inter-frame time inside a mini-burst and the inter-mini-burst time inside a burst depend on the end-station characteristics (hardware, software and protocols). Moreover, the number of frames by mini-burst is related to the flow control window size used by the protocols.

We define below the various parameters of the model :

- L : the random variable (r.v.) associated to the frame length (in bits). One assumes that L is given by a uniform distribution on the integer interval $[\mathcal{L}_{min}, \mathcal{L}_{max}]$.
- T_b : the r.v. associated to the silence period between two successive bursts. One assumes that T_b is a Poisson process of parameter $\mathcal{T}_b$.
- N_m : the r.v. associated to the number of mini-bursts in a burst. One assumes that N_m is given by a uniform probability on the integer interval $[1, \mathcal{N}_m]$.
- N_t : the r.v. associated to the number of frames in a mini-burst. One assumes that N_t is given by a uniform probability on the integer interval $[1, \mathcal{N}_t]$.
- T_t : the r.v. associated to the delay between the beginning of two successive frames in a mini-burst. One assumes that T_t is a Poisson process of parameter $\mathcal{T}_t$.
- T_m : the r.v. associated to the delay between the beginning of the last frame in a mini-burst and the beginning of the first frame in the next mini-burst. The r.v. T_m is the sum of two independant r.v. T_r and T_R.
- T_r : the r.v. associated to the response time of an end-user. One assumes that T_r is a Poisson process of parameter $\mathcal{T}_r$.
- T_R : the r.v. associated to the network response time.

2.1.2. Adaptation of the packet trains model to LAN interconnection

We have seen that the packet trains model is characterized by several parameters, each of them having a network significance in the case of a LAN traffic. When the packet trains model is applied to a LAN interconnection traffic, no any additional parameter is needed. However, some parameters such as the network response time T_R must be re-interpreted according to the number of frame relay switches considered along the permanent virtual circuit (PVC). Indeed, frames exchanged between two end-users connected on distinct LANs travel across bridge/routers and switches. The inter-miniburst time T_m has not only a LAN component but also a frame relay network component.

Compared to the LAN traffic model described in the previous section, a new factor makes the Poisson arrival assumption inadapted to our network configuration. Indeed, the intermediate nodes (the frame relay switches) have to manage simultaneously several PVCs. Depending on the load of the input VCs, an intermediate node may be in a high load or in a low load activity period. For instance, a switch routes packets with longer delays if it is in a high load period than if it were in a low load period. In other words, packet interdeparture delays at the output of a switch cannot be assumed as independent. The traffic pattern between two end-users A and B on distinct LANs is described in Figure 4.

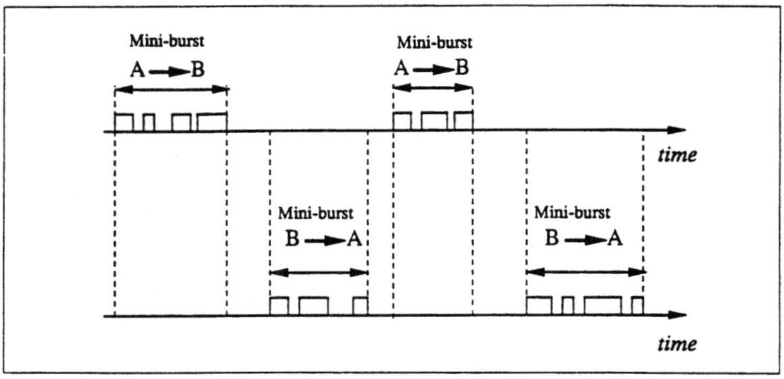

Figure 4. Traffic pattern between an end-user pair on distinct LANs

The values of the packet trains model's parameters are determined from experimental frame interarrival monitoring. We take some of these values in our simulations and we make arbitrary choices for other values and for the distribution laws which are application and protocol-dependent (e.g. distribution of the frame size). Table 1 summarizes the values and the distribution laws of the parameters adopted to model the traffic between two end-users connected on distant LANs.

Table 1
Traffic parameters between an end-user pair on distinct LANs

Parameter	*Choice*
Frame size L	Uniform distribution, [500 bits, 12500 bits]
Mini-bursts/burst (N_m)	Uniform distribution [1, $\mathcal{N}_m$], $\mathcal{N}_m$ = 3, 8, 16, 17
Frames/mini-burst (N_t)	Uniform distribution [1, $\mathcal{N}_t$], $\mathcal{N}_t$ = 1, 3, 8
Inter-burst time (T_b)	Exponential distribution [20 s]
Inter-miniburst time (T_m)	$T_r + T_R$
Inter-frame time (T_t)	Exponential distribution [35 ms]
Response time of a station (T_r)	Exponential distribution [65 ms]
Propagation time across the backbone (T_R)	backbone dependent

2.2. Network modeling

Our network has a linear topology. It is 900 kms long and is composed by 2 end-systems (ES) and a variable number of intermediate systems (IS). The ES and IS are connected by 2 Mbit/s links (Figure 5). The ES are bridge/routers and the IS are frame relay switches. Distance between a bridge/router and its access switch is negligible. The intermediate systems are assumed to be equidistant along the data path.

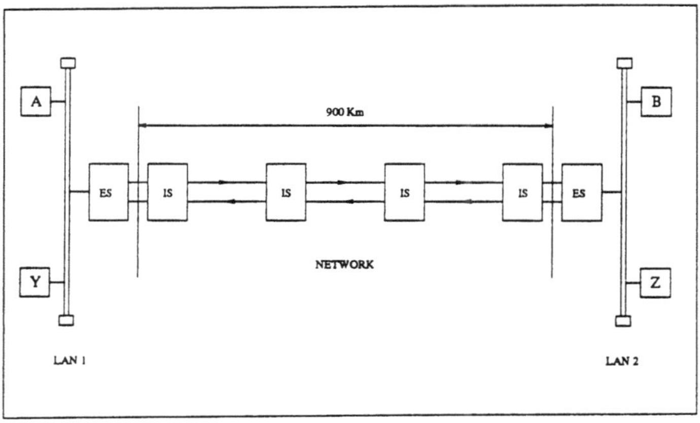

Figure 5. Network topology

N traffic sources of the type of those described above deliver frames to each bridge/router at a peak rate of 10 Mbit/s. According to the packet trains model, the upstream and downstream flows are synchronized between each end-user pair (Figure 6). Each bridge/router has an unlimited buffering capacity and its frame processing time T_p is 0.07 ms. The switches are modeled according to the characteristics of the TRT CP200 switch. This equipment exists with different firmwares. The firmware used in this study ignores the congestion bits (FECN, BECN, DE). Each switch has 1000 buffers of 128 bytes. Its buffer allocation policy is of the "full sharing" type. In this allocation buffer policy, all the output queues share the same buffers pool which minimizes the frame loss rate but does not allow fairness ([14]). The mean switching time is set to 0.5 ms.

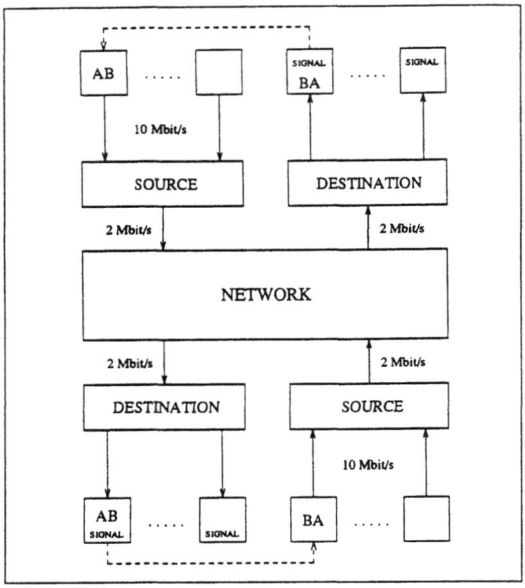

Figure 6. Synchronized data flow between two end-users

3. PERFORMANCE EVALUATION

3.1. Simulation results

Simulations are performed with 2 possible error bit rates (0 and 10^{-6}), various numbers of switches (2-4) and 4 types of traffic sources (Table 2). We investigate performance of the network (frame relay backbone) in terms of average throughput, frame loss rate and response time. We define response time as the elapsed time between the instant when a frame enters the access switch and the instant when the last bit of this frame leaves the last switch. An increase of the offered load is obtained by enlarging the number N of traffic sources entering the bridge/routers.

Table 2
Types of traffic sources

Type	$\mathcal{N}_t$	$\mathcal{N}_m$	*Average number of frames/burst*
1	1	17	9
2	3	8	9
3	8	3	9
4	3	16	18

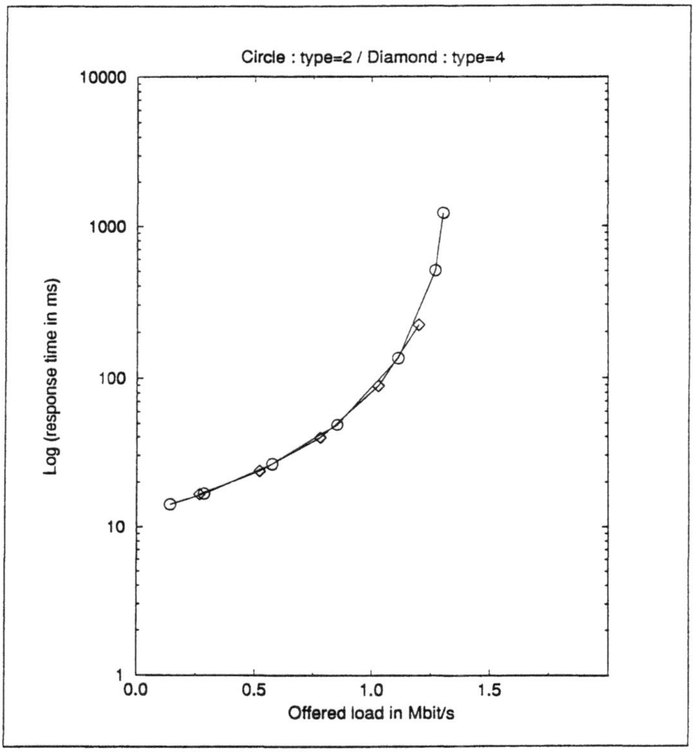

Figure 7. Influence of the average number of frames per burst on frame response time

In all our simulations, it appears that the frame loss ratio due to buffer overflow is negligible. This result suggests an accurate buffer dimensioning of the TRT CP200 switches With an error bit rate of 10^{-6}, the frame loss rate is 2.8 percent. Afterwards, only results with null error bit rate and 4 switches are presented. Let us investigate the influence of the type of traffic sources on network performance. In a first step, we compare traffic types 2 and 4. These traffic types have the same $\mathcal{N}_t$ value but different $\mathcal{N}_m$ values. Traffic 2 and 4 only differ by their average number of frames per burst (Table 2). A burst of type 4 contains twice more frames than a burst of type 2. Figure 7 shows that for a given offered load, no difference occurs on the frame response time between traffic types 2 and 4.

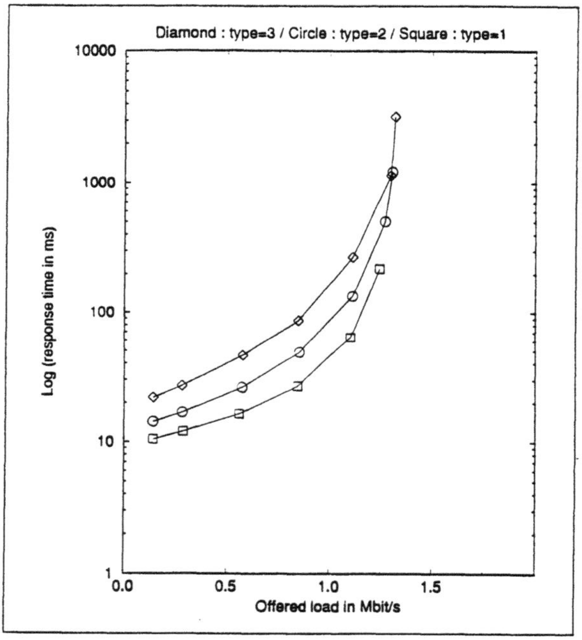

Figure 8. Influence of the window size on the response time

In a second step, we compare traffic types 1, 2 and 3. These traffic types have different $\mathcal{N}_t$ and $\mathcal{N}_m$ values, the average number of frames per burst remaining unchanged. These three kinds of traffic only differ by the window size of the stations (Table 2). For a given offered load, Figure 8 shows significant differences on the frame response time. The larger is the station window size, the longer is the frame response time. These results can be interpreted in considering the burstiness of the traffic outcoming a bridge/router on the 2 Mbit/s link. If several traffic sources of type 2 or if several traffic sources of type 4 are multiplexed in the bridge/router, the resulting traffic has roughly the same shape for a given offered load. This is why in that case, network performance remains the same.

At the opposite, when traffic sources of type 1, or of type 2, or of type 3 are multiplexed in the bridge/router, the resulting traffic does not have any longer the same shape. This is why in this second case, different network performance is observed for a given global offered load. As it is illustrated by Figure 9, the window size of the stations influences compactness of the bursts leaving the bridge/router.

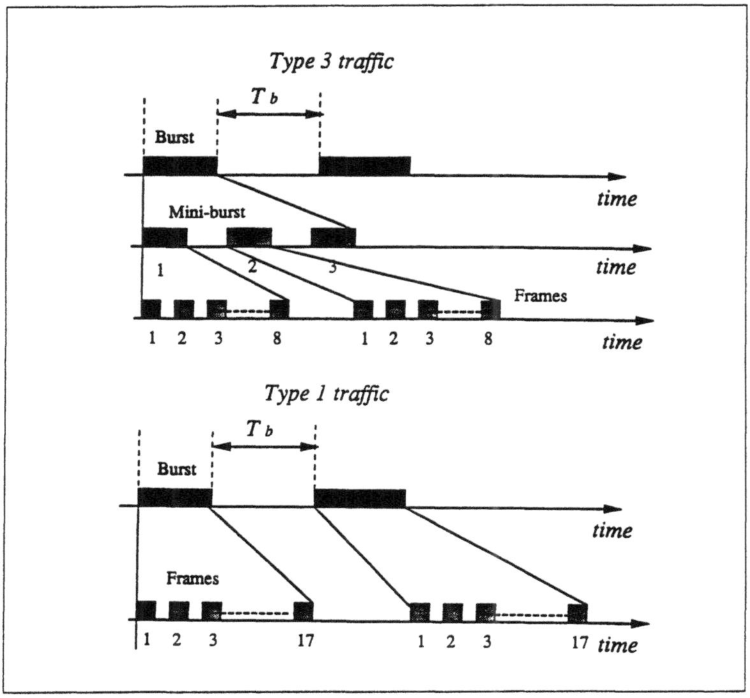

Figure 9. Time structure of traffic types 1 and 3

3.2. Evaluation of the global offered load

Let X_m and X_b be respectively the random variables associated to the mini-burst duration and to the burst duration. The mean frame length is given by :

$$E[L] = \frac{(\mathcal{L}_{min} + \mathcal{L}_{max})}{2} \tag{1}$$

The average number of mini-bursts in a burst is :

$$E[N_m] = \sum_{i=1}^{\mathcal{N}_m} \frac{i}{\mathcal{N}_m} = \frac{(\mathcal{N}_m + 1)}{2} \tag{2}$$

and the average number of frames in a mini-burst is :

$$E[N_t] = \sum_{i=1}^{\mathcal{N}_t} \frac{i}{\mathcal{N}_t} = \frac{(\mathcal{N}_t + 1)}{2} \tag{3}$$

The mean mini-burst interarrival delay and the mean frame interarrival delay are respectively :

$$E[T_r] = \frac{1}{\mathcal{T}_r}$$
$$E[T_t] = \frac{1}{\mathcal{T}_t} \tag{4}$$

The offered load λ by the N sources is given by the ratio of the average amount of bits transfered per second. From Figure 3, it appears that λ is given by :

$$\lambda = N\frac{\{E[N_t]E[L]E[N_m]\}}{\{E[T_b] + E[X_b]\}} \tag{5}$$

Knowing that the expectation of T_b is $\mathcal{T}_b^{-1}$ and using equations (1), (2), (3) and (4), one gets :

$$\lambda = N\frac{\{(\mathcal{N}_t + 1)(\mathcal{L}_{min} + \mathcal{L}_{max})(\mathcal{N}_m + 1)\}}{2^3\left\{\mathcal{T}_b^{-1} + E[X_b]\right\}} \tag{6}$$

Let ϕ be the average throughput of each of the N sources. The expectation of the burst duration X_b is given by :

$$E[X_b] = (E[N_m]-1)\{(E[N_t]-1)E[T_t]+E[T_m]\}+E[X_m] \quad (7)$$

where $E[T_m]$ is given by :

$$E[T_m] = E[T_r]+E[T_R] = \mathcal{T}_r^{-1}+E[T_R] \quad (8)$$

and $E[X_m]$ is given by :

$$E[X_m] = (E[N_t]-1)E[T_t]+\frac{E[L]}{\phi} \quad (9)$$

Including equations (8) and (9) in equation (7), one gets :

$$E[X_b] = \frac{(\mathcal{N}_t-1)\mathcal{T}_t^{-1}(\mathcal{N}_m+1)}{4}+\frac{(\mathcal{N}_m-1)}{2}\left\{\mathcal{T}_r^{-1}+E[T_R]\right\}+\frac{(\mathcal{L}_{max}+\mathcal{L}_{min})}{2\phi} \quad (10)$$

One notices that $E[X_b]$ depends on the mean station response time $E[T_r]$. The station response time depends itself on the upper layers protocols (transport and above) implemented in each end-user equipment. The workload corresponding to each of the N sources is described by a Markov Modulated Poisson Process (MMPP). Not any exact analytical description of the three levels packet trains model is available in the literature. The solution to this problem will be presented in our next publication. If source modeling is based on the packet trains model, an analytical resolve of the average frame response time through a frame relay network is quite complex. Indeed, the first difficulty consists in determining the analytical model associated to the superimposition of N packet trains sources. Such an approach seems tractable only if the network response time $E[T_R]$ is upper-bounded by a given value $\mathcal{T}_{Rmax}$.

4. CONCLUSION

In this paper, we have investigated frame relay network performance in using a source modeling based on the packet trains model. First of all, we have underlined why neither a Poisson process nor a Poisson with bulk arrival process were a good description of LAN traffic. The packet trains model allows probably the most faithful description of LAN traffic. We have chosen to adapt this model to the case of a frame relay environment. For that purpose, one has assimilated in our simulator the traffic submitted at the input of a bridge/router to the superimposition of several packet trains sources. So as to consider a realistic network configuration, we have also included in our simulation model the charateristics of the TRT CP200 full sharing packet switch. For a 900 kilometers long and 2 Mbit/s permanent virtual circuit, it appears that the maximum allowable global load on a physical link is about 1.4 Mbit/s. Under this load threshold, buffer overflow is negligible. When traffic sources only differ by their mean number of mini-bursts per burst, the mean end-to-end frame reponse time remains unchanged for a given global offered load. This result may be interpreted by the fact that the average number of mini-bursts per burst (the N_m parameter) does not sensibly modify the shape of the multiplexed flow. At the opposite, for a given global offered load, the mean end-to-end frame response time is strongly modified if both the average number of frames per burst and the average number of mini-bursts per burst fluctuate. This second result may be interpreted by the fact that, in that case, burstiness of the multiplexed flow is sensibly modified. The higher is the sporadicity, the worst is the mean end-to-end frame response time. If the notion of mini-burst is neglected and if frame length is assumed constant, the behaviour of a source is very similar to that of a packetized voice. One of our coming papers will deal with this aspect. As mentionned in section 3.2, we propose in our next paper an exact analytical modeling of the packet trains model based on the MMPP theory. Some of our coming studies will also deal with source policing and traffic shaping techniques applied to frame relay networks.

REFERENCES

1. J. P. Cavanagh, "Applying the Frame Relay Interface to private Networks", IEEE Communications Magazine, pp. 48-64, March 1992.
2. H. Santoso and S. Fdida, "Frame relay : a solution for high bandwidth networking", Computer Communications magazine, Vol. 16, N7, pp. 432-439, July 1993.
3. J. Lamont and M. H. Hui, "Some Experience with LAN Interconnection via Frame Relay", IEEE Communications Magazine, pp. 21-24, September 1989.
4. S. A. Taylor, "Frame Transport Systems", IEEE Communications Magazine, pp. 66-70, March 1992.
5. "DSS1 - Core aspects of Frame Protocol for use with Frame Relay Bearer Service", ANSI T1S1/90-214R1, November 30, 1990.
6. C. L. Bottomley *Al.*, "Traffic Measurements on a working wide area network", Proceedings of ITC'91, 1991.
7. R. Gusella, "A Measurement Study of Diskless Workstation Traffic on an Ethernet", IEEE Transactions on communications, September 1990.
8. R. Jain and S. A. Routhier, "Packet Trains - measurements and a new model for computer network traffic", IEEE Journal on Selected Areas in Communications, Vol. SAC-4, N6, September 1986.
9. B-G Cambier, "Modélisation du trafic sur un réseau local Ethernet", Rapport de stage ENST au CNET, Octobre 1993.
10. "Addendum to T1.606 - Frame Relay Bearer Service - Architectural Framework and Service Description", ANSI T1S1/90-175R4, 1990.

11. "Frame Mode Bearer Service - FMBS", CCITT I.233, Geneva, 1991.
12. H.J. Fowler and W.E. Leland, "LAN traffic characteristics with implications for broadband network congestion management", IEEE Journal of Selected Areas in Communications, Vol.9, no.7, pp. 1139-1149,1991.
13. K. J. Chen and K. M. Rege, "A comparative performance study of various congestion control for ISDN Frame Relay network", Proceedings IEEE Infocom'89, pp. 674-675, 1989.
14. F. A. Tobagi, "Fast packet switch architecture for broadband integrated services digital networks", Proceedings of the IEEE, pp. 133-167, 78(1) 1990.

Broadband Communications, II (C-24)
S. Tohmé and A. Casaca (Editors)
Elsevier Science B.V. (North-Holland)

TRIBUNE: On the Leading Edge of Broadband Testing

J. Scarr, G. Bosveld, A. Doelman, J. de Bie[a]

[a]PTT Research, P.O. Box 421, 2260 AK Leidschendam, The Netherlands

Abstract

Facilities for testing implementations of Broadband ISDN are currently being developed by the RACE project TRIBUNE. One of the primary objectives of this project is to provide a test-bed, with a standardised Broadband User-Network Interface (B-UNI), which can be used by other European projects. The test-bed is being designed to incorporate the latest and most stable ITU-TS recommendations. This will insure that equipment which is suitable to be tested using the test-bed is also suitable for subsequent commercial exploitation. The projects which will use the test-bed are developing equipment which contain a user side implementation of the B-UNI, for example, a broadband terminal or a broadband Customer Premises Network. The TRIBUNE project will provide testing facilities for the user projects which are well in advance of commercially available test equipment. The test-bed and associated test tools will enable all protocol layers at the B-UNI, including access signalling, to be tested. To increase the speed and thoroughness with which new prototypes can be tested, an innovative approach is being taken to testing and many new techniques are being applied during the project. In 1994 the test-bed will be operational and the users of the network will come to the test-bed site in Leidschendam, The Netherlands, to test their equipment. It is the intention of the project partners to continue to seek contact with potential users of the network.

Keyword Codes: C.2.2
Keywords: Network Protocols, ATM (Asynchronous Transfer Mode) networks, Protocol specification, testing and verification

1. INTRODUCTION

Broadband ISDN and the associated technology Asynchronous Transfer Mode (ATM) have been acclaimed to be the answer to everybodies communications requirements. From the network operators point of view B-ISDN offers the chance to provide the customer with what he requires in a flexible manner, it is also possible to replace the proliferation of existing networks with a single universal network. Many advantages can also be seen for the users of public networks; a single network offering all the required services is the ideal situation. However in order to successfully introduce a new technology, thorough and efficient testing of all the network elements and terminals is required. This paper presents the work being done in the RACE project TRIBUNE (Testing Ratification and Interoperability of the Broadband User-Network Interface), which will enable emerging implementations of B-ISDN equipment to be tested. TRIBUNE is specialised in specifying, implementing and testing the protocols of the Broadband User-Network Interface (B-UNI).

Within TRIBUNE work is being done on many aspects of the B-UNI, informal and formal specifications are being produced, test tools and test cases are under development and implementations of both the user and the network side are being built. It is imperative that the B-UNI is completely and correctly specified, principally because it marks the boundary between equipment of the network operator and the equipment of the customer. The protocols specified at the B-UNI must offer the flexibility which is required if all current and many future services are to be supported. Also by ensuring that the B-UNI is well specified this boundary is opened up for competition, equipment from a number of vendors can be interconnected and should interoperate correctly.

Implementations of the latest international standards for the B-UNI will be integrated in TRIBUNE to form an operational network. The project is implementing the B-UNI recommendations which meet the ITU-TS release 1 requirements. The network will function as a reference implementation which can be used for testing and demonstration purposes. To increase the value of the project within the RACE programme, the network is available to other European projects which require broadband testing facilities. These projects are generally referred to as the TRIBUNE user projects. A number of candidates have already made plans for testing using the test-bed. For example, the RACE project MIMIS will use the network to test and demonstrate an advanced broadband teleconferencing application. ATM trials of MPEG II video and audio, being developed by the EUREKA project VADIS, will be carried out using TRIBUNE. Several other projects have also announced plans for using the network.

There are three important factors which have been instrumental in insuring that the test-bed is fit for purpose:

- The network has been designed to take into account the requirements of user projects. All types of user equipment can be accommodated, it is however a prerequisite that the equipment has a B-UNI for connection to the test-bed.
- The specification of the B-UNI has been based on the emerging ITU-TS specifications and every effort is being made to remain in line with changes as they occur. Users of the network are therefore developing equipment which is in line with the latest standards and consequently close to commercial exploitation.
- A testing strategy has been produced which is intended to enable all elements of the network to be efficiently and thoroughly tested. Complete testing is imperative if the network is to function as the reference it is intended to be.

User projects are encouraged to follow the TRIBUNE testing strategy as closely as possible.

These factors, design, specification and testing, which are of great importance in the project, will be elaborated upon in this paper.

2. TRIBUNE TEST-BED DESIGN

The TRIBUNE project will result in the realisation of a broadband test-bed. This test-bed will offer a standardised B-UNI for the connection of user equipment. This section describes the overall design of the TRIBUNE test-bed.

Figure 1 shows the design of the test-bed. The test-bed equipment is centered around the B-UNI and can be divided into the following five modules: the ATM switch, the SDH switch, the TRIBUNE Control Network, equipment for interconnection to external networks and the user side equipment (terminals). Each of these modules will be elaborated upon.

The heart of the network is formed by the ATM switch which offers a number of standardised B-UNIs for the connection of user equipment. In order to meet the objective of supplying a test-bed for third-parties a number of possible physical interfaces are supported by the switch. These are: eight SDH based interfaces (all electrical) and four cell based interfaces (which can be configured as either electrical or optical). The bit rate supported by these interfaces is 155 Mbit/s. Operations and Maintenance (OAM) flows are only supported by the SDH framed interfaces. An overview of the physical interfaces supported by the switch is given in table 1.

The switch is able to route ATM cells from any input to any output port. Routing takes place based on the Virtual Path Identifier (VPI) or the Virtual Connection Identifier (VCI) in an ATM cell. The switch is therefore able to function as a VP cross connect and a VC switch. The switch can perform both broadcasting and multi-casting functions. In broadcast mode cells received via a particular port are duplicated and sent to all other ports; in multi-casting mode cells received via a particular input are duplicated and sent to selected output ports. The broadcast and multi-casting mode of the ATM switch will be used by the RACE II project MIMIS to demonstrate their point-to-multipoint broadband tele-conferencing application.

The ATM switch is controlled by the ATM switch controller. This controller is able to access and make available a number of parameters regarding the current status of the switch and the active connections (i.e. cell deletion due to policing, total number of cells per interface per time interval, ATM header single bit error, ATM header multiple bit errors and buffer overflow conditions). These parameters are available to user projects to facilitate for example performance testing scenarios. For projects interested in testing their equipment when cell losses occur, the switch is able to operate in tromboning mode. This means that the load of the switch can be increased to its limits thereby forcing cell loss due to buffer overflow.

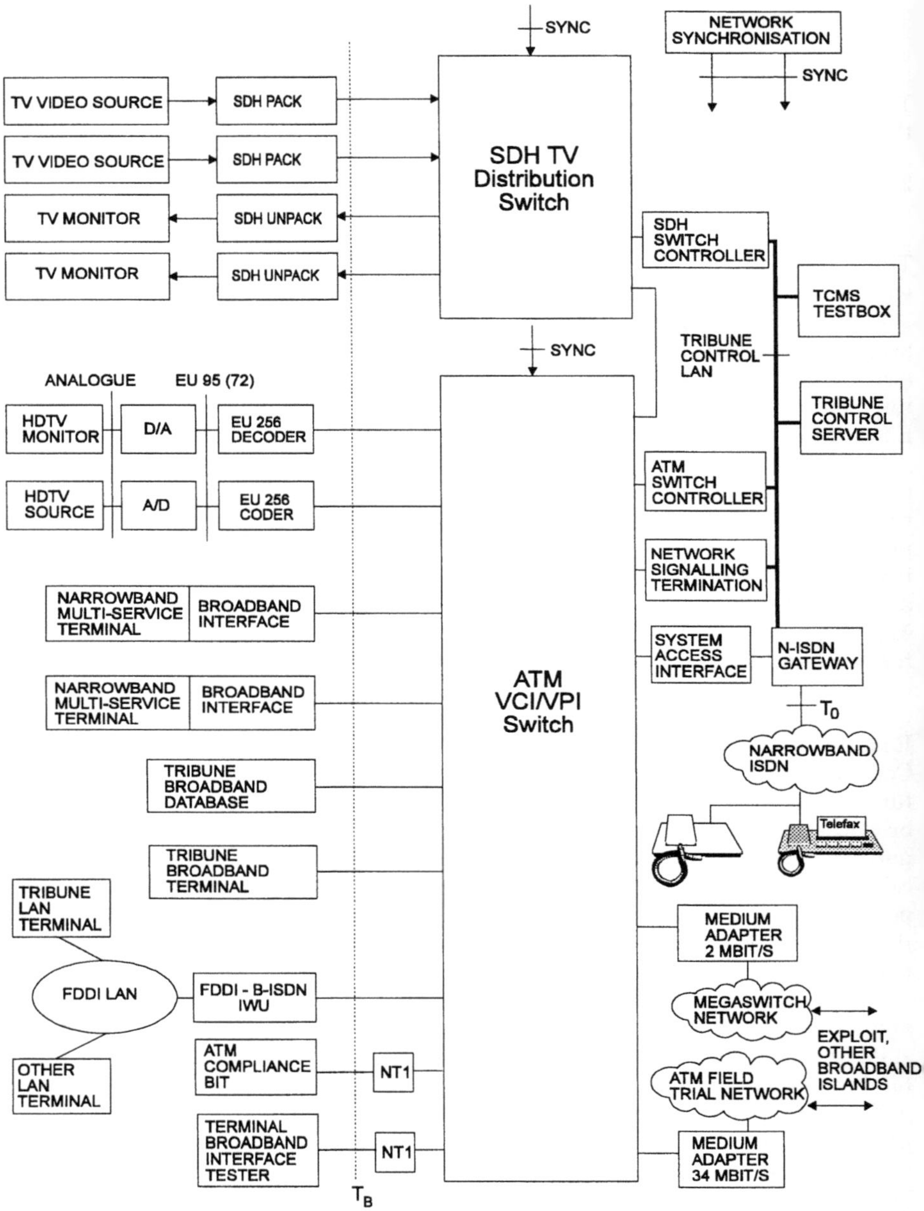

Figure 1. TRIBUNE Test-bed Design

Table 1
Interfaces supported by the TRIBUNE ATM switch

Interface type	Physical Medium	TC sub-layer	ATM Layer
Cell based	Optical according to G.957 Bit timing Physical medium	According to I.432 Cell rate decoupling HEC sequence generation/verification Cell delineation and scrambling	According to I.361 (deviations from I.361 are documented in [2])
Cell based	Optical according to G.703 Bit timing Physical medium	According to I.432 Cell rate decoupling HEC sequence generation/verification Cell delineation and scrambling	According to I.361 (deviations from I.361 are documented in [2])
SDH based	Electrical according to G.703 Bit Timing Physical medium	According to I.432 Cell rate decoupling HEC sequence generation/verification Cell delineation and scrambling Transmission frame adaption Transmission frame generation/recovery	According to I.361 (deviations from I.361 are documented in [2])

Connection Admission Control (CAC) and User Parameter Control (UPC, also referred to as policing) are both supported by the switch based on the ATM peak user cell rate. CAC guarantees a users' Quality of Service. This means that the switch is able to transfer the cells within the stipulated limits for cell loss. The maximum delay and also the delay jitter are taken into account when the addition of a new connection is being decided. Policing consists of monitoring traffic belonging to a particular connection, to determine whether or not the characteristics of the connection are those agreed when the connection was set up. Policing also means taking certain action if the connection exceeds these agreed characteristics. In the TRIBUNE ATM switch this action will be the deletion of cells belonging to the connection. The police function board of the switch is able to police cell streams with a speed of 0.35 cell/s (135bit/s) up to the full link rate of 155 Mbit/s.

The ATM switch is connected by a single bi-directional connection to the distributive SDH (Synchronous Digital Hierarchy) switch. In TRIBUNE this switch will be used as a distributive switch for demonstrating TV services and will be included in performance testing scenarios. The switch has sixteen uni-directional data inputs and sixteen uni-directional data outputs each with a bitrate of 155 Mbit/s. All interfaces are electrical with SDH framing. No OAM flows are supported by the switch. The switch is capable of distributing sixteen STM-1 (Synchronous Transport Module-1) channels without affecting the payload of the carrier. It should be noted that the SDH switch routes the entire payload of all STM-1 frames received on an interface to the selected outputs. In contrast, the ATM switch unpacks the SDH frame and routes the payload as individual ATM cells. The SDH switch for example can be used by user projects to enlarge the end-to-end delay of ATM cell streams passing the ATM switch, thereby altering some performance characteristics of a connection.

A separate network has been designed to perform the functions required for controlling the network and performing the network side call control functions. This network, the TRIBUNE Control Network (TCN), is based on an Ethernet Local Area Network (LAN). To enable the afore mentioned functions to be carried out a number of pieces of equipment are connected to the LAN. These are: the two switch controllers, the TRIBUNE Control Server (where the call control is implemented), the Network Signalling Termination (where the network signalling is terminated) and the N-ISDN Gateway (see figure 1).

The TRIBUNE Control Server (TCS) is the central co-ordination point of the TCN. It provides the following functions: call control, switch control and some administrative functions. The TCS communicates with the other TCN modules via specially designed messages, the TRIBUNE Control Messages (TCMs).

Termination of the signalling protocols on the network side of the B-UNI is performed by the Network Signalling Termination (NST). In TRIBUNE Meta-Signalling is not implemented as only point-to-point scenarios are foreseen at the B-UNI. On each interface of the ATM switch signalling is performed using the pre-assigned signalling VCI. The ATM switch performs translation of this pre-assigned VCI so that the switch is able to distinguish various pieces of user side equipment. After this translation all signalling cells are routed to the NST. The NST sends the primitives which emerge from Access Signalling via a TCM to the TCS. When the NST receives messages from the TCS, these are passed as primitives to the Access Signalling protocol.

The TCS, the NST and both switch controllers provide extensive monitoring capabilities of the call and connection control activities in the test-bed. This enables user projects to access a lot of diagnostic information concerning call and connection problems.

Three modules for interconnection to other networks are attached to the ATM switch: an N-ISDN Gateway, a Medium Adapter 2 Mbit/s and a Medium Adapter 34 Mbit/s. The N-ISDN Gateway will provide access to the N-ISDN network from the network and vice versa. The N-ISDN Gateway will provide three basic rate interfaces to the Dutch N-ISDN network (ETSI ISDN). Two Medium Adapters are connected to the ATM switch. Both Medium Adapters perform mapping of ATM cells into the PDH (Plesiochronous Digital Hierarchy) frame structure. The Medium Adapter 2 Mbit/s enables the TRIBUNE network to be connected to the switched 2 Mbit/s network. The Medium Adapter 34 Mbit/s will enable the TRIBUNE network to be connected to the European ATM Field Trial network. A 34 Mbit/s link will connect the test-bed with the international ATM pilot node of

The Netherlands, which is located in Amsterdam. Via the Field Trial Network the test-bed will be connected with other RACE networks, for example, the EXPLOIT network in Basel. The interconnected test-beds can be used to demonstrate broadband services and perform more extensive end-to-end testing. Users of these experimental services will be able to gain a realistic feeling for future broadband services. The European ATM Pilot Network does not support signalling so management procedures are required to set up connections. Both the N-ISDN Gateway and the Medium Adapter are available to user projects who wish to to connect to other networks.

The ATM switch, SDH switch, TCN, N-ISDN Gateway and Medium Adapters basically form the network which is available for the connection of user equipment. However, besides making a test-bed and test tools available to user projects the TRIBUNE project has some additional sub-goals. These are the demonstration of:

- the distribution of broadband TV and HDTV signals;
- the interworking between N-ISDN and B-ISDN;
- the interworking between a broadband LAN (FDDI) and an ATM network;
- the interworking and interconnection between different broadband islands.

The user side equipment which is rendered in figure 1 each address one of the above mentioned objectives. Each piece of equipment will now be elaborated upon.

The distribution of television signals is one of the services which can be demonstrated with the network. A number of PAL sources are available. The PAL signal is digitally encoded at 108 Mbit/s. After being encoded in digital form the signal will be packed into an STM-1 frame. The resulting SDH signal is transmitted to the SDH switch. Two SDH pack modules are available. Equipment to perform the reverse process is also available, i.e. to unpack the SDH frame and to reconstitute the analogue TV signal. TV monitors are available to display the TV signal.

To enable HDTV signals to be transported by an ATM network operating at 155 Mbit/s, the raw HDTV signal must be encoded before it can be transported across the network. The EU-256 codec, which has been developed as part of the Eureka programme, has the required functionality to enable digital video signals to be interfaced to a B-UNI and on the reception side to restore the coded signals to the required digital video standard. The functionality of the codec system will be enhanced in TRIBUNE to implement the signalling protocols of the B-UNI. Both an HDTV source and an HDTV monitor are available at the TRIBUNE site.

A number of equipment modules are being developed to demonstrate the interworking between a broadband LAN (FDDI) and a broadband ISDN network [1]. These modules are: an ATM-FDDI interworking unit, a broadband terminal and a broadband database (see figure 1). The interworking scenario aims to demonstrate the retrieval of multimedia broadband data from the broadband database via either the broadband terminal attached to the ATM network or an FDDI terminal attached to the FDDI network. The application which has been chosen for the demonstration is a hotel information system which allows the user to view holiday venues and select hotel accomodation. The information about the hotels is offered to the user as text, graphics, still picture, video and audio. In this demonstration the MPEG II codec developed within the EUREKA VADIS project is used.

Besides being used to demonstrate the interworking between FDDI and an ATM network, the FDDI network is available for the connection of user projects' equipment. Either single attachment or dual attachment stations may be connected to the FDDI network. The network protocols used are TCP/IP. These protocols have been chosen due to the availability of many commercial implementations. The use of XTP (eXpress Transport Protocol) may be investigated later in the project if a suitable implementation of XTP is available.

Two codecs are included in the project to provide a demonstration of the different broadband services which can be expected. The HDTV codec demonstrates a very high bandwidth service giving a very high quality picture. The MPEG II codec illustrates image services using a very high compression ratio. MPEG II codecs will typically be used for interactive image services such as video on demand.

Two Multi-Service Terminals will be developed for connection to the test-bed. They will enable the demonstration of interworking between B-ISDN and N-ISDN. The terminals will support a number of narrowband services (teletex, telex, telephone, telefax group III and IV, file transfer and video services). Although supporting N-ISDN services the terminals will contain a broadband interface. The two terminals are able to communicate with each other as well as with terminals connected to the public N-ISDN network. The connection to the public network will be realised via the N-ISDN Gateway.

3. TRIBUNE TEST-BED CONTROL

A data path through the test-bed can be provided in either of two ways: manual configuration of the ATM and/or SDH switch or automatic configuration of these switches. Automatic configuration is provided if broadband signalling is implemented in user equipment. The control of the test-bed will be addressed in this section.

Manual configuration of the test-bed is done by means of the TCS. The TCS operator uses the Human Machine Interface of the TCS to assemble the control message needed to configure the appropriate switch and containing the required connection type. With this degree of manual control it is possible, for example, to establish virtual paths through the network, or to setup virtual connections with or without a policing function. For the manual configuration of the SDH switch the input port and output port of the switch have to be provided. For configuration of the ATM switch the TCS operator has to provide the input port number, the output port number, the VPI/VCI value, the peak cell rate, the direction and the policing choice.

Manual control of the test-bed is necessary for those user projects which are not planning to implement signalling in their B-UNI. These user projects only require a data path through the network. The network will assign these users a VPI/VCI combination.

For user projects which plan to implement signalling the TRIBUNE test-bed supports automatic configuration of the ATM switch. User equipment will be assigned a directory number based on the port number of the ATM switch to which it is connected. The equipment modules which provide the automatic configuration of the switch are the NST (termination of network side signalling), the TCS (call control) and the controllers of the ATM and SDH switch (connection control). In order to appreciate this split of functions, consider the setup of a call across the network.

When a user terminal wishes to establish a call, a signalling message is sent over the B-UNI. This message contains, amongst other elements, the Directory Number of the called party, and a description of the type of connection required. This signalling message is transferred across the B-UNI using the signalling VCI 5. The ATM switch is preconfigured to route such cells to the NST. The access signalling function in the NST will check the signalling message and notify the TCS. The TCS will check that the calling user is allowed to make the call. If so, the routing algorithm is run. This will return the ATM switch port number which corresponds to the Directory Number of the called party which was contained in the signalling message received. Together with the ATM switch port number of the calling user and the connection type wanted, this information is passed to the ATM switch. After performing CAC the ATM switch passes the result (connection accepted or rejected) to the TCS. Based on this result the TCS will initiate additional signalling to the called party or terminate the signalling with the calling party (via the NST).

4. THE TRIBUNE B-UNI SPECIFICATION

The TRIBUNE B-UNI specification will follow as closely as possible the recommendations being produced by ETSI and the ITU-TS. These recommendations describe all the layers and sub-layers of the B-UNI. TRIBUNE will produce a specification which is according to the ITU-TS release 1 requirements.

The TRIBUNE B-UNI specification specifies what will be supported by the test-bed. In this respect it is directed towards implementors of the specification, it does not contain all the options which are possible at the B-UNI. When necessary, additional information has been added to the specification to complete it and tailor it to the requirements of the TRIBUNE project. Within TRIBUNE every effort will be made to ensure that the specification is clear and unambiguous and that consequently the undesirable situation of non-interworking equipment will be avoided. Formal specification of the ATM

Adaptation Layer and above using SDL'88 are part of this approach. All implementations are based heavily on this SDL specification and are partly generated automatically from this specification.

The following subsections identify the origin of each part of the specification and give an outline of any modifications which have been made (some of these modifications have been contributed to the standardisation bodies). More detailed information can be found in the Initial TRIBUNE B-UNI Specification [2].

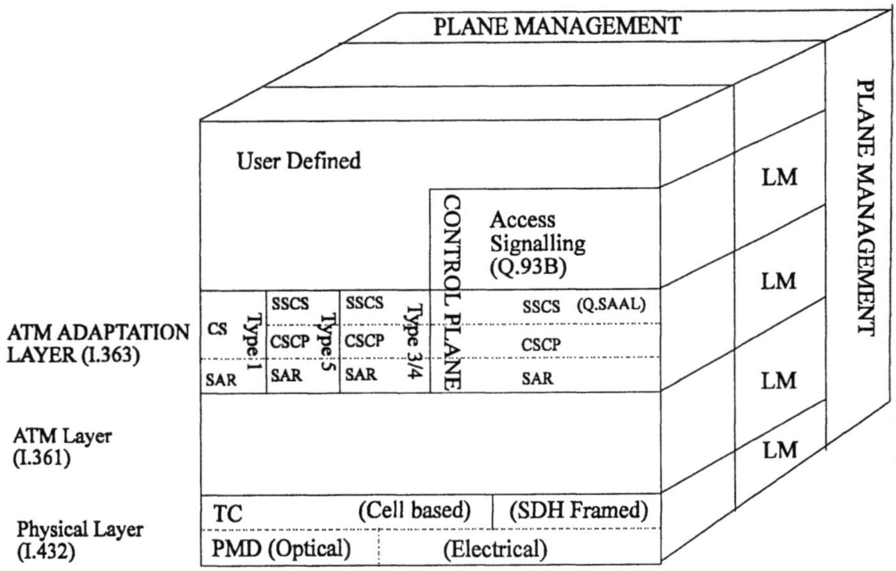

Figure 2. TRIBUNE B-UNI Protocol Reference Model

4.1 The B-UNI Protocol Reference Model

The Protocol Reference Model used for the specification is based on ITU-TS Draft Recommendation I.321. It is illustrated in figure 2. The model is divided into a Control Plane, a User Plane and a Management Plane:

- The Control Plane performs call and connection control functions.
- The User Plane provides user information transfer.
- The Management Plane provides functions to manage the entire protocol stack. Plane Management provides co-ordination between each of the layers and planes, and performs management functions related to the overall

system. A separate Layer Management (LM) entity is associated with each layer, and performs management of resources specific to its associated layer.

4.2 The Physical Layer

The specification for the Physical Layer is based on ITU-TS Recommendation I.432 (June 1992). Additional information for the OAM aspects has been taken from ITU-TS Recommendation I.610.

The Physical Layer specification is divided into two sub-layers, the specification of the Physical Medium Dependent (PMD) sub-layer, and the specification of the Transmission Convergence (TC) sub-layer. The PMD sub-layer provides a bit transmission capability. The TC sub-layer performs the functions necessary to structure the ATM cell stream for transmission and recovery at the receiving end. Three possible combinations of PMD and TC will be offered by the TRIBUNE network (see also table 1):

- TC based on SDH framing and PMD based on electrical transmission at 155.52 Mbit/s.
- TC based on pure ATM cell based transmission and PMD based on electrical transmission at 155.52 Mbit/s.
- TC based on pure ATM cell based transmission and PMD based on optical transmission at 155.52 Mbit/s.

4.3 The ATM Layer

The ATM Layer specification is conformant with ITU-TS Recommendation I.361, with some simplifications because of the limitations of the TRIBUNE ATM switch. The most important limitations are:

- No use of Cell Loss Priority bit.
- No Meta-Signalling (since only point-to-point signalling connections are supported in TRIBUNE).
- No traffic and congestion control functions.

4.4 The ATM Adaptation Layer

The core parts of the ATM Adaptation Layer (AAL) are derived from the ITU-TS recommendation of the AAL (ITU-TS Recommendation I.363). Some simplifications have been made because of the limitations of the TRIBUNE switch on congestion control and cell loss priority handling.

The AAL to be used for the transport of signalling information shall be AAL type 5. This AAL type 5 can be divided into three sub-layers: the Segmentation and Reassembly sub-layer (SAR), the Common Part Convergence Sub-layer (CPCS) and the Service Specific Convergence Sub-layer (SSCS). The SSCS should enhance the service of the Common Part of the AAL to the actual needs of Access Signalling.

In the Initial TRIBUNE B-UNI specification [2] no SSCS is specified. The SSCS will be included in the Enhanced TRIBUNE B-UNI specification which will be issued in December 1993. This SSCS will be completely conformant to the Q.SAAL recommendation of ITU-TS. There are also several AAL types specified within the TRIBUNE project that can be used for the transport of user data. These AAL types include AAL type 1, AAL type 3/4 and AAL type 5. It should be noted that user projects are free to implement their own AALs (provided they do not need the N-ISDN Gateway or the LAN Interworking Unit).

4.5 Access Signalling

The specification for Access Signalling (AS) in the Initial TRIBUNE B-UNI specification [2] is based on ITU-TS Draft Recommendation Q.93B of September 1992. Modifications to the specification to comply with agreements made by ITU-TS since September 1992 are being included in the living TRIBUNE B-UNI Specification.

The main assumptions made in the tailoring of the specification to the TRIBUNE application were as follows:

- All messages will use the en-bloc sending mode.
- A point-to-point physical interface configuration at the B-UNI is assumed.
- Enhancement of the specification for control of distributive calls (point-multipoint calls).

Furthermore some procedures that were not fully developed (such as some of the negotiation procedures) have been left out.

4.6 Availability of TRIBUNE B-UNI Specification

The Initial TRIBUNE B-UNI Specification [2] is available for user projects which are interested in implementing the TRIBUNE B-UNI. For implementors of the TRIBUNE B-UNI a living specification of the TRIBUNE B-UNI is available which incorporates the latest changes in the standard. Usually the formal SDL specification of AAL type 5, SSCS and AS can also be made available to user projects.

5. TRIBUNE TESTING STRATEGY

It is the main objective of the TRIBUNE project to offer to users a reliable and robust test-bed with a standardised B-UNI. To meet this objective a thorough testing strategy has been developed. This testing strategy will be used during the integration and operational phase of the test-bed. The testing strategy is heavily based on experience gained during the integration phase of the RACE BUNI Demonstrator project (R.1081). New state-of-the-art techniques have been introduced in this strategy, a glass box approach is used and the ferry clip testing method applied. Furthermore formal specifications of protocols and test cases are made in TRIBUNE. All these features of the testing strategy will be addressed in this section.

From the start of the project it was clear that specifying, developing and integrating a complex test-bed would not prove easy to achieve in only three years by a consortium of at least eleven different companies. The aim of testing therefore should not be to burden, but rather to facilitate this whole process of specification, development and integration of the test-bed. A glass box approach will make this possible.

A glass box approach to testing aims at better systems and a reduced testing phase in the system life cycle. According to the glass box concept this can be achieved with the following principles [3]:

- Specifications must have a high quality. A system cannot be better than its specifications.
- Specifications must be provided to the supplier in time to allow for sufficient time for system development.
- The supplier must demonstrate that the process of system development is carried out according to good engineering practice. This includes testing practices.
- The supplier is responsible to demonstrate that the system conforms to what has been agreed in the contracts.
- The supplier must report progress at regular intervals, to avoid surprises with respect to the final delivery date of the product.

Although not all principles are applicable to the TRIBUNE project the general guidelines of the glass box approach are adopted in the project. Basically in TRIBUNE testing is made an integral part of the life-cycle of a piece of equipment. Testing should start in the earliest possible phase of the life cycle of an implementation and be visible not only to the implementor, but also to other members of the project.

The first step in the TRIBUNE glass box approach is the specification of parts of the control plane of the B-UNI (Common Part AAL, SSCS and AS) in the formal specification language SDL'88. This will allow validation of the specification, using validation tools. Furthermore code generators can be used to make an implementation of the B-UNI. Based on the commercial availability of these tools and the fact that SDL'88 is standardised by the ITU-TS the choice for SDL'88 was made. Formal specification of protocols also reduces the effort necessary for interoperability testing if communicating equipment are all based on the same formal description. This could lead to significant time savings in the integration phase of the test-bed.

A second step into the TRIBUNE glass box approach is the adoptation of the ferry clip testing method, introduced by Zeng and Rayner [4]. This method is designed to alleviate the difficulties of the well known distributed test architecture [5]. The approach can be succesful due to the introduction of a direct test access inside a B-UNI implementation. This requires a relatively small adaptation to the B-UNI implementation. The ferry clip method will be elaborated upon in the next subsection. The testing facilities which the project offers to user projects, to a certain extent depend on the implementation of the ferry clip by these user projects.

A last step in the TRIBUNE glass box approach is the formal description and automatic generation of test cases. As manual derivation and maintenance of test cases is a very time consuming process, in TRIBUNE a strategy is adopted to derive the test cases from the formal description of the different protocol layers of the B-UNI. The description language used for the test cases is the Tree and Tabular Combined Notation (TTCN), as it is the test notation language standardised by ISO [5]. Using TTCN as a description language aids in readability of the test cases and should lead to repeatable results. A TTCN-to-C compiler will be developed to facilitate the transformation of the test cases to executable format. Using the formal specification of the different protocol layers for the automatic derivation of test cases also helps in the maintainabilty of the generated test suites. As the formal specifications are updated on new released ITU-TS recommendations, less effort is needed to generate new conformant test suites.

The TRIBUNE test cases will be executed on the Broadband Interface Tester (BIT), which will be available to user projects at the TRIBUNE site. The BIT will be further elaborated upon in the test tools subsection.

5.1 The Ferry Clip Testing Method

As part of the TRIBUNE glass box approach the ferry clip testing method is adopted by the project. The ferry clip method allows a (sub-)layer of a B-UNI implementation to be tested in complete isolation from his environment. This is very useful in diagnostic testing during the development phase of a B-UNI implementation since it can be used to locate errors before the complete system has been integrated or finalised. The ferry clip method will be further elaborated upon in this subsection.

The ferry clip principle was introduced by Zeng and Rayner [4]. This principle defines a way to have direct access to a single layer of an implemented protocol stack. The ferry principle aims to use the power of the local test method [5], while reducing the amount of test functionality inside the implementation. In the case where it is possible to access both the upper and lower service boundaries of the Implementation Under Test (IUT, i.e. a single protocol layer) this new test method is referred to as the ferry clip.

To test an IUT, two ferry entities are implemented: the Active Ferry Clip (AFC) in the test system and the Passive Ferry Clip (PFC) in the System Under Test (SUT). This is illustrated in figure 3. The arms of the PFC are attached to the upper and lower service boundaries of the IUT to provide direct control and observation of each boundary. The arms of the AFC are attached to the Lower Tester (LT) and the Upper Tester (UT) in the test system. Running the ferry control protocol, the two clips provide a transparent data transfer service which allow their users (IUT, UT and LT) to exchange test data. This service enables the UT and the LT, external to the SUT, to function as if they resided in the

SUT. In order to transfer ferry data units, a ferry transport service is needed. This service can be an existing data transfer service or a user defined one. Within TRIBUNE the ferry transfer service will be provided by TCP/IP running over Ethernet. The test manager in the test system is responsible for operator event processing, test cases selection and ferry channel management. The lower tester and upper tester are test case executors and test event handlers.

In order to have access to the upper and lower boundary of a protocol layer (the IUT) of a B-UNI implementation test boundaries are defined in the SUT (rendered as Interface Adapters (IA) in figure 3). These test boundaries can be viewed upon as Points of Control and Observation (PCOs). The test boundaries are defined in such a manner that data can be monitored, inserted at, or extracted from these boundaries. The TRIBUNE project has specified a flexible ferry control protocol that can configure PCOs (test boundaries) located in the SUT to perform one of these tasks.

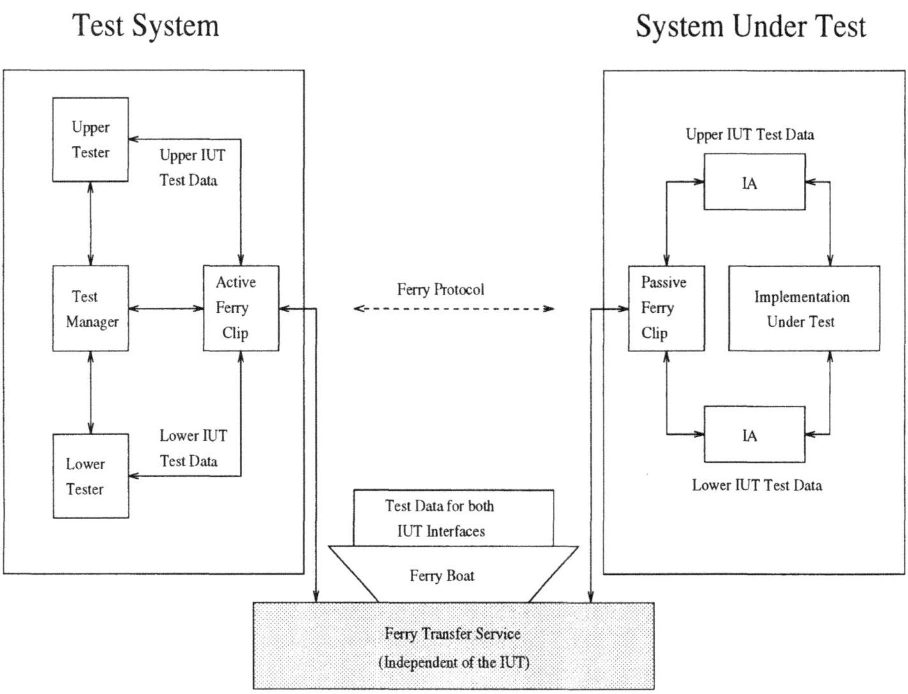

Figure 3. The Ferry Clip Testing Approach

The usage of the ferry clip testing method requires extra test functionality inside a B-UNI implementation. Compared to the total effort of implementing a B-UNI this effort is minimal. The main effort is the development of some software routines. User projects are encouraged to implement this functionality as a number of the available TRIBUNE test tools are based on the ferry clip testing method. A formal description of the ferry control protocol usually can be made available to user projects. Although advised, implementation of the ferry clip is not mandatory for user projects. Test tools are available at the TRIBUNE site to support the testing of equipment without implemented ferry clip. However test scenarios for such equipment are less extensive.

For an extensive overview of the usage of the ferry clip method in the TRIBUNE project please refer to [6].

Within the project the ferry clip testing method is not only used in stand-alone testing of terminal equipment, but also in the integration and operational phase of the test-bed. The ferry clip specification used in TRIBUNE allows for the usage of two or more PCOs situated in different implementations. This enables the usage of the ferry clip for layer-to-layer interoperability testing. This will not further be elaborated upon in this paper.

5.2 Test Tools

Protocol conformance testing encompasses both formal conformance testing carried out at test laboratories for certification purposes and diagnostic testing carried out by protocol implementors during development. In TRIBUNE a number of test tools are built to assist in testing implementations of protocols of the B-UNI on the conformance to their specification. The term compliance testing has been adopted to address this diagnostic part of protocol conformance testing. This section addresses the different test tools which will be available to user projects to test the compliance of their protocol implementations. Two of these test tools rely on the ferry clip approach described in the previous section.

In TRIBUNE emphasis is placed on compliance testing the Control Plane of the B-UNI, as from a network point of view this is the vital part of the B-UNI. As the type of traffic over a user channel can be very diverse, (as are the number of possible protocols in the User Plane) compliance testing of protocols in the User Plane is left to the responsibility of the user projects and is not supported by any TRIBUNE test tools.

A number of measures have been taken to ensure a thorough compliance testing of the Control Plane part of the B-UNI. The B-UNI has been logically split into four different parts, in the SUT between these parts PCOs are defined, the PFC is implemented in the SUT and loop back functionality for testing the lower layers is defined. The division of the B-UNI into separate groups is used as a basis for the need of test equipment. This resulted in the following test tools:

The ATM compliance BIT

Testing the PMD, TC and ATM layers of a B-UNI implementation is done with the ATM compliance BIT. The ATM compliance BIT is not based on the ferry clip testing method. As the PMD, TC and ATM layers (and Common Part AAL) reside in hardware it is difficult to define and implement a PCO on the boundary between the AAL and the ATM layer. To resolve the problem of testing these layers loop back functionality is defined in the ATM layer. By use of an OAM cell the SUT can be configured to loop back the contents of the ATM cell to the tester over a specific Virtual Channel. User projects are encouraged to implement this loop back functionality.

The AAL compliance BIT

The SAR sublayer and CPCS in the Control Plane of the B-UNI are tested with the AAL compliance BIT. The AAL compliance BIT relies on the implementation of a PFC and one PCO in the SUT. This PCO is defined on the boundary between CPCS and SSCS in the Control Plane. This PCO can be reached via the PFC. In testing the SAR and CPCS layers two channels between the AAL compliance BIT and the SUT are used: the ferry channel (TCP/IP running over Ethernet) and the B-UNI.

The Q.93B/Q.SAAL compliance BIT

The SSCS and AS layers in the Control Plane of the B-UNI are tested with the Q.93B/Q.SAAL compliance BIT. This compliance BIT relies on the implementation of a PFC and three PCOs in the SUT. These PCOs are defined between CPCS and SSCS, between SSCS and AS and between AS and Call Control. This allows the SSCS and AS layer to be single layer tested thereby largely reducing the complexity of the test cases. The Q.93B/Q.SAAL compliance BIT is connected to the SUT via the ferry channel (TCP/IP running over Ethernet). The Q.93B/Q.SAAL compliance BIT executes test cases written in TTCN.

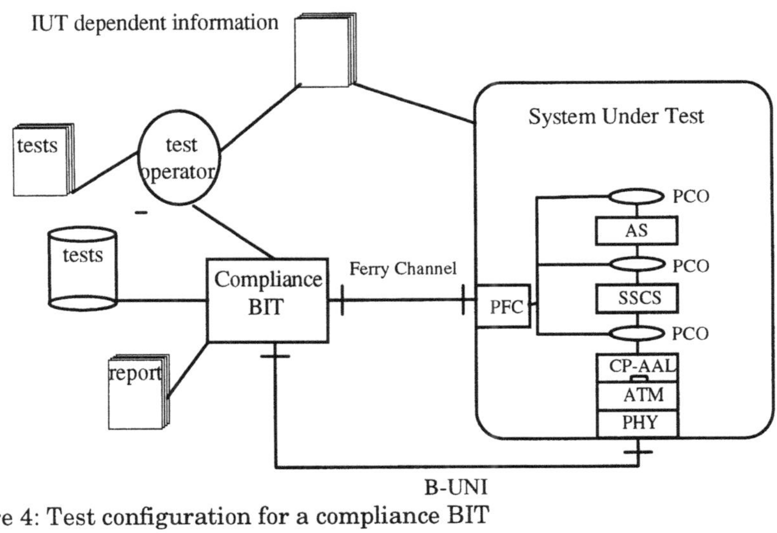

Figure 4: Test configuration for a compliance BIT

The PCOs which are defined in the B-UNI by means of the ferry control protocol can be enabled to monitor, insert or extract data in either direction (up or down) or disabled to act as a transparant passage for data. This means that apart from testing purposes these PCOs can be used to monitor network traffic during the operational phase of the test-bed thereby providing extensive monitoring capabilities.

In figure 4 the general configuration for a compliance BIT is given. The compliance BIT is attached to the SUT via the B-UNI interface and/or via the ferry channel. The compliance BIT is operated by the test operator who has a paper version of the tests executed by the compliance BIT available. Usually the first task of the operator in executing these tests is adjusting the tests to the IUT: some IUT dependent information has to be provided. This information typically concerns options implemented in the IUT. After the test operator has physically connected the SUT to the compliance BIT and parameterised the tests execution can start. After performing a number of tests the compliance BIT generates a test report. The report is then analysed by a protocol expert.

Besides the three configurations of the BIT which are used for compliance testing, two other BIT configurations have been identified in the TRIBUNE project. These are:

The Terminal BIT

The Terminal BIT plays an important role when the TRIBUNE network is being integrated. It is used to check that the AS and Call Control functions of the network and possible user terminals are functioning correctly. Further it is used in interoperability testing scenarios. The Terminal BIT is available to user projects to integrate and test their terminals.

The Performance BIT

The performance BIT can be used when the network has been fully integrated and performance characteristics of the network are measured.

6. TIME SCALES AND FUTURE PLANS

The TRIBUNE project started in January 1992, the first two years of the project have been used predominantly for specification and design work which has led to the implementation of the various network elements. The first half of 1994 will be used to integrate and test the network. The latter part of 1994 can be used to demonstrate the network and associated terminals and to test and evaluate the performance of the network.

For user projects the network will be available for the beginning of 1994 for performing testing up to the ATM layer. As the year progresses the complexity of what can be tested at the TRIBUNE site will increase. By the middle of 1994 all the testing facilities, including those for testing the access signalling will be complete.

TRIBUNE is one of the minority of RACE projects which has been selected by independent evaluators to be extended until the end of 1995. The TRIBUNE extension will be used to exploit the results of the project in terms of demonstrations and maintaining the testing facilities. The extra year will also be used to interconnect the TRIBUNE network with the EXPLOIT network and in this way to contribute to the ATM Field Trial experiments.

More information

The TRIBUNE test-bed will be located at the premises of PTT Research in Leidschendam, The Netherlands. The test-bed facilities are available to all projects which have a European character. Any project interested in using the test-bed should contact PTT Research, the Prime Contractor of the project. The authors of this paper would be pleased to answer any questions and supply any additional information to potential users of the test-bed.

ABBREVIATIONS

AAL	ATM Adaptation Layer	AFC	Active Ferry Clip
AS	Access Signalling	ATM	Asynchronous Transfer Mode
BIT	Broadband Interface Tester	B-UNI	Broadband User Network Interface
CAC	Connection Admission Control	CPCS	Common Part Convergence Sublayer
CS	Convergence Sub-layer	FDDI	Fiber Distributed Data Interface
ISDN	Integrated Services Digital Netw.	IUT	Implementation Under Test
LAN	Local Area Network	LT	Lower Tester
LM	Layer Management	NST	Network Signalling Termination
OAM	Operations and Maintenance	PCO	Point of Control and Observation
PDH	Plesiochronous Digital Hierarchy	PFC	Passive Ferry Clip
PMD	Physical Medium Dependent	SAAL	Signalling ATM Adaptation Layer
SAR	Segmentation and Reassembly	SDH	Synchronous Digital Hierarchy
SDL	Specification Description Lang.	SSCS	Service Specific Convergence Sub
SUT	System Under Test	STM	Synchronous Transport Module
TC	Transmission Convergence	TCM	TRIBUNE Control Message
TCN	TRIBUNE Control Network	TCS	TRIBUNE Control Server
TTCN	Tree and Tabular Combined Not.	UPC	User Parameter Control
UT	Upper Tester	VCI	Virtual Channel Identifier
VPI	Virtual Path Identifier	XTP	eXpress Transport Protocol

REFERENCES

1. Bie, J. de, ATM-FDDI Interworking in TRIBUNE, First Symposium on Interworking, November 1992, Bern, Switzerland.
2. Initial TRIBUNE B-UNI Specification, TRIBUNE RACE Deliverable, Nr S/P/5.2/03, March 1993, Issue 1.
3. Slijkhuis, R., Hendrich, H., Coenen, H., Hoogervorst, M., Verbetering Ontwikkelproces en Software Kwaliteit AXE, KPN NWB Report Nr 91-128, 1991 (in Dutch).

4. Zeng, H.X. and Rayner, D., The impact of the Ferry Concept on Protocol Testing. Protocol Specification, Testing and Verification V, North-Holland publishers, 1986, pp 533-544.
5. ISO/IEC 9646: Information Technology, OSI Conformance Testing Methodology and Framework, July 1991, Part 1, 2, 3, 4, 5.
6. Witteman, M. and Wuijtswinkel, R. van, ATM broadband network testing using the Ferry Principle, Sixth International Workshop on Protocol Test Systems, July 1993, Pau, France.

Broadband Communications, II (C-24)
S. Tohmé and A. Casaca (Editors)
Elsevier Science B.V. (North-Holland)

Implementation of an ATM Network Emulator

Alejandro Alonso Hernán, Luis A. Merayo Fernández, Luisa Mola Morales,

Hernán Salas Asencios

Telefónica Investigación y Desarrollo

Emilio Vargas 6, 28043 Madrid (SPAIN)

Tel.: +34 1 337 46 35

Fax.: + 34 1 337 45 02

e.mail: vax047@beta.tid.es

Abstract

The development and use of Asynchronous Transfer Mode (ATM) technology has experienced a great advance on the last years. At present, it existits a real need for ATM test tools. These test tools have to verify the proper operation of the ATM networks components and to evaluate the global network performance.

This paper describes an ATM Network Emulator (ANE), developed by Telefónica I+D, and capable of emulating the behavior of a real ATM network by means of controlling the main parameters that characterize these kind of networks (constant and variable cell delay, cell insertions, cell losses, cell errors and bit errors). With this tool it is possible to verify the performance of the AAL without the availability of the ATM network, or to emulate loaded conditions by means of controlling the network parameters in the existing unloaded one.

1.-Introduction

On the last years, B-ISDN has experienced a great advance with the development and use of Asynchronous Transfer Mode (ATM) technology.

The test of the network components under a wide range of operating conditions, and the network global performance evaluation is a real need for the ATM projects and experiments.

ATM generators/analyzers have been developed providing testing capabilities for the network lowest layers (physical, ATM). It is possible to measure cell errors, losses, missinsertions and transfer delays using these tools, but none of them take into account the requirements for the ATM Adaptation Layer (AAL) testing. Therefore, a tool to simulate certain network performance parameters in a controlled way is required. Such a tool should transform an incoming ATM flow into an outcoming one as if the original flow had crossed a real ATM network, with special stressing on performance parameters selected by the users.

Telefónica I+D, conscious of the above mentioned need, has developed an ATM Network Emulator (ANE). It is able to emulate the behavior of a real ATM network controlling the main parameters that characterize it (constant and variable cell delay, cell insertion rate, cell loss rate, cell error rate, bit error rate). In this way it is possible to verify the performance of the AAL without the availability of the ATM network, or to emulate loaded conditions by means of controlling the network parameters in the existing unloaded one. Following, some examples of applications scenarios of the ANE are given:

- **Simulation without any network equipment**. In this configuration the ANE is connected to two terminal equipments in order to test the AAL implemented in each one. Note that the ANE does not yet include either signaling nor OAM functions, so this configuration may require some additional network equipment.
- **Simulation with network equipment**. The characteristics of the ANE make it possible to be inserted on any of the network interfaces (SB, TB,

UB or NNI), enabling controlled load simulations in an unloaded network.

- **Other applications** for the ANE are also possible: e.g. it can be used as a Variable Bit Rate (VBR) generator from a Constant Bit Rate (CBR) source when using the variable delay generator block.

This paper is organized as follows: in Section 2 a general architectural description of the ANE is presented, Sections 3, 4, 5 and 6 describe the ANE main components (Line Interface, Core, Control and Operation Unit), finally in Section 7 a summary of the ANE main characteristics is included.

2.-ANE general description

As it has been above mentioned, the main objective of the ANE is to emulate the behavior of a real ATM network through a controlled manipulation of the parameters that characterize it: variable and constant cell delay, cell insertion rate, cell loss rate, cell error rate and bit error rate. As these parameters are associated to the physical and ATM layers, the network emulator has been designed taking into account this multilayer operation. In order to guarantee the layer independence, the physical layer functions should be independent of the ATM layer ones, allowing to associate different physical layers to the same ATM layer emulator.

Other desirable characteristic for the ANE should be to expand this layer independence into a function independence, providing that all functions associated to the ATM layer (cell variable delay, cell insertion rate, cell loss rate and cell constant delay) may be programmed, activated and deactivated independently.

Taking into account the above mentioned requirements, the selected architecture for the ANE is indicated in figure 1:

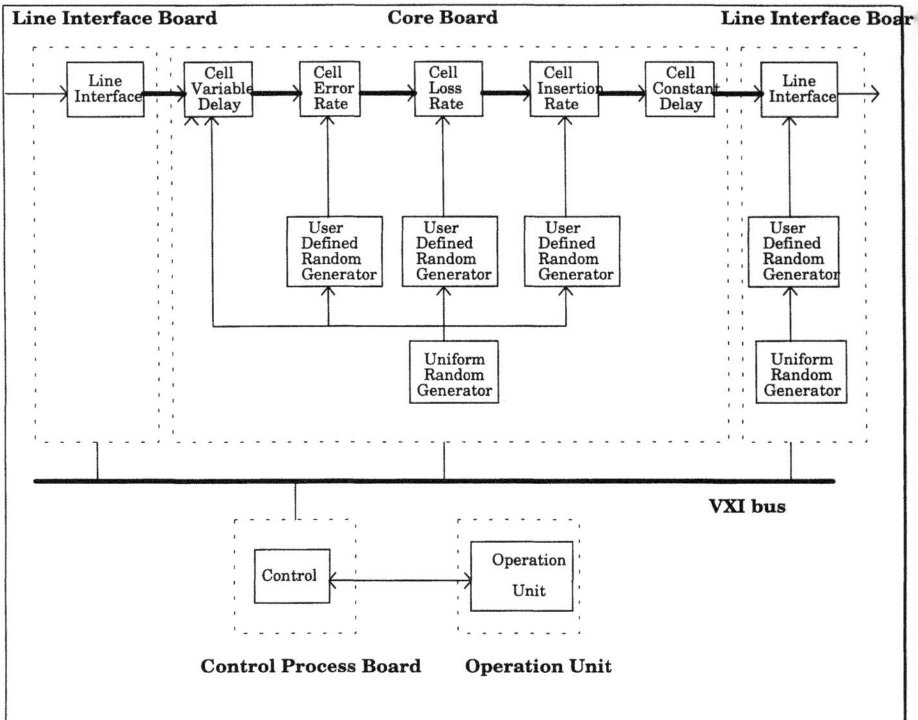

Figure 1. ANE architecture

Physically, the ANE modules are implemented in a standard VXI C-size system that guarantees expansion capability. In the present configuration, a complete system is based on the following elements (see figure 2):

- Line Interface Board with one bidirectional 155.52 Mb/s ATM-STM1 optical interface. In addition to the ATM-SDH and electro-optical conversion, the transmitted SDH frame can be affected by a user controlled Bit Error Rate function.
- Core Board that, operating over the incoming ATM cells, produces constant and variable cell delays, cell losses, cell insertions and cell errors in a controlled way according to the user specifications.
- Control Board that controls and configures Line Interfaces and Cores.
- Operation Unit (486 PC) that provides a user graphical interface (Microsoft Windows based).

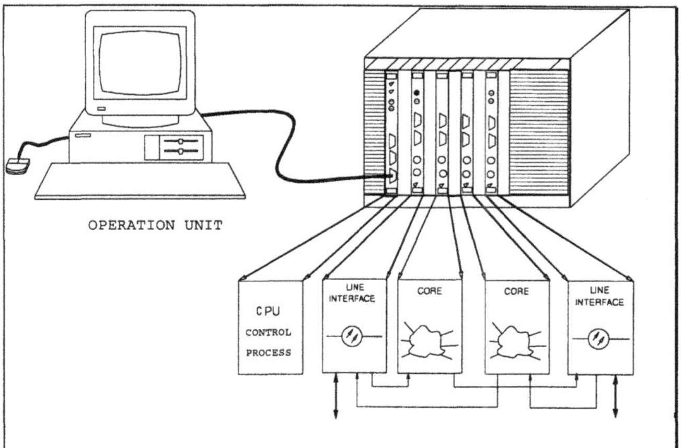

Figure 2. ANE physical structure

3.-Line Interface Board

This board, that incorporates transmitter and receiver functionalities, is responsible for the physical layer functions and simulates the Bit Error Rate (BER) of the transmission line. The Line Interface block diagram is shown in figure 3.

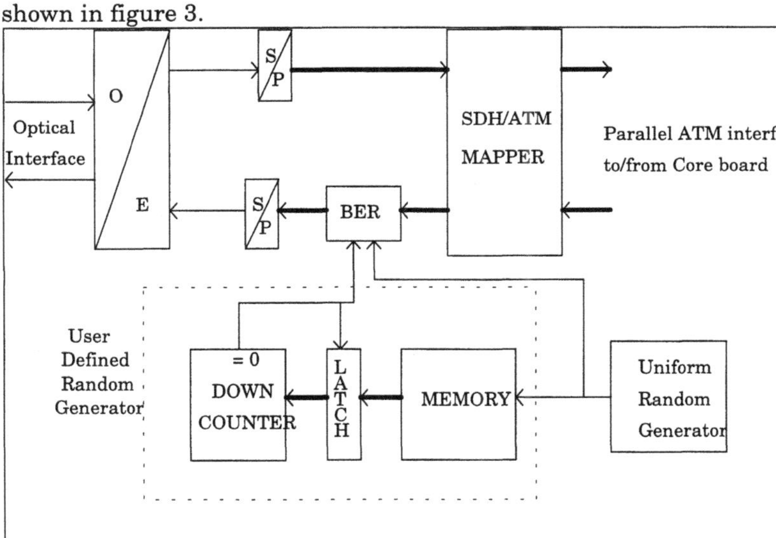

Figure 3. Line Interface block diagram

The board is composed by the following blocks:

- **Optical/Electrical converter (O/E)**.
- **Serial/Parallel (S/P)**.

- **SDH/ATM Mapper**. This block is based on a single ASIC that simultaneously and continuously maps an ATM cell stream to the physical SDH level and vice versa, as per CCITT recommendations I.432, G.707, G.708 and G.709. It generates and detects the physical layer path and section alarms (AIS and RAI) and also provides quality of service statistics in both SDH and ATM links.
- **Bit Error Rate (BER) generator**. This block modifies the SDH frame produced by the SDH/ATM Mapper introducing bit errors that simulate the BER of the transmission line. Given that this type of errors are, generally, bursty, this block is capable of emulating this behavior by generating up to 4 consecutive bit errors in a selected octet. The octet affected by the error/s is selected according to a user defined probability distribution. The probability of having 1, 2, 3 or 4 consecutive erroneous bits is also defined by a user defined probability distribution statistically independent to the distribution used to select the erroneous octet. The BER block can be enabled or disabled with a range (when activated) between 10^{-4} and 10^{-12}.
- **User Defined Random Generator**. This block selects the erroneous byte in the SDH frame based on a user defined distribution implemented by means of the Montecarlo method. The implementation is based in a memory (4Kx40, $2^{40} \approx 10^{12}$) where the user writes the statistical distribution to be generated. This memory is continuously accessed with a random uniform address (generated by the Uniform Random Generator). The data read from the memory is loaded into a down counter. Every time the counter reaches 0, it generates an indication of byte to be modified by the BER block.
- **Uniform Random Generator**. This block generates the uniform pseudo random sequences used in the board.

4.-Core board.

The core board is responsible for emulating the impairments corresponding to the ATM layer. These impairments are variable and constant cell delays, lost cells, inserted cells and errored cells.

The block diagram of the board is depicted in figure 1 (section 2). The main characteristics can be summarized in the following list:

- Possibility to connect several Cores in cascade.
- Parallel ATM interface with the Line Interface Board.
- VXI interface with the Control Process Board.
- Possibility to enable/disable independently each one of the modules associated to the Core.

The remaining part of the chapter describes in detail each one of the modules that constitute the board.

4.1.-Variable delay module.

One of the new aspects that has to be considered when introducing ATM are the impairments introduced by the different transit delay related to each one of the cells associated to a connection. This variable delay is mainly due to the use of buffers in the ATM networks. By this reason, the traffic pattern of the received side is different from the traffic pattern sent by the source. As a consequence of this, it is very important to know the effect of the variable delay on the receiver side. Therefore, it would be of great interest to be able to control this parameter. The network emulator offers a very valuable tool to control the variable cell delay in a very flexible way.

It is necessary to fulfill the next expression because the received cells must arrive in the same order as they are transmitted:

$$VD(t_{n+1}) > VD(t_n) - T \qquad (1)$$

being T the inter departure time of two consecutive cells and $VD(t_i)$ the variable delay associated with cell i.

Other characteristic that has to be taken into account when designing the ANE is the distribution probability function of the variable cell delay. This variable cell delay is mainly due to the use of buffers along the network. Taking into account that a cell, in general, goes though several buffers, and based on the central limit theorem, it is possible to claim that the global variable delay is quite similar to a normal distribution. However, the ATM networks work in a discrete cell time basis instead of a continuous time. Therefore, the variable cell delay shall correspond to a discrete distribution, as similar to the normal one as possible. With these assumptions, the most correct solution is to consider the binomial distribution as the best approach to the variable cell delay, because the similarities between the two mentioned distributions.

The next step is to find a method that complies with the two former constraints (cell sequence and binomial distribution). In the network emulator this is achieved by means of a birth-death process as depicted in figure 4. Selecting the corresponding parameters for birth rate (λ_i) and death rate (μ_i) it is possible to achieve the binomial distribution [1][2] (in these expressions λ and μ are arbitrary constants) :

$$\lambda_i = (N-i)\,\lambda$$

$$\mu_i = (i+1)\,\mu$$

Moreover, a large range of distributions can be easily achieved changing λ_i and μ_i parameters (geometric, Poisson, Engset, Erlang-C and Erlang-B formulas...). The user of the network emulator can provide these parameters in order to achieve one of the former distributions.

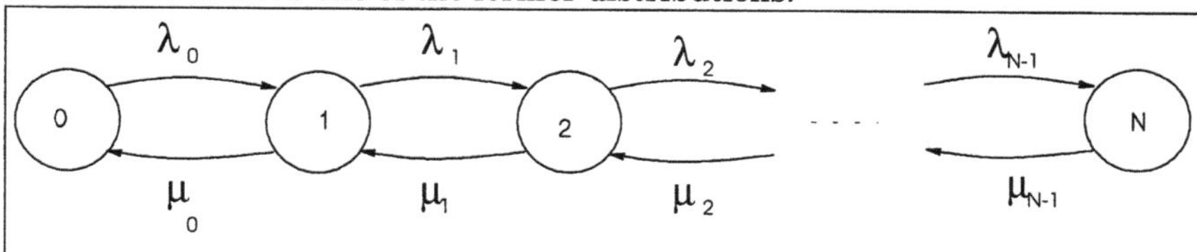

FIGURE 4.-Variable Delay implementation

In order to meet the cell sequence integrity condition, the chain can only change its state when an unassigned cell (see ITU-T Recommendation I.361) arrives at the emulator. In this case, if the chain is in state "i" the next state may be "i-1", "i" or "i+1". The probabilities associated to each of these possibilities are dependent of the λ_i , μ_i parameters [3]. When an assigned cell arrives at the emulator the state of the chain never changes; in this case a variable delay equivalent to the present value of the chain (in cell units) is applied. These actions are summarized in the next table :

Cell type	**Change of state**	**Variable delay**
Unassigned cell	YES	NOT APPLICABLE
Assigned cell	NO	APPLICABLE

The next example clarifies this process.

A cell stream formed by the following data arrives at the network emulator :

A_1 U_1 U_2 U_3 .... U_K A_2 ... (A=Assigned cell, U=Unasigned cell).

If the chain of the network emulator, when the cell A_1 arrives, is in state "n", the following cell delay shall be applicable :

Cell	State	Feasible next states[1]
A_1	n	n
U_1	n	n-1, n, n+1
......	.....	......
U_K	n-(K-1)...,n,...n+(K-1)	n-K,...,n,...n+K
A_2	n-K,...,n,...n+K	n-K,...,n,...n+K

In this example, the inter departure time between two consecutive assigned cells (A_1 and A_2) is T=K+1 (expressed in cell units). As it can be seen, the condition for the arrival in sequence (see expression 1) is always fulfilled :

Variable Delay of $A_2 \geq$ Variable Delay of A_1 - K (see table above)

Variable Delay of $A_2 >$ Variable Delay of A_1 - (K+1)

[1]This indicate the delay (in cell units) applied to the cell. This delay is only applied to assigned cells.

Variable Delay of A_2 > Variable Delay of A_1 - T (see expression 1)

This implementation also imposes the next limitation :

Variable Delay of A_2 < Variable Delay of A_1 + T

Or, in general :

Variable Delay of A_{N+1} < Variable Delay of A_N + T

This implies that the emulator, from the variable delay point of view, supposes that the changes in the network load are not very abrupt. This is equivalent to claim that the variable cell delay of two consecutive cells is always bounded. This constraint is not significant, specially when the links capacity is large (as it is usually the case of ATM networks).

The variable cell delay module of the network emulator implements a chain as the one described above. All the assigned cells are written in a FIFO, being the waiting time of each one exactly the same as the state of the chain in the instant of the arrival. A maximum state value of 256 is allowed in the implementation. This value is equivalent to a varaible cell delay of 724 μsec, obtained by multiplying 256 by the time duration of a cell transmited at 149.76 Mbs (bit rate of the payload of a STM1 frame). Larger figures could be achieved by using more than one Core.

4.2.-Cell error module

This module is quite similar to the Bit Error Rate module implemented in the Line Interface board. The difference is focused in that in this case it is possible to operate over the bits of the payload of the cell in a selective way, instead of operating in the whole SDH frame. It is also important to note that only assigned cells are taken into account, because errors in the unassigned cell payload are not meaningful (the payload of this type of cells is irrelevant). Finally, it is convenient to remark the bursty nature of the bit error in the transmission line. As a consequence of this, it should be convenient to implement in the ANE the possibility of consecutive bit errors. Taken into account all the above comments, the main characteristics of this module are :

- Random generation of bit error in the assigned cell payload, according to a distribution defined by the user.
- Possibility to generate up to four consecutive bits errors, all belonging to the same octet.
- Possibility to select the VPI/VCI of the cells to be errored.
- The cell error rate values lie in the range 10^{-3} to 10^{-10}.

The hardware implementation of this module is depicted in figure 5.

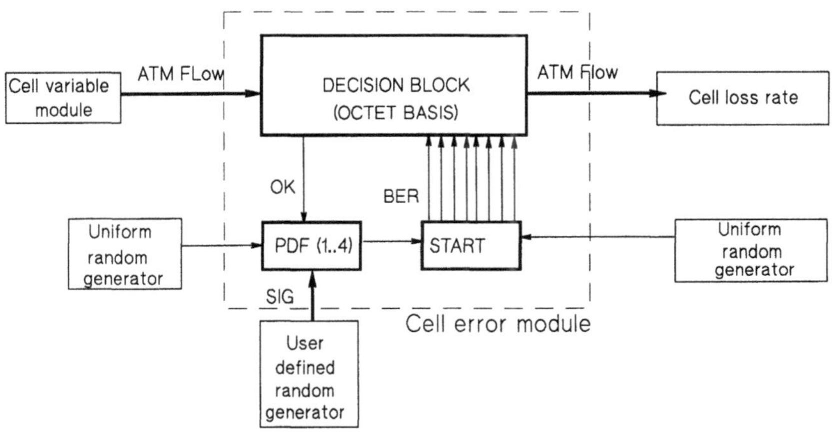

FIGURE 5.-Cell error module

The PDF block decides, in a random way, the number of consecutive bits that have to be changed. All these bits belong to the same octet (the octet is selected by the decision block as described below). For this action it is necessary that both SIG (coming from the user defined random generator, see section 3) and OK (coming from the decision block) signals be active. SIG signal shall be activated in a random way, according to the set of parameters selected by the user. OK signal shall be activate by the decision block (figure 5). The condition for the activation of the OK signal is described below.The number of consecutive errored bits are loaded into the start module. This module decides which bits of the selected octet have to be changed through the activation of the corresponding BER signals (see figure 5). The BER signals that are not activated correspond to bits that have not to be changed. Afterwards, all BER signals are deactivated till the next SIG activation.

The decision block works in an octet basis and activates the OK signals when the incoming octet belongs to the payload of an assigned cell (note that the assigned cell has to belong to one of the VPI/VCIs selected by the user). This block also changes the bits associated to the activated BER signals, coming from the start block.

4.3.-Cell loss module

Cell loss outcomes in ATM networks are mainly due to two reasons :

- Undetected and no corrected bit errors in the ATM header (because of bit errors in transmission lines).
- Buffer overflow.

From these assumptions, the ANE has to fulfill the following constraints:

- The existing bit errors in transmission lines are very low and bursty, therefore, a certain probability of consecutive lost cells arises. A similar effect occurs when buffer overflows. In this case, all cells arriving at the buffer during the overflow state shall be lost. As a consequence, the network emulator has to be capable of generating consecutive cell lost outcomes.
- On the other hand, the type of probability distribution is not yet clearly defined in case of cell lost outcomes. This implies that the cell loss outcomes generated by the emulator must to cover a very spread range of

distributions. The network emulator obtains this distribution from the User Defined Random Variable Generator module, which has been described previously in chapter 3.

- The CLP (Cell Lost Priority) bit included in the ATM header makes it possible to set a cell loss probability as a function of the VPI/VCI values. The network emulator must be able to select a VPI/VCI value (or a set of them) affected by the cell loss. Cells not belonging to this set of VPI/VCIs are never lost.
- Finally, the range of values associated to the cell loss has to be very large. The ANE is able (when the cell loss function is enabled) of handling values of cell loss probabilities which lie in the range from 10^{-2} to 10^{-10}.

The hardware implementation of this module, covering all the above constraints, is depicted in figure 6. The ATM flow entering to the decision block arrives from the Cell error module. The SIG signal comes from the User Defined Random Generator block. This block is similar to the one used in the Line Interface (see chapter 3), so it is not necessary to describe it again.

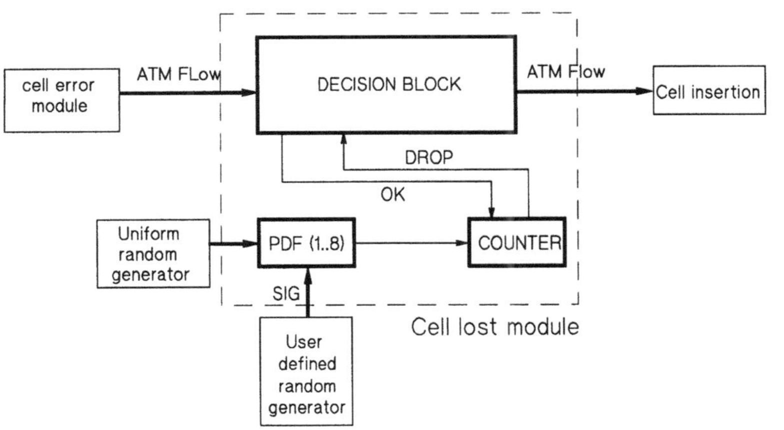

FIGURE 6.-Cell loss module

When the SIG signal is activated (according to a distribution defined by the user), the PDF block randomly determines the number of consecutive cells that have to be lost. This number can vary from 1 to 8, being the probability of each option defined by the user. The number of consecutive lost cells is loaded into a 3-bit counter. This counter is decreased with the activation of the signal OK. This signal is asserted by the decision module when an assigned cell which belongs to the programmed VPI/VCIs set is received through the incoming ATM flow. Every time the counter is decreased, it activates the signal DROP indicating to the decision block that the cell has to be extracted from the ATM flow. This cell is replaced by an unassigned cell, or by an assigned cell with a header previously defined by the user, in order to identify them easier. Finally, when the counter reaches the value 0, this process is stopped till the next SIG signal activation.

4.4.- Cell insertion module

Cell insertion outcomes are mainly due to the non corrected and undetected errors in the ATM header. Furthermore, and in order to produce an inserted cell, it is also necessary that the new errored cell belongs to one existing connection. This is equivalent to claim that the cell insertion probability is very low, being also very low (practically null) the probability of having several consecutive insertions. The final effect of the insertion cell is the existence of new cells belonging to different connections, as if unassigned cells were turned onto assigned cells. Because of these reasons, the main characteristics of this module are :

- Random insertion of assigned cells, according to a distribution defined by the user.
- The assigned cell inserted (header and payload) is also defined by the user.

- No consecutive cell insertion is envisaged.
- Only unassigned cells are replaced by the inserted user defined assigned cell.
- The ranges associated to the cell insertion (when the module is enabled) vary from 10^{-3} to 10^{-12}.

The hardware implementation of this module is depicted in figure 7.

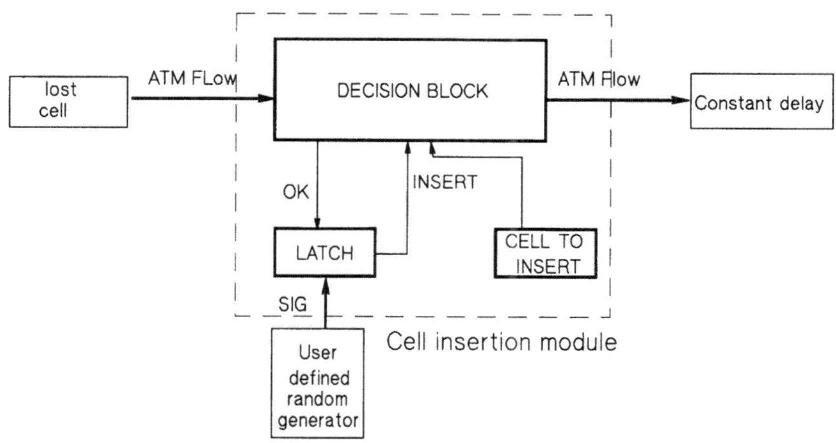

FIGURE 7.-Insertion cell module

The decision block receives the ATM flow coming from the previous module. This decision block activates the OK signal whenever an unassigned cell is received. On the other hand, the latch registers the activation of the SIG signal. This signal comes from the user defined random module (similar to the one described in section 3). Therefore, the activation of this signal follows a distribution probability previously defined by the user. The latch activates the INSERT signal when the SIG signal has been previously registered and the OK signal is activated. When the INSERT signal is activated, the decision block replaces the incoming unassigned cell by the user programmed assigned cell. No new insertions are produced till the next SIG activation.

4.5.-Cell constant delay module

In an ATM network, the constant delay is due to the propagation time of the transmission lines and also due to the latency of the switching and transmission equipment. This delay is deterministic, not being necessary the use of any random generator. On the other hand, one of the features that are desirable for the emulator is the possibility of simulate differential delays between different components belonging to the same call (for instance, simulate the differential delay between voice and video in a video telephony call). With these assumptions, the main characteristics of the constant delay module are the following :

- Delay generation for all the cells associated to a set of VPI/VCIs previously programmed by the user. The rest of VPI/VCI shall have a null constant delay. It is possible to select the delay generation independently of the VPI/VCI. In this case all cells shall be affected with the programmed constant delay.
- In case of selecting a set of VPI/VCIs for constant delay, the emulator does not generate a real constant delay. It only assures a minimum delay associated to these cells. This minimum delay is equivalent to the constant delay requested by the user. The cells which do not belong to the set of VPI/VCI shall be applied a null delay.
- The maximum constant delay per core is 54 msec. Longer constant delay values can be achieved by using more than one Core.

The hardware implementation associated to this module is depicted in figure 8. The ATM flow arrives at the write block. This block reads the VPI/VCI and, depending on the VPI/VCI value, it writes the incoming cell into the null delay block or into the minimum delay block. The null delay block is a FIFO with a capacity of 1 ATM cell. The minimum delay block is a RAM memory. The address in which the cell is written depends on the constant delay requested by the user. The read block is a server of the null and minimum delay modules. The read block shall extract cells from the null delay block, and only when there is not any cell in this block shall extract from the minimum delay block. As it can be seen, the delay of a cell written into the minimum block shall be equivalent to CD + K, being CD the constant delay requested by the user and K the number of cells written into the null block between the entry time of the cell in the minimum block and its departure time.

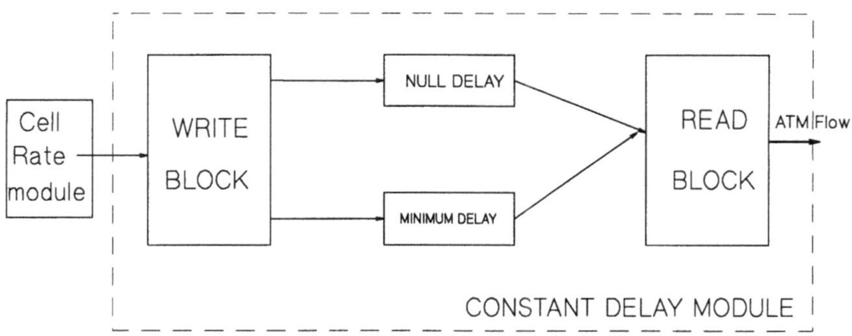

FIGURE 8.-Constant Delay module

Note that when the delay is independent of the VPI/VCI cell value, no cells are written into the null delay. In this case the delay applied to the cells is exactly the constant delay requested by the user (K=0).

5.-Control Process Board

The main functionality of this board, based on a M680x0 processor, is the communication between the Operation Unit and the Line Interface and Core boards. For this purpose, this board has two interfaces:

- A standard RS232 serial interface, controlled by a software protocol, to communicate with the Operation Unit.
- A VXI interface allowing the control and configuration of the remainder system elements (Line Interfaces and Cores).

6.-Operation Unit

The Operation Unit (based on a 486 PC) basic functions are:

- To offer a friendly and easy-to-use way to control the ATM Network Emulator.
- To elaborate the parameters required by the hardware to program Line Interfaces and Cores according to the user selections.
- To inform the user about the state of each element.

The software has been developed as a Microsoft Windows application, and has been designed following the philosophy of those applications: menu driven, easy-to-use/easy-to-learn, etc. .It allows:

- To activate or deactivate each module independently from any other.
- To select the final value or statistical distribution of the related parameter (constant and variable delay, cell error rate, cell loss rate, cell misinsertion rate or bit error rate). If the parameter can be applied to some specific cells depending on their header, it is possible to select the header that will be compared with the input cells and a mask. The mask selects some bits as "don't care" so they will not be used during the comparison.

Once the user requirements have been fixed, the software transforms the user values into the parameters required by the hardware and send them to the Line Interface and Core boards through the Control Process that verifies that the parameters have been correctly written onto the hardware.

7.-ANE features summary

In this paper an ATM Network Emulator developed by Telefónica I+D has been described. The main features of the current configuration (release 1.0) are summarized in the following:

- SDH optical interfaces at 155.52 Mbit/s.
- Possibility of working with a data recovered clock, internal clock or external 128K/2.048M Hz clock. It also provides a 2.048 Mhz clock.
- Generation of cell variable delays, cell losses, cell misinsertions, errored cells and cell constant delays over the incoming ATM cells.
- Possibility to apply a controlled bit error rate over the outgoing SDH frame.
- Each function can be activated or deactivated independently from the others.
- VXI C-size bus oriented system to interconnect the hardware components.
- Full modular design to facilitate future enhancements.
- Windows based man-machine interface.
- Parameter variability (when the associated module is enabled):

Parameter	minimum	maximum
Bit error ratio	10^{-12}	10^{-4}
Variable cell delay	0	722 μs
Cell lost ratio	10^{-10}	10^{-2}
Cell insertion ratio	10^{-12}	10^{-3}
Cell error ratio	10^{-10}	10^{-3}
Constant delay	3 ms.	54 ms

The range of these values can be modified connecting two or more Core boards in cascade.

The modular design of the ANE facilitates the system growth. It is possible to design some new elements like:

- New PDH 2, 34 or 140 Mbits/s Line Interfaces.
- New Core board to modify any other parameter not included in the present system.
- New boards to provide signaling and OAM functionalities.

8.-References

[1] L. Kleinrock. *Queueing Systems, Vol. 1. Chapter 3: Birth-Death Queueing Systems in Equilibrium.* John Willey & Sons Ed, 1975

[2] B. Magliaris et al. *Performance Models of Statistical Multiplexing in Packet Video Communications.* IEEE Trans. on Communications, Vol. 36, No. 7, July 1988.

[3] S. M. Ross. *Introduction to Probability Models. Chapter 6: Continuous-Time Markov Chains.* Academic Press, 1985.

SWITCHING

Broadband Communications, II (C-24)
S. Tohmé and A. Casaca (Editors)
Elsevier Science B.V. (North-Holland)

A Growable Folded ATM Switching Fabric Architecture

Andrzej Jajszczyk

EFP - The Franco-Polish School of New Information and Communication Technologies, ul. Mansfelda 4, 60-854 Poznań, Poland

Abstract

In this paper a growable folded ATM or fast packet switching fabric architecture is proposed. The size of the fabric can be increased modularly in a broad range. The fabric is expanded by connecting additional basic fabrics to the existing structure by some interconnecting modules. The expansion process does not disturb the existing cabling. The growth of the fabric size does not increase the number of required switching stages. The hardware complexity of the new fabric is moderate and is discussed in the paper. Some important implementation issues such as the physical design and reliability are also presented.

Keywords: Switching Techniques; Network Architecture and Design, ATM (Asynchronous Transfer Mode) Networks

1. INTRODUCTION

One of the important issues associated with the design process of an ATM switching system is the selection of such a switching fabric architecture that allows manufacturing nodes of various sizes. Moreover, the system should have the ability to grow smoothly during its service life. The simplest solution could be based on the use of the same fabric of the maximum target size for all systems, but equipped with different number of ports. Such a solution is not economical for all but very small ATM switching systems. In practice, the switching fabric should have a modular structure. Moreover, extension procedures should not significantly disturb the existing wiring between modules.

In this paper a new, growable ATM architecture is proposed. This architecture makes it possible to expand the size of a folded ATM fabric in a smooth way, preserving its number of switching stages. The growth does not involve any changes of the existing cables.

In Section 2 a brief survey of known growable folded fabrics is given. The next section presents the expansion procedure and discusses the hardware complexity of the proposed growable fabric. Section 4 deals with important implementation issues such as the physical design and fabric reliability.

2. GROWABILITY OF FOLDED ARCHITECTURES

Numerous papers have been devoted to growable ATM or fast packet architectures. The variety of the proposed solutions results from different approaches to ATM switching as well as differences in available technology. Here, the focus is put on the growability of folded architectures. In such architectures some parts of the fabric transport information in one direction, and the same parts transport other information in the opposite direction [1]. An example of a folded two-stage fabric is shown in Fig. 1 [2]. In such a fabric cells can be transferred between terminals belonging to the same first-stage switching element without using the second stage. The concept of the size expansion in such architectures is illustrated by an example shown in Fig. 2 (the simplified method of presentation, as illustrated by Fig. 1b, was used). Let us assume that the basic ATM switching element has the capacity of 16 bidirectional links. If only a few ports are needed, the fabric consists of a single switching element (denoted by *A*). To achieve a higher capacity than 8 ports the second stage should be added and the number of first-stage switching elements has to be increased (up to 8). Thus, we obtain fabric *B*, containing at most 64 ports. The required throughput is achieved by adding an appropriate number of second-stage switching elements. The fabric of the maximum capacity, denoted by *C*, is obtained analogously. If we assume that the ultimate fabric consists of at most three stages, we can use the total capacity (i.e., 16 ports) of the third stage elements to connect the two-stage modules, without leaving some ports for the connections to the next stage. We can note that the presented method of expansion does not disturb already existing links.

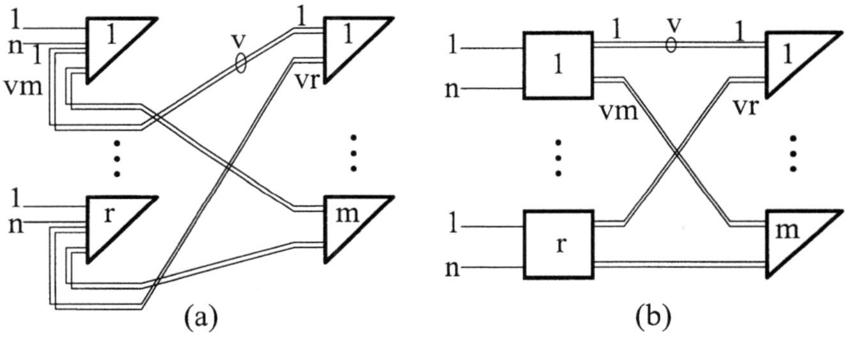

Figure 1. Two-stage folded fabric; (a) structure of the fabric; (b) simplified method of presentation

Cells or packets in a folded fabric are routed by different number of switching elements, depending on the addresses of input and output ports. For example, if a cell is to be sent between ports belonging to the same switching element, only this element is involved in the switching process. A communication between ports of the same two-stage module but different switching elements requires routing to the second stage where cells are reflected back to the first stage. The intermodule communication moves the reflection point to the third stage. In the latter case a cell traverses five switching elements between the input and the output port.

The concept described above is used in various ATM or fast packet architectures, although some of them use a different interconnection pattern between stages. However, as has been shown, most of these fabrics are isomorphic [3]. In the *Athena* switching fabric, proposed by de Prycker *et al.* [1], [4], a 16×16 switching element based on the output queueing principle is used. Henrion *et al.* have described a three-stage folded ATM fabric, later referred to as

Roxanne [1], [5]. The switching fabric is composed of standard switch module boards of 128 × 128 inlets/outlets of 150 Mbit/s each. Physically, four links of 150 Mbit/s are multiplexed to 600 Mbit/s, resulting in a 32 × 32 physical switch module board. Each board is to be composed of a number of shared buffer 32 × 32 integrated switching elements. Technological constraints, at present, resulted in an interim solution based on 16 × 16 integrated switching elements and 64 × 64 board size [5], [6]. Similar, three-stage folded ATM or fast packet growable architectures have been also described by Obara and Okamoto [7], as well as Tanabe and Ohtsuki [8].

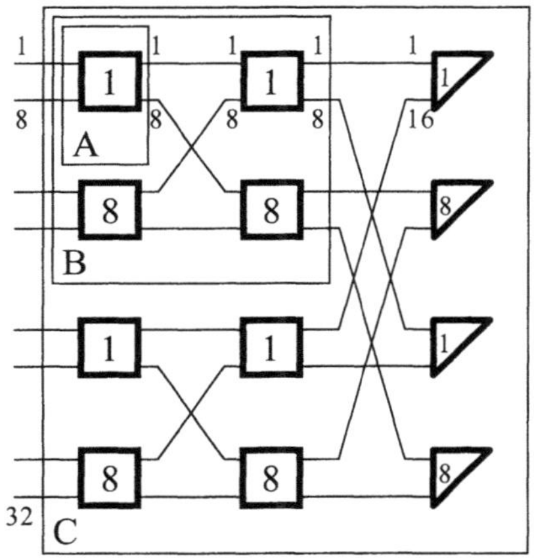

Figure 2. Growability of a conventional folded switching fabric

Fischer, Fundneider, Goeldner and Lutz discussed a two-stage folded ATM architecture using 16 × 8 shared buffer switching elements [9]. They proposed a detailed procedure for the fabric growth which ensures that none of the existing ATM connections is disrupted. A similar folded fabric implemented with 8 × 8 input queue space switches has been described by Stevens [10].

3. PROPOSED GROWABLE FABRIC

The basic folded fabric, which size is then expanded, contains N ATM lines and is composed of s stages ($s \geq 2$). To increase the number of lines two times we add a second identical fabric to the existing basic structure and connect them by additional modules, as shown schematically in Fig. 3. These modules have the same structure as the s_2-stage part of the basic fabric. We can note that the s_1- and s_2-stage modules are interconnected by using bidirectional ATM links. Such a bidirectional link contains one incoming and one outgoing link. For clarity of presentation, incoming and outgoing links for each module are depicted by lines of different thickness.

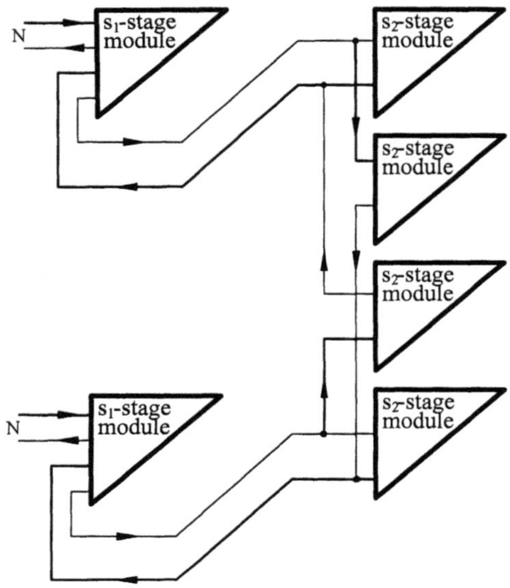

Figure 3. Doubling the size of the basic fabric

To increase the size of the fabric by additional N lines we add the next basic fabric and connect it to the existing structure by using the s_2-stage connecting modules. In general, by expanding the size of the fabric k times we obtain the structure shown in Fig. 4. It contains k identical basic fabrics and $k(k-1)$ modules interconnecting these basic fabrics. To increase the fabric's size from $(k - 1)N$ to kN ATM lines one basic fabric of N ports and $2(k - 1)$ interconnecting modules are added. Note that each line in Fig. 4 represents a group of lines incoming to or outgoing from switching modules. Fig. 5 illustrates the growth of a three-stage fabric from 128 to 256 lines. It is assumed that the basic switching element contains 16 bidirectional ports.

We should note that since a cell always "sees" each path as a path in a basic fabric, the fabric growth does not influence the throughput significantly. The throughput of the fabric can be adjusted by changing the ratio between the number of first-stage switch lines which serve as fabric's terminals and the number of lines which interconnect it to the next stage.

The number of switching elements in the proposed growable fabric can be expressed by the following formula

$$C_{kN} = kC_N + k(k-1)C_{N,m}\,, \qquad (1)$$

where:

C_{kN} is the number of switching elements in a fabric containing kN ATM lines,

C_N is the number of switching elements in a basic fabric,

$C_{N,m}$ is the number of switching elements in the s_2-stage module (see Fig. 3).

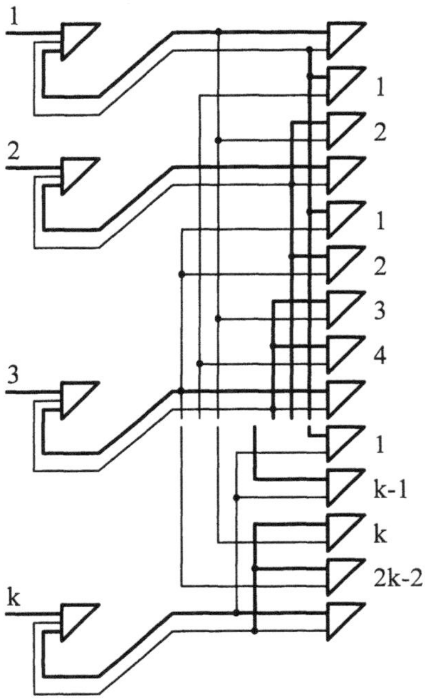

Figure 4. Principle of the fabric growth

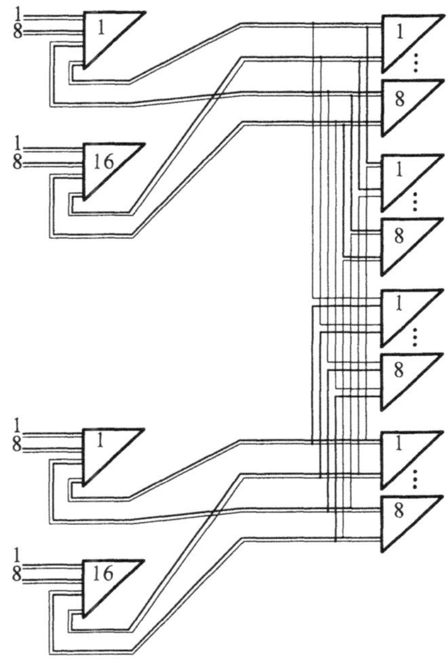

Figure 5. Example of fabric growth from 128 to 256 lines

Since the number of interconnecting modules grows as fast as $k(k - 1)$ it has a significant influence upon the fabric's total cost. Therefore, it is desirable to use simple interconnecting modules, containing as few switching elements as possible.

To illustrate the dependence between the hardware cost and the fabric's size the number of switching elements, each of the maximum capacity of 8 ATM lines, required in a folded growable fabric of various sizes is shown in Fig. 6. The basic fabric has the following parameters: $m = n = 4$, $r = 8$, and $v = 1$. For comparison, the numbers of switching elements in conventional folded Clos-like fabrics without compression or expansion in the first stage are shown. All numbers of switching elements relate to fabrics optimal in the sense of the minimum number of required switching elements and have been calculated by using the dynamic programming method [11]. An analogous comparison for fabrics containing switching elements of 16 ATM lines each is shown in Fig. 7. In this case we compare the number of required switching elements for two different initial basic fabrics. One containing 128 terminals ($m = n = 8$, $r = 16$, $v = 1$) and the other having 64 terminals ($m = 4$, $n = r = 8$, $v = 2$). We can see that the latter contains more switching elements than the former, although the latter allows for more smooth growth.

We can see that the cost of the proposed growable fabrics exceeds, at some point, the cost of the conventional folded Clos-like structure. However, conventional fabrics often require more than two stages which results in longer signal delays and requires more complex routing procedures. The numbers of required switching stages in the conventional fabrics are marked in Figs 6 and 7. The proposed growable fabrics of our examples always have only two stages. We can note a significant "jump" in the cost of the conventional fabrics associated with the change of the number of stages.

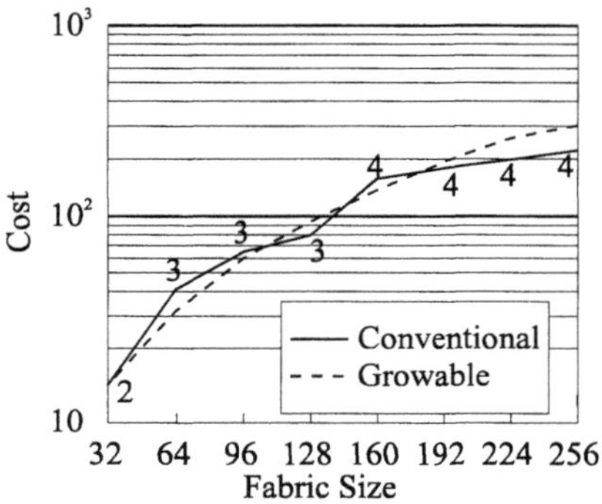

Figure 6. Number of switching elements with 8 ATM lines in fabrics of various sizes

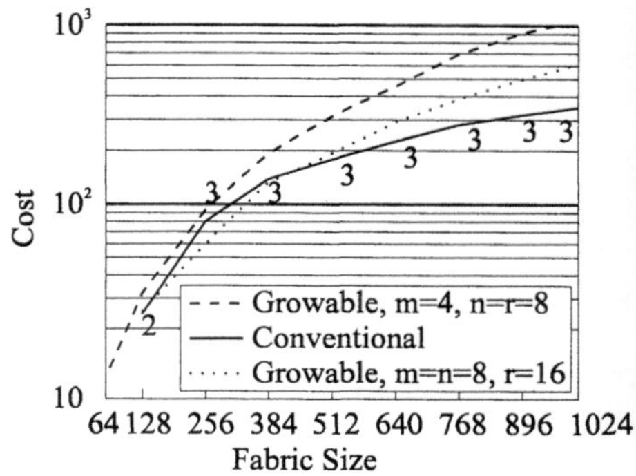

Figure 7. Number of switching elements with 16 ATM lines in fabrics of various sizes

The issue of the number of stages is illustrated in greater detail in Fig. 8, for fabrics containing switching elements of 8 and 16 ATM lines.

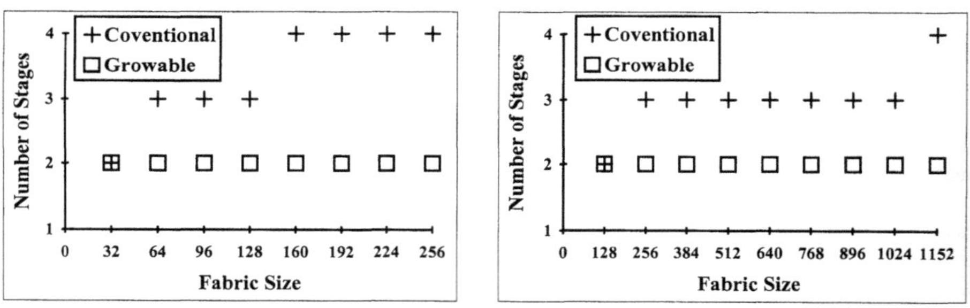

(a) (b)

Figure 8. Number of stages in fabrics of various sizes; (a) switching elements of 8 ATM lines, (b) switching elements of 16 ATM lines

4. IMPLEMENTATION ISSUES

4.1. Branching Points

One of the important implementation issues associated with the proposed growable fabrics concerns branching points denoted by dots in Figs 3 to 5. Such points, along with complicated wiring and high speed ATM signals, may lead to great technological difficulties due to electrical signal reflections. In some systems, simple passive branching gives satisfactory results [9]. In the case of passive branching, the switching elements should have the capability of accepting only calls with the appropriate values in the self-routing headers.

Another option is to apply demultiplexers and multiplexers in splitting and combining points, respectively. Demultiplexers can be controlled by bits in a cell header.

More complex functions can be performed by the multiplexers. These functions can include the prevention of the internal congestion when two or more cells compete for the access to the same multiplexer output. This can be accomplished by backpressure. The multiplexers can be kept simple if an appropriate backpressure mechanism is provided between switching elements in consecutive stages. Kozaki *et al.* used standard 32 × 32 ATM switching elements as 16 two input - one output multiplexers [12]. In another solution, the function of the input and the output branching points (as those shown in Fig. 3a) was performed by header translators [13].

The immunity to electromagnetic interference and the large bandwidth make the optical solutions to wiring and cabling problems a very attractive option. When the interconnects are based on optical fibres, branching points may be either passive (splitters, combiners) or active (e.g., guided-wave-based switches). In the case of passive branching, a significant loss of optical signal power may require optical amplification. A free-space optical interconnect seems also to be a viable solution to the wiring problem.

4.2. Physical Design

The basis of any physical design is the system partitioning [14]. At the beginning of the design process the size and the structure of the basic fabric have to be chosen. There is a trade-off between the initial cost, the maximum target capacity, and the growth increment (see, for example, Fig. 7). The unification of board, or chip, types is also of a primary importance. These board types highly depend on the available technology as well as the basic fabric structure. An example of a possible fabric partition into several standard board types is shown in Fig. 9. Separate boards are used here for the first- and the second-stage modules, as well as demultiplexers and multiplexers which serve as the branching points. The demultiplexers and multiplexers in the initial fabric should have a spare capacity to facilitate growth. Letters *A*, *B*, and *C* mark fabric modules used during the growth from N, through $2N$, to $3N$ lines, respectively. We can note that the switching modules, denoted by rectangles in the figure, contain a multiplicity of elementary switches (see, for example, Fig. 5).

4.3. Reliability

The proposed growable architecture makes it possible to attain high reliability standards required in switching systems. In practice, various methods are used to meet these standards, e.g., complete or partial fabric duplication. Such a duplication results in a significant cost penalty as well as complicates the whole system. High reliability is an inherent feature of the proposed architecture [15], [16]. Since each basic fabric forming a complete fabric of any capacity greater than N can operate independently of the other basic fabrics, then a fault in any basic fabric does not affect the remaining parts of the complete structure. If each line interface of an ATM switch is connected to at least two basic fabrics, then a fault in one basic fabric eliminates only ATM links associated with it, and the whole traffic will be served by ATM links connecting the line interface to other basic fabrics. A temporary degradation of the ATM switch throughput is usually accepted in practice. We can note that by using appropriate fault recovery procedures it is also possible to block only ATM links connected to faulty first-stage switching elements, not the whole basic fabric.

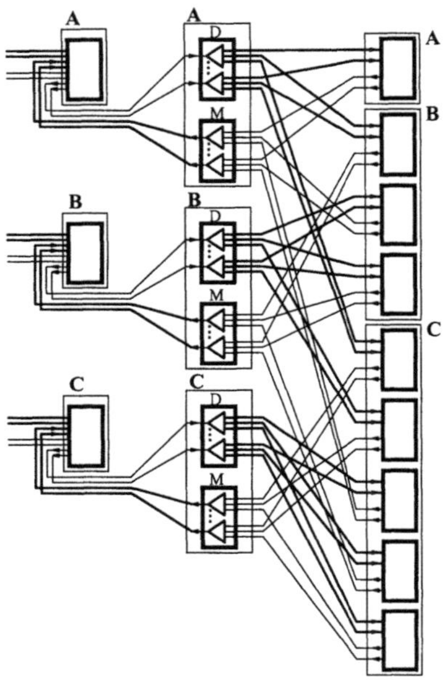

Figure 9. 3*N* fabric structure with possible partition into boards

5. CONCLUSIONS

We have proposed a new, growable folded architecture for ATM switching fabrics. The architecture is best suited to memory-based switching elements and has the following advantages:

- growability,
- moderate hardware complexity,
- fixed and small number of switching stages,
- high reliability.

The size of the fabric can grow smoothly over a wide range. The growth does not disturb the existing wiring of the fabric. The architecture allows also for throughput adjustment, provided the appropriate initial design. The number of switching elements required in the proposed fabric is comparable to the analogous number for conventional Clos-like ATM architectures. The important feature of the new fabric is the fact that the growth does not increase the number of switching stages which is always equal to the number of stages in the basic fabric. The modularity of the proposed structure makes it possible to attain high system reliability only by appropriate connections of line interfaces to fabric inputs and outputs.

Several important issues, such as contention resolution or multicasting capabilities, have not been addressed in this paper. However, since the growable fabric is seen by an input/output pair as a conventional one, the contention and multicasting problems can be solved in a similar way as in the conventional fabrics. Some traffic studies and a quantitative reliability analysis of the proposed architecture will be required in future.

REFERENCES

1. M. de Prycker, *Asynchronous Transfer Mode: Solution for Broadband ISDN*. Chichester, England: Ellis Horwood, 1991.
2. A. Jajszczyk, "One-sided switching networks composed of digital switching matrices", *IEEE Trans. Commun.*, vol. COM-35, pp. 1383-1384, Dec. 1987.
3. A. Jajszczyk, "On combinatorial properties of broadband time-division switching networks", *Comp. Networks and ISDN Syst.*, vol. 20, pp. 337-382, Dec. 1990.
4. M. de Prycker *et al.*, "An ATM switching architecture with intrinsic multicast capabilities for the Belgian broadband experiment", in *Proc. XIII Int. Switching Symp.*, Stockholm, Sweden, May/June 1990, vol. V, pp. 111-118.
5. M.A. Henrion *et al.*, "Switching network architecture for ATM based broadband communications," in *Proc. XIII Int. Switching Symp.*, Stockholm, Sweden, May/June 1990, vol. V, pp. 1-8.
6. T.R. Banniza, G.J. Eilenberger, B. Pauwels, and Y. Therasse, "Design and technology aspects of VLSI's for ATM switches", *IEEE J. Select. Areas Commun.*, vol. 9, pp. 1255-1264, Oct. 1991.
7. H. Obara and S. Okamoto, "Self-routing fast packet switch with in-service modular growth," *Electron. Lett.*, vol. 26, pp. 1286-1287, Aug. 1990.
8. S. Tanabe and K. Ohtsuki, "A hyper-distributed switching system architecture for B-ISDN", in *ICC '91 Conf. Rec.*, Denver, CO, June 1991, pp. 1431-1435.
9. W. Fischer, O. Fundneider, E.-H. Goeldner, and K.A. Lutz, "A scalable ATM switching system architecture", *IEEE J. Select. Areas Commun.*, vol. 9, pp. 1299-1307, Oct. 1991.
10. P. Stevens, "The Siemens AG ATM laboratory model", in *Proc. 4th Fast packet Switching Research Workshop*, Sydney, Australia, May 1989, paper no. 25.
11. A. Jajszczyk, "A dynamic programming approach to optimization of switching networks composed of digital switching matrices", *IEEE Trans. Commun.*, vol. COM-35, pp. 1342-1346, Dec. 1987.
12. T. Kozaki *et al.*, "32 × 32 shared buffer type ATM switch VLSI's for B-ISDN", *IEEE J. Select. Areas Commun.*, vol. 9, pp. 1239-1247, Oct. 1991.
13. A. Itoh *et al.*, "Practical implementation and packaging technologies for a large-scale ATM switching system", *IEEE J. Select. Areas Commun.*, vol. 9, pp. 1280-1288, Oct. 1991.
14. T.C. Banwell *et al.*, "Physical design issues for very large ATM switching systems", *IEEE J. Select. Areas Commun.*, vol. 9, pp. 1227-1238, Oct. 1991.
15. A. Jajszczyk and W. Kabaciński, "Reliable digital switching networks", *Electron. Lett.*, vol. 22, pp. 93-94, Jan. 1986.
16. A. Jajszczyk and W. Kabaciński, "Multistage digital switching network", *US Patent*, no. 4,956,840, Sept. 1990.

Broadband Communications, II (C-24)
S. Tohmé and A. Casaca (Editors)
Elsevier Science B.V. (North-Holland)

Performance analysis of an ATM switch Based on a three-stage Clos interconnection network under non-uniform traffic patterns*

André-Luc Beylot, Iosefina Kohlenberg and Monique Becker

Département Informatique, Institut National des Télécommunications
9, rue Charles Fourier 91011 Evry Cedex - FRANCE
and
Laboratoire MASI
5, place Jussieu 75230 Paris Cedex - FRANCE

Abstract
A model is designed for the performance evaluation of an ATM switch based on a three-stage Clos interconnection network under non uniform traffic pattern. Two dissymetrical traffics are presented : Single Source to Single Destination (SSSD) traffic and Multi-Hot-Spot traffic (MHS). An exact model of a switching element is presented : the cell delay and the cell loss probability are derived for the aggregate input flow and for each individual flow. An exact model is presented for the first stage and an approximate model is proposed for the second and the third stages. The performance criteria are derived for each kind of traffic. It can be applied to lots of traffic patterns. Discrete events simulations are used to validate our model.

Keyword Codes : C.2.1; I.6.4
Keywords : Computer Communication Networks, Network Architecture and Design, ATM (Asynchronous Transfer Mode) networks, Performance evaluation, Simulation and Modeling, Model Validation and Analysis

1. INTRODUCTION

ATM (Asynchronous Transfer Mode) is the technique recommended by the CCITT for broadband network [1]. The information flow is sent in fixed length packets called "cells". ATM can transmit many kinds of traffics through the same transmission channel. Performance of ATM networks will depend on transmission and switching performance. Many ATM switch designs have been proposed. A multi-path switch based on a three-stage Clos interconnection network [2-4] is studied. In this paper, non-uniform traffic patterns are considered. Several models have been proposed for interconnection networks, especially for Banyan Networks. Those studies were not considering the case of losses. The study of switching elements with output dedicated queues has extensively been done [5-6]. Several models have been reported in the literature which include a non-uniform traffic.
The examples of non-uniform traffic patterns can be classified as follows :
<u>Non-uniform destination distribution</u>
• SSSD type : each input port sends most or all of its cells to one output port [7-8]
• Hot-Spot Traffic pattern : one (or several) output has a higher access load [7,9].
<u>Non-uniform input process</u>
In [8], one input port sends its cells to one output port and receives a heavier load.
In [10-11], the input processes are IBP type; they differ by the squared coefficient of variation.

---

* This work was supported by a CNET grant

We shall study two cases of non-uniform destination distribution:
- one with only one hot-spot output port corresponding to one heavy load input port. We shall call this: SSSD case.
- one with several hot spot output ports receiving traffic from several input port with high load, we shall call this: MHS case (Multi Hot-Spots).

The problem of modeling input traffic is not simple. Assumptions can be done about sources [12-14], measurements have been performed [15], but input traffic in a switch will be mostly output traffic from other switches. We are working on this problem. In this paper we shall assume that the number of sources is large enough for the input traffic to be approximated by a Bernoulli traffic.

We propose an exact model of a switching element in the Switching Clos architecture under Bernoulli arrivals (the parameter of the Bernoulli law depends on the input port) when the choice of the output port depends on the input port. For each ouput queue the mean response time and the cell loss probability can be computed. The performance criteria are then derived for each input flow. The quality of service must be guaranteed for each kind of services.

An approximate model is then proposed for the switch. We will study the most important parameters of ATM switches: the cell delay, the cell loss probability for each input/ouput port.

Part 2 will present the interconnection network. Part 3 will introduce the model, the hypotheses and approximations. Part 4 will derive the exact solution of the first stage of the switch and the approximate solution of the switch. Part 5 will present some results of the analytical model that are validated by simulation; the influence of dimensioning parameters will be emphasized.

2. MULTI-STAGE INTERCONNECTION NETWORK

Our switch is based on a three-stage Clos interconnection network with a number **N** of input ports equal to the number **O** of output ports. The switching elements are supposed to be without internal blocking and it is assumed that each cell arriving on each of the inputs may be switched in one slot. Several cells may have the same output link destination. So buffers have to be placed at the output ports of the switching elements. Let us consider a **C(N,a,b)** Clos Interconnection network where **N** is the total number of input ports, **a** and **b** the number of input and output ports for each switching element of the first stage.

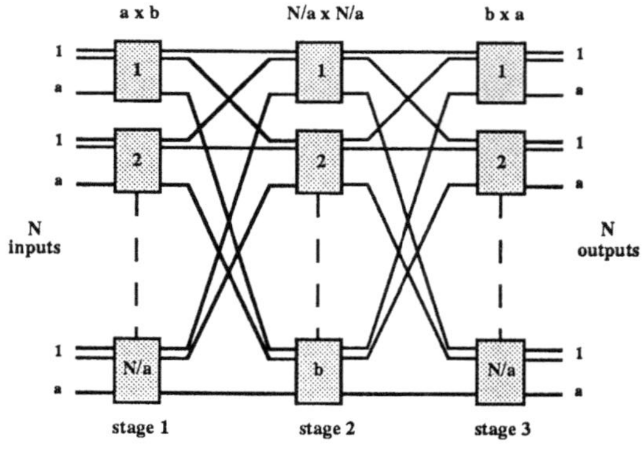

Figure 1. C(N,a,b) three-stage Clos network

The main characteristics of this switch are:

- There are b eventual paths from one input to one output

- Switching elements at stage k are identical
- Each switching element includes output FIFO queues with finite capacity
- No backpressure signals are exchanged between adjacent stages [16]
- When a new cell arrives in a full buffer, it is lost.
- A physical path is assigned for each call during call set-up.

In order to obtain the performance of this switch, we will study the cell delay and the cell loss probability. An analytical model of a Clos network will be presented in section 3.

3. ANALYTICAL MODEL

3.1 Hypotheses

The set of eventual parameter values is very large. Two examples of dissymetric traffic are studied. The proposed method may be extended to a large set of possible configurations.

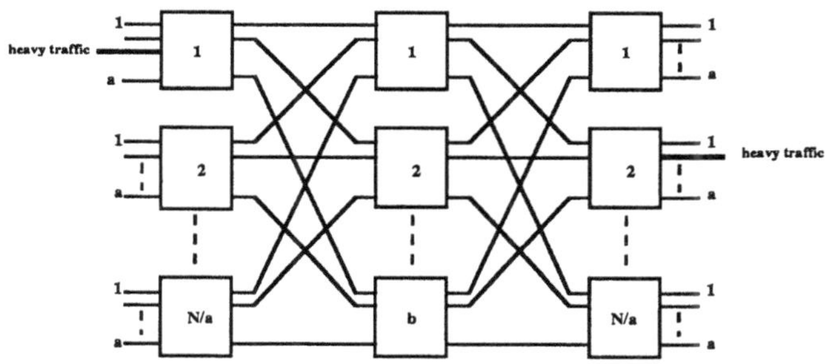

Figure 2. SSSD traffic

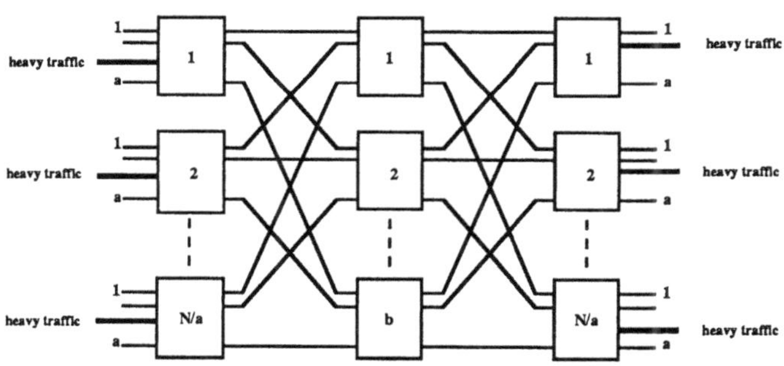

Figure 3. MHS traffic

SSSD case (Fig.2): we assume that only one input link receives heavy load traffic and the destination of this traffic is only one output. Other input links receive a lower traffic. Their cell destinations are uniformly distributed on all the switch outputs, including the heavy traffic output (SSSD path imbedded in a uniform traffic pattern). Consequently, the output chosen by the heavy traffic is a hot-spot. It is assumed that the call number is large enough for heavy load as well as for low load, for the path choice to be equidistributed. We assume that when the

number of calls is large enough, the choice of the matrix of the second stage is uniformly (and randomly) done. This hypothesis is an approximation.
MHS case (Fig.3): we assume that several input links (one per switching element of the first stage) receive heavy traffic and send their cells to several output ports (one per switching element of the third stage). The choice of the output port is random and equidistributed. The other input links receive lower traffic. The cells are sent on the other output ports (and not on the heavy traffic ports). The choice is random and equidistributed.

3.2 Modeling and approximations

As there are no backpressure signals, there is independence between the behavior of each stage k and all stages i with i greater than k. The first stage will first be studied and the results will be used for the study of the second stage and so on. The switching elements at any stage k are assumed to be statistically independent. As sources are supposed to be independent, this assumption is exact for the first stage. We shall consider the same approximations as in a previous work [3-4]: flows between stages are Bernoulli processes. The parameter of this law is determined by the expected value of the output load of the previous stage. The first stage model is exact. The main difference between the work presented here and the previous work is the assumption that traffics are dissymmetric. Consequently, the switching elements of a same stage have not the same performance.

Let us describe in detail for the SSSD case the cell classes and the different queues in each stage of the switching node. We shall then quickly indicate the modeling method for the MHS case.

1st stage SSSD case :
On the first stage there are two types of switching elements: the switching element where there is one heavy traffic input link (type A) and the switching elements where the traffic is low on every input link (type B). Only one queue has to be studied in each of the two types of switching elements, because input traffics are assumed to be independent and the choice of the second stage switching element is random and equidistributed. Then expected values of delay and loss probability may be computed for each traffic types. The output process on each output link of the first stage switching elements will be approximated by a geometrical process with an exact throughput (the loss probabilities will be taken into consideration).

2nd stage SSSD case :
The hypotheses and approximations leads to only one type of switching element. Each of them receives on its input links independent traffics coming from distinct (i.e. independent) switching elements of the first stage. There are two types of input traffic :
- on one input link receiving traffic from a type A switching element, the traffic will include a part of the heavy traffic and a part of the low traffic (both of them come from the same switching element). Low traffic cells will go to a random and equidistributed output of the second stage switching element. Heavy traffic goes to only one output link corresponding to a given output switching element of the third stage.
- on the other input links receiving low traffic from a type B switching element of the first stage, cells will go to a random and equidistributed output. So, there will be two types of output traffics: type A' traffic goes to the switching element which contains the heavy traffic output, and type B' traffic goes to the other switching elements. The output processes will be approximated by a geometric process taking into account the loss probability, for each traffic.

3rd stage SSSD case :
On the third stage there are two types of switching elements: one switching element (type A') contains the heavy traffic output, the others are type B' elements. In type B' elements there is only one type of output, while in type A' elements there is one output receiving heavy traffic and the others. On type A' elements, input processes are composed of a part of heavy load traffic going to only one output and a part of low traffic equidistributed among the outputs. On

type B' elements, input processes are only low traffic processes and they are randomly spread among the outputs; the outputs are equidistributed. It is not easy to solve the third stage, because input processes are not independent. Low traffic outputs receive input traffics that are composed of superposition of independent low traffics; these processes will be approximated by geometrical processes having the same throughput. On the contrary, heavy traffic comes from only one input. So it cannot be considered as being composed of superposition of independent traffics: in the type A' switching element, traffic going to the heavy traffic output will be the whole heavy traffic and a part of the low traffic, for this part the independence assumption holds and the geometrical approximation is valid.

First stage, MHS case:
On the first stage there is one heavy traffic in each of the switching element. It will be necessary to study only one switching element and one output queue.
Second stage, MHS case:
On the second stage, all the input traffics are identical and independent. The choice of the output link is equidistributed. Again, only one output queue has to be studied.
Third stage, MHS case:
On the third stage input processes include one part of heavy traffic which goes to only one output and low traffic which is spread among the other outputs. Let us assume that heavy traffic and low traffics are independent (there are for both traffic many input processes which have the same behaviour). Two queues will be studied: heavy traffic output queue and one of the others.

Let us solve now a switching element receiving a dissymetric traffic. For each of the outputs, average performance criteria will be derived in the case of the infinite capacity queue and in the case of finite capacity queue, without taking into account the different classes of cells. In the case when there are two types of cells, the criteria for each of the classes will then be obtained. (The heavy traffic case may be solved directly and the low traffic performance is deduced from the average performance and the low traffic performance). This method may be easily extended to the case when there is a larger number of traffic classes.
Let us first evaluate the performance of a switching element with **a** inputs and **s** outputs. Then we shall consider the Clos network with the chosen hypotheses and approximations.

4. PERFORMANCE EVALUATION OF A NON-SYMMETRIC SWITCHING ELEMENT UNDER DISSYMETRIC TRAFFIC

4.1 Introduction and notations

Several works have been done about the performance of symmetric elements with infinite output queues [5,6,17]. Our work is an extension of [5] but we consider the state of the queues before the head cell leaves. The finite capacity case has been presented in [18] for a uniform traffic. Let us derive the performance of a switching element.

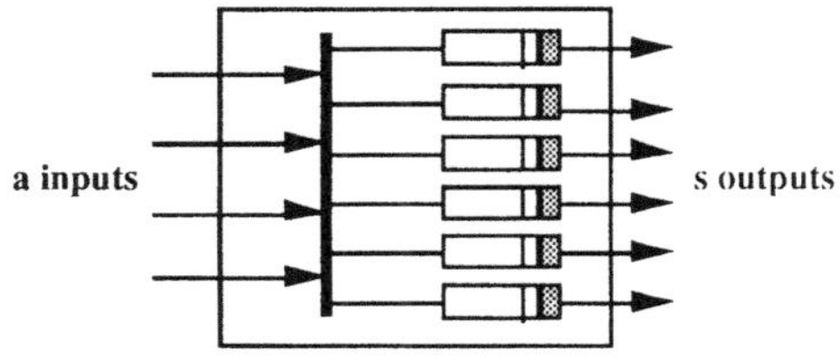

Figure 4. A switching element

Let us consider a switching element with **a** inputs and **s** outputs. Buffers are on output links. Time is discrete. Departures are treated before arrivals. Throughput is the same on input links as on output links. Switching is the following. In one slot:

- In each queue the head cell gets out of the switching element.
- Cells arrive on each input and are put on the output queue they choose. When several cells choose the same output during the same slot, they will be buffered in the queue, back of the cells that were already there, in a random order.
- Cells may be switched in one slot and several of them to the same output.
- Departures are taken into account before arrivals. Consequently, a cell arriving in an empty queue will get out at the beginning of the next slot.

The two main assumptions concerning the input traffic are:

- Independent traffic arrives on each input i; this input process is assumed to be a simple Bernoulli process (parameter p_i).
- Each incoming cell at input i will choose the output j with probability q_{ij}.

Let us note, at time t:

$L_j(t)$: number of cells in the output queue j (just before departure)
$R_j(t)$: response time of output queue j
$N_j(t)$: number of cells in the output queue j (just after departure)
$A_j(t)$: number of cells that want to join the queue
$B_j(t)$: number of cells that join the queue during the slot t.
$\pi_j(t)$: number of cells lost because output queue is full
M : capacity of the buffers
Cj(t) : size of the block of cells that join the queue with a given cell **c**.
λ_j : input load, μ_j : service rate, Λ_j : output load, ρ_j : utilization factor,
Π_j : cell loss probability

Index i will be added to refer to class i customers. Let us note X(z) the z-transform of random variable X.

4.2 Infinite capacity queue

To get an approximation of the delay let us consider first the case of an infinite capacity queue. Let us study the mean number of cells in each output queue j and the mean response time.
Let us consider that switching one cell and have a cell leaving the switching element needs at least one slot (a cell will be taken into account at the beginning of slot t and can be at least delivered at the beginning of slot t+1) [6]. It follows :

$$L_j(t+1)=L_j(t)+A_j(t)-\delta_{L_j(t)}$$

where $\delta_{L_j(t)}=1$ if $L_j(t)>0$ (If there is one cell at slot t, it will have left the switch at time t+1)
$A_j(t)$ is independent of t and will be noted A_j. $L_j(t)$ is a Markov process which evolution can be represented by the following Markov chain (let us note $b_i = \Pr[A_j(t)=i]$).
The computation of $L_j(z)$ looks like the computation of the z-transform of the number of customers in an M/G/1 queue [19].

$$L_j(z) = \frac{A_j(z)\,(1-\rho_j)\,(1-z)}{A_j(z)-z}$$ where ρ_j is the utilization factor.

λ_{ij} is the probability that a cell arrives at input i and will join queue j: $\lambda_{ij} = p_i\, q_{ij}$.

The input load λ_j is equal to $\lambda_j = \sum_{i=1}^{a} p_i\, q_{ij}$. The service time is equal to one slot time.

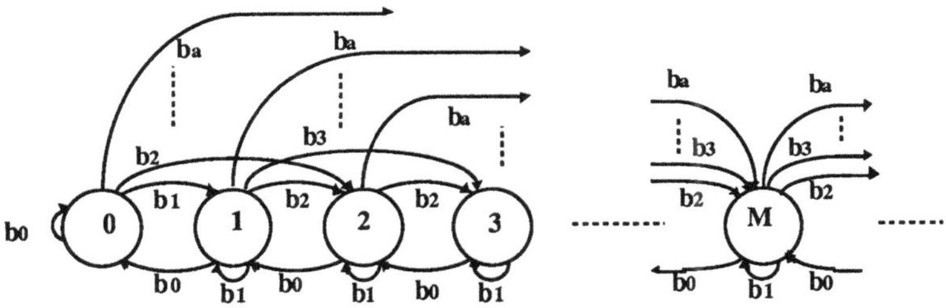

Figure 5. The Markov chain representing the number of cells in the output queue

We thus obtain $\rho_j = \sum_{i=1}^{a} p_i \, q_{ij}$

The z-transform of incoming cells traffic is equal to

$$A_j(z) = 1 + \sum_{k=1}^{a} \left[\sum_{1 \le \alpha_1 < \ldots < \alpha_k \le a} \prod_{m=1}^{k} \lambda_{\alpha_m j} \right] (z-1)^k$$

The mean number of customers in the output queue can then be computed using L'Hospital rule. The mean response time of output queue j can then be derived :

Let us note $\tau_j = \sum_{i=1}^{a-1} \lambda_{ij} \sum_{k=i+1}^{a} \lambda_{kj}$

$$E[L_j] = \lambda_j + \frac{\tau_j}{1 - \lambda_j} \qquad E[R_j] = \frac{E[L_j]}{\Lambda_j} = 1 + \frac{\tau_j}{\lambda_j (1 - \lambda_j)}$$

This value is an upper bound of the mean response time of the finite output buffer j.

4.3 Finite Buffers

Let us compute $L_j(t)$ when buffers are finite to obtain the cell loss probability. Let us study the Markov chain related to the finite buffer. Now:

$$L_j(t+1) = \text{Min}\,(M, L_j(t) + A_j(t) - \delta_{L_j(t)})$$

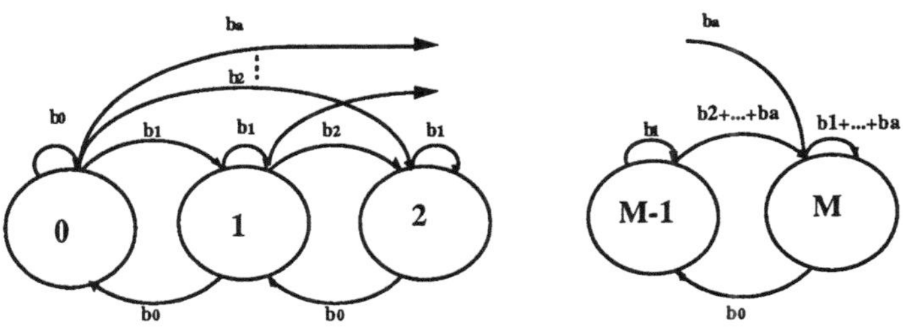

Figure 6. Markov Chain - Finite Buffer

The mean number of cells can be derived and Little's formula can be applied to compute the mean response time :

$$E[L_j] = \sum_{k=0}^{M} k \, Pr[L_j=k] \qquad\qquad E[R_j] = \frac{E[L_j]}{1 - Pr[L_j=0]}$$

The mean cell loss will be equal to :

$$Loss = \frac{\lambda_j - \Lambda_j}{\lambda_j} = 1 - \frac{1}{\lambda_j}(1-Pr[L_j=0])$$

The steady state of the Markov chain is easily computed. One Markov chain has to be solved for each output port.

4.4 Performance criteria of each individual flow

4.4.1 Cell loss probability

Notations will be the same as in the beginning of part 4. The service time is constant and equal to one slot time : $1/\mu_j = 1$. Since $\mu_j=1$,

$$\Pi_j = 1 - \frac{1 - Pr[L_j=0]}{\lambda_j}$$

For flow i : $\Pi_{ij} = 1 - \frac{\Lambda_{ij}}{\lambda_{ij}}$. Consequently, $\Pi_j = \Sigma_i \frac{\lambda_{ij}}{\lambda_j} \Pi_{ij}$. Let us get the Π_{ij}.

Let us study the case when, one input is heavily loaded and the others have a low load. Let us assume that each of the others has the same average input load and cells choose random equidistributed outputs. This study can easily be extended to a more general case.
Let us use index 1 for cells which belong to heavy traffic and 2 for the others. It follows :

$$\Pi_j = \frac{\lambda_{1j}}{\lambda_{1j} + (a-1)\lambda_{2j}} \Pi_{1j} + \frac{(a-1)\,\lambda_{2j}}{\lambda_{1j} + (a-1)\lambda_{2j}} \Pi_{2j}$$

Let us compute Π_{1j}. The loss rate for cells arriving on input 1 and going to output j is equal to the average number of lost cells arriving on input 1 and going to output j per time unit, divided by the arrival rate on input 1 of cells going to output j.

$$Pr[\pi_{1j}=1] \quad = \Sigma_{h,k} Pr[\pi_{1j}=1 \text{ and } A_{1j}=1 \text{ and } A_{2j}=h \text{ and } L_j=k]\ , \quad \Pi_{1j} = \frac{E[\pi_{1j}]}{\lambda_{1j}}$$

All the states of the Markov chain for which a cell may be lost, have to be considered, taking into account other arrivals of cells going to the same output.
If at time t, there are k cells in the queue and h cells that arrive from inputs, a cell will be lost if $(h + k - \delta k + 1) > M$. In that case $(h + k - \delta k + 1 - M)$ cells will be lost.

The loss probability for cells arriving from input 1 is: $\frac{h + 1 + k - \delta k - M}{h + 1}$

The states corresponding to L(t) cells in the output queue, for which cells may be lost are the states for which: $L(t) \geq 0$ if $a>M$ and $L(t) \geq M-a+2$ if $a \leq M$.

It corresponds to values of k such that: $h + k - \delta k + 1 \geq M + 1$ with $h \leq a-1$

It follows: $M - k + \delta k \leq h \leq a-1$

Let us consider the steady state of the Markov chain.

• $a \leq M$

$$\Pr[\pi_{1j}=1] = \sum_{k=M-a+2}^{M} \sum_{h=M-k+1}^{a-1} \Pr[\pi_{1j}=1 \text{ and } A_{1j}=1 \text{ and } A_{2j}=h \text{ and } L_j=k]$$

$$= \sum_{k=M-a+2}^{M} \sum_{h=M-k+1}^{a-1} \frac{k+h-M}{h+1} \Pr[A_{1j}=1] \Pr[A_{2j}=h] \Pr[L_j=k]$$

$\Pr[A_{1j}=1] = \lambda_{1j}$ and $\Pr[A_{2j}=h] = \binom{a-1}{h} \lambda_{2j}{}^{h} (1-\lambda_{2j})^{a-1-h}$

$\Pr[L_j=k]$ is computed by solving the Markov chain of the finite capacity queue.

• $a > M$

$$\Pr[\pi_{1j}=1] = \Pr[L_j=0] \sum_{h=M}^{a-1} \Pr[A_{1j}=1] \Pr[A_{2j}=h] \frac{h+1-M}{h+1}$$
$$+ \sum_{k=1}^{M} \Pr[L_j=k] \sum_{h=M-k+1}^{a-1} \frac{k+h-M}{h+1} \Pr[A_{1j}=1] \Pr[A_{2j}=h]$$

So the loss probability is derived from :

• $a \leq M$

$$\Pi_{1j} = \sum_{k=M-a+2}^{M} \Pr[L_j=k] \sum_{h=M-k+1}^{a-1} \frac{k+h-M}{h+1} \binom{a-1}{h} \lambda_{2j}{}^{h} (1-\lambda_{2j})^{a-1-h}$$

• $a > M$

$$\Pi_{1j} = \Pr[L_j=0] \sum_{h=M}^{a-1} \binom{a-1}{h} \lambda_{2j}{}^{h} (1-\lambda_{2j})^{a-1-h} \frac{h+1-M}{h+1}$$
$$+ \sum_{k=1}^{M} \Pr[L_j=k] \sum_{h=M-k+1}^{a-1} \frac{k+h-M}{h+1} \binom{a-1}{h} \lambda_{2j}{}^{h} (1-\lambda_{2j})^{a-1-h}$$

4.4.2 Response time with infinite buffers

The average delay was calculated for every cell, in the infinite capacity case. It was shown that:

$$E[R_j] = \frac{E[L_j]}{\Lambda_j} = 1 + \frac{\tau_j}{\lambda_j (1-\lambda_j)} \quad \text{with } \tau_j = \sum_{i=1}^{a-1} \lambda_{ij} \sum_{k=i+1}^{a} \lambda_{kj} \quad \text{and} \quad \lambda_j = \sum_{i=1}^{a} \lambda_{ij}$$

In the case we are interested in, it follows: $\lambda_j = \lambda_{1j} + (a-1)\lambda_{2j}$

$$\tau_j = \sum_{i=1}^{a-1} \lambda_{ij} \sum_{k=i+1}^{a} \lambda_{kj} = (a-1)\lambda_{1j}\lambda_{2j} + \frac{(a-1)(a-2)}{2} \lambda_{2j}^{2}$$

The response time of cell **c** is made up of two parts: the first part corresponds to the number of cells that cell **c** sees when it arrives into the queue; the second part corresponds to the cells which arrive at the same time as cell **c** and which happen to be placed ahead of cell **c** (including the cell **c** itself).

$R_{ij} = R'_{ij} + R''_{ij}$

The computation of R'_{ij} is derived from the number of cells after departure.

$Pr[R'_{ij} = k] = Pr\,[L_j - \delta_{L_j} = k]$

Consequently,

$$E[R'_{ij}] = \sum_{k=1}^{+\infty} k\ Pr[\,R'_{ij} = k\,] = \sum_{k=1}^{+\infty} k\ Pr[L_j = k+1]$$

$$E[R'_{ij}] = E[L_j] - (1 - Pr[L_j=0]) = E[L_j] - \rho_j$$

R'_{ij} does not depend on the class of cell **c**. The computation of R''_{ij} depends on the class of cell **c**. Let us compute the response time when **c** is a heavy traffic cell (type 1 cell). The response time R''_{1j} depends on the number of type 1 and type 2 customers which arrived at the same instant as the considered type 1 cell. Let us consider the case when this cell arrived at the same time as (h-1) other type 1 cells and m other type 2 cells.

$Pr[R'_{ij}=k] = \Sigma_{h,m}\, Pr[R''_{ij}=k\ ,\ C_{1j}=h\ A_{2j}=m$ and cell **c** is placed in kth place$]$

As $Pr[C_{1j}=1] = 1$, cell **c** may be placed in kth place in the queue if **m** is included between (k-1) and (a-1).

$$Pr[R''_{1j}=k] = \sum_{m=k-1}^{a-1} Pr[A_{2j}=m \text{ and cell } \mathbf{c}\text{, which is of type 1 is placed in kth place}]$$

$$= \sum_{m=k-1}^{a-1} \frac{Pr[A_{2j}=m]}{m+1}$$

$$E[R''_{1j}] = 1 + \frac{a-1}{2}\lambda_{2j}$$

$$E[R_{1j}] = E[R'_{1j}] + E[R''_{1j}] = E[L_j] - (\lambda_{1j} + (a-1)\,\lambda_{2j}) + 1 + \frac{a-1}{2}\lambda_{2j}$$

It is also possible to compute the response time of cell **c** if it belongs to low traffic :

$$E[R_{2j}] = \frac{\lambda_j}{(a-1)\,\lambda_{2j}} E[R_j] - \frac{\lambda_{1j}}{(a-1)\,\lambda_{2j}} E[R_{1j}]$$

4.4.3 Response time with finite buffers

In case of a finite capacity queue, the average response time (for a cell of type 1 or 2) is:

$$E[R_j] = \frac{E[L_j]}{\Lambda_j} = \frac{E[L_j]}{1-Pr[L_j=0]} \quad \text{and} \quad R_j = R'_j + R''_j$$

But, those variables are not independent, since the number of cells that will enter the queue depends on the number of cells that were there when they arrived, because of the losses. The computation of R'_j is the same as in the infinite capacity queue, but it is included between 0 and M-1 (M is the queue capacity) and does not depend on the class of the cell **c**.

$E[R'_j] = E[L_j] - (1 - Pr[L_j=0]) = E[L_j] - \rho_j$

It remains to compute R''_{1j}

$$E[R''_{1j}] = \Sigma_{b,b1,k}\ E[R''_{1j}/N_j=k,C_{1j}=b_1,A_{2j}=b]\ Pr[N_j=k,C_{1j}=b_1,A_{2j}=b]$$
$$= \frac{\Sigma_{b,b1,k}\ b_1\ E[R''_{1j}/N_j=k,B_{1j}=b_1,A_{2j}=b]\ Pr[N_j=k,B_{1j}=b_1,A_{2j}=b]}{E[B_{1j}]}$$

This formula gets simpler since $B_{1j}(t)$ can only be 0 or 1.

$$E[R''_{1j}] = \frac{\Sigma_{b,k}\ E[R''_{1j}/N_j=k,B_{1j}=1,A_{2j}=b]\ Pr[N_j=k,B_{1j}=1,A_{2j}=b]}{Pr[B_{1j}=1]}$$

As far as $(b+k+1) \leq M$ no cell is lost. In this case, $Pr[B_{1j}=1/N_j=k \text{ et } A_{2j}=b] = 1$

$$E[R''_{1j}/N_j=k,B_{1j}=1,A_{2j}=b] = 1 + \frac{b}{2}\ .$$

When $(b+k+1) > M$, among the $(b+1)$ cells which have arrived, $(b+k+1)- M$ cells are lost.
In this case, $Pr[B_{1j}=1/N_j=k \text{ and } A_{2j}=b] = \frac{M-k}{b+1}$

$$E[R''_{1j}/N_j=k,B_{1j}=1,A_{2j}=b] = \frac{M-k+1}{2}$$

Two cases have to be considered, when cells may or maynot be lost, from an empty queue :

• $M<a$

$$E[R''_{1j}] = \frac{\sum_{k=0}^{M-1} Pr[N_j=k] \left[\sum_{b=0}^{M-k-1} (1+\frac{b}{2}) Pr[A_{2j}=b] + \sum_{b=0}^{M-k-1} \frac{M-k+1}{2} \frac{M-k}{b+1} Pr[A_{2j}=b] \right]}{\sum_{k=0}^{M-1} Pr[N_j=k] \left[\sum_{b=0}^{M-k-1} Pr[A_{2j}=b] + \sum_{b=0}^{M-k-1} \frac{M-k}{b+1} Pr[A_{2j}=b] \right]}$$

• $M \geq a$

$$E[R''_{1j}] = \frac{\sum_{k=0}^{M-a} Pr[N_j=k] \sum_{b=0}^{a-1} (1+\frac{b}{2}) Pr[A_{2j}=b] + \sum_{k=M-a+1}^{M-1} Pr[N_j=k] \left[\sum_{b=0}^{M-k-1} (1+\frac{b}{2}) Pr[A_{2j}=b] + \sum_{b=0}^{M-k-1} \frac{M-k+1}{2} \frac{M-k}{b+1} Pr[A_{2j}=b] \right]}{\sum_{k=0}^{M-a} Pr[N_j=k] + \sum_{k=M-a+1}^{M-1} Pr[N_j=k] \left[\sum_{b=0}^{M-k-1} Pr[A_{2j}=b] + \sum_{b=0}^{M-k-1} \frac{M-k}{b+1} Pr[A_{2j}=b] \right]}$$

4.5 Approximate algorithm

Let us indicate the use of the solutions derived in part 4.4 for the SSSD case. The same kind of algorithm has been derived for MHS case. It was explained in part 3.2 that, for the SSSD case, in the first stage, type A switching elements (with one heavy traffic input) and type B switching elements (with no heavy traffic) have to be solved. For type A switching elements, there are two types of inputs: one input receives heavy load and the others receive low load. For type B switching elements there is only one type of inputs, they receive low load. The solution derived in 4.3 is valid for the two types of switching elements; it is exact for the first stage.
Let p_h be the load of the heavy load and p_l the low load.
For type A switching elements: $\lambda_{1j} = \frac{p_h}{b}$ and $\lambda_{2j} = \frac{p_l}{b}$. For type B switching element: $\lambda_j = a\frac{p_l}{b}$

For the second stage there is only one type of switching element but there are two types of input traffic coming from the two types of first stage switching elements and two types of output queues will have to be solved: type A' and type B' queues. The solution derived in 4.3 may be applied for the second stage, but it is approximate.

For type A' queue: $\lambda_{1j} = \frac{p_h}{b}(1-\tau_1) + (a-1)\frac{p_l}{b}(1-\tau_2) * \frac{a}{N}$ and $\lambda_{2j} = a\frac{p_l}{b}(1-\tau'_2) * \frac{a}{N}$

For type B' queues: $\lambda_{1j} = (a-1)\frac{p_l}{b}(1-\tau_2) * \frac{a}{N}$ and $\lambda_{2j} = a\frac{p_l}{b}(1-\tau'_2) * \frac{a}{N}$

where τ_i are the cell loss probabilities of the first stage of class i cells.

For the third stage there are 3 types of output queues. First there is the queue with heavy traffic, there are two types of inputs: one input receives heavy traffic and a part of the low one, the other inputs receive only low traffic. For heavy traffic input processes might be approximated by Bernoulli processes of parameter ϕ but those processes are not independent since the heavy traffic comes from one input of the first stage. Nevertheless during a short time period only one input receives heavy traffic. The heavy traffic was approximated by one Bernoulli process with parameter $b\phi$ on one only input. (The input with heavy traffic is not always the same, when the time goes). The second type queues are the low traffic output queues inside an A' switching element. The third type queues are every queue inside a type B' switching element. For the second and the third type queues there is only one type of input traffic. The solution derived in 4.3 may be applied for the third stage, but it is approximate.

For the first type queue: $\lambda_{1j} = p_h(1-\gamma_1) + \frac{N-1}{N}\frac{p_l}{b}(1-\gamma_2)$ and $\lambda_{2j} = \frac{N-1}{N}\frac{p_l}{b}(1-\gamma_2)$ where γ_i are the cell loss probabilities of the first two stage.

For the second and the third type queue: $\lambda_j = \frac{N-1}{N} p_l (1-\gamma'_2)$

5. RESULTS

5.1 Introduction

We intend to **dimension the switch**. It is not easy to do it from simulations: the simulations for low losses would deal with rare events and the simulations for exhaustive values of the parameters would be too long. So, in fact it is first necessary to **validate the analytical model** and then to use it to dimension the switch. The points in the analytical model were validated by discrete events simulations. The confidence intervals were around 10-20%. (For cell loss probability of 10^{-7} they are 20%. For cell loss probability higher than 10^{-6} they are around 10%). There are many parameters. The switch architecture, the memory sizes on each stage may vary, the input load of the heavy and the low traffics may also vary.

Eventually we obtain many classes of customers. The results are presented for the heavy traffic and for the low traffic : the low traffics were aggregated (in SSSD case, there were 6 classes of low traffic and it is very difficult to obtain a good estimation for the cell loss probability for each class). Then the switch may be dimensioned in order to give the best ratio performance/cost either for the average traffic or for one particular traffic.

Several parameters were chosen. The total number **N** of input links is 128 and the number **a** of input links of first stage switching elements is 4. Let us describe this configuration by **C(128,4,b)**. The **b** parameter will vary. We intend to chose the best value for b and the buffer sizes on each stage.

The method will be the following: we shall compare the performance of different choices of the buffer sizes **for a given global memory size M**. Let **M(k)** be the memory size on stage k.

$$M = N\frac{b}{a}M(1) + N\frac{b}{a}M(2) + N\,M(3)$$

The best repartition is the one that leads to the smallest value of the cell loss probability. The global memory size is fixed (equal to 48x128 for SSSD case and to 72x128 for MHS case). The cell loss probability and the delay are presented as functions of the memory size on the first stage for the best choice for the buffer size on the second stage (then the size on the third stage is determined).

5.2 Results SSSD case

The load of the heavy traffic is 0.6 and the load of the low traffic is 0.3. So the load on the heavily loaded output is: Θ_h=0.897 and it is Θ_l=0.297 on the others. The performance criteria will be presented as a function of the memory size on the first stage and as a function of one of the traffics, when the other is chosen.

5.2.1 Results as a function of the memory size of the first stage

Fig. 7,9 and 11 represent the cell loss probability; Fig. 8, 10 and 12 the cell delay.
Validation of the analytical model: the analytical model gives results which are very near the simulation results. From the cell loss probability as a function of the memory size in the first stage, it appears that the analytical model is better for the small memory size in the first stage, because if the memory size is smaller then the cell loss is larger in the first stage where the analytical model is exact.

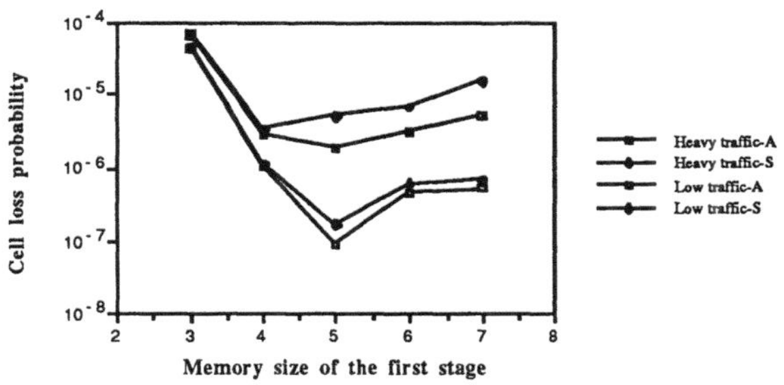

Figure 7. Cell loss probability as a function of the memory size of the first stage
SSSD case, C(128, 4, 8), Θ_h=0.9 and Θ_l=0.3

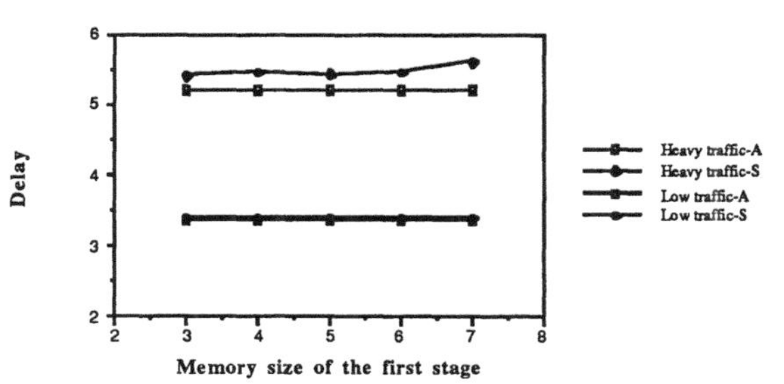

Figure 8. Cell delay as a function of the memory size of the first stage
SSSD case, C(128, 4, 8), Θ_h=0.9 and Θ_l=0.3

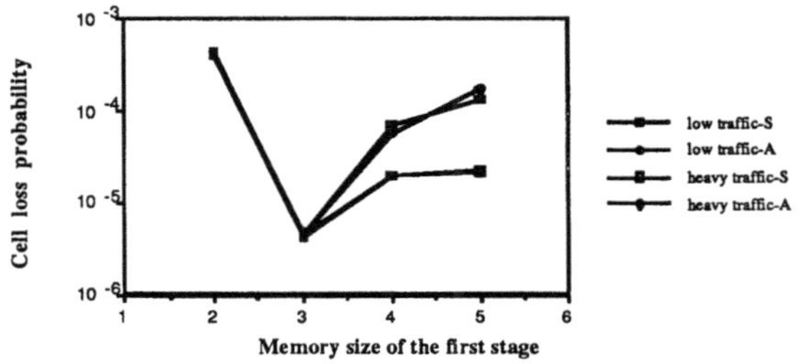

Figure 9. Cell loss probability as a function of the memory size of the first stage SSSD case, C(128, 4, 16), Θ_h=0.9 and Θ_l=0.3

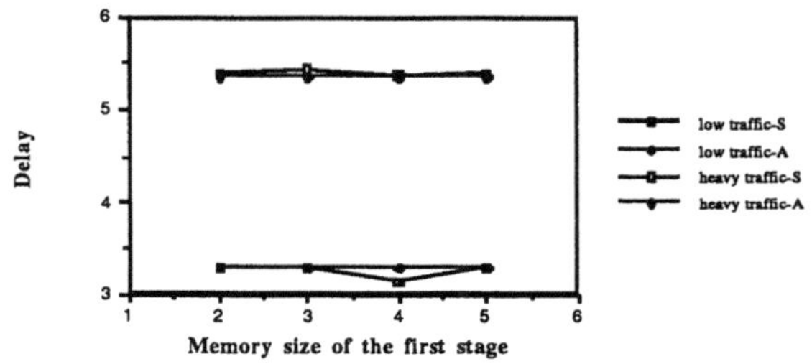

Figure 10. Cell delay as a function of the memory size of the first stage SSSD case, C(128, 4, 16), Θ_h=0.9 and Θ_l=0.3

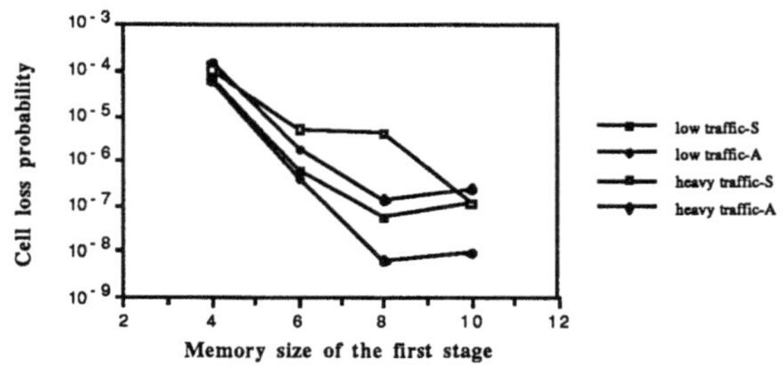

Figure 11. Cell loss probability as a function of the memory size of the first stage SSSD case, C(128, 4, 4), Θ_h=0.9 and Θ_l=0.3

The influence of the memory size change on the cell delay is negligible. The cell delay of the heavy traffic is around 5.5 slots and the cell delay of the low traffic is around 3 to 3.5 slots. The cell loss probability as a function of the memory size of the first stage shows an optimal

value for the memory size: it is 5 for C(128, 4, 8) architecture (Fig. 7), 3 for C(128, 4, 16) architecture Fig. 9), 8 for C(128, 4, 4) architecture (Fig. 11).
The cell loss for the heavy traffic is larger than for the low traffic. These values are comparable for the small values of memory size of the first stage. For larger memory sizes of the first stage, the cell loss for the heavy traffic is more important that the cell loss of the low traffic. For three architectures the best results for the cell loss probability as a function of the memory size of the first stage were obtained for C(128, 4, 4) where the cell loss is about 10^{-9}. For this value, the simulation results are not good enough after one week simulation! A good confidence interval would need a 2-3 month simulation.

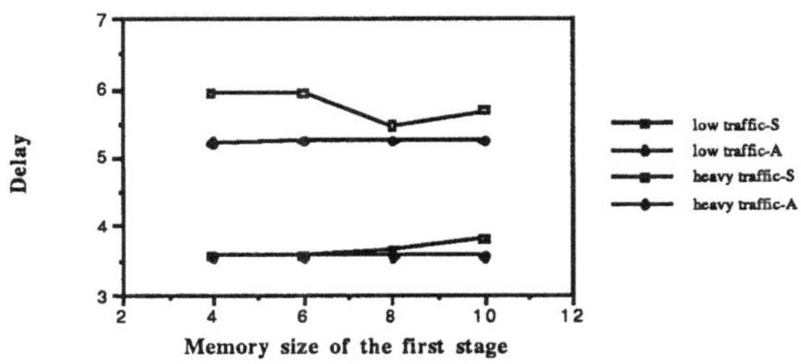

Figure 12. Cell delay as a function of the memory size of the first stage
SSSD case, C(128, 4, 4), Θ_h=0.9 and Θ_l=0.3

Table 1 shows the best memory sizes for a global memory of 48x128, for output links loads Θ_h=0.9 and Θ_l=0.3 for heavy and low traffic, induced by input load p_h=0.6 and p_l=0.3, and presents the values of the loss probabilities for heavy and low traffic and of total delays for heavy and low traffic. The confidence intervals are presented for the performance parameters.

Table 1 : SSSD traffic: performance results for several architectures and the best memory repartition.

Archit.	Θ_h	Θ_l	p_h	p_l	M(1)	M(2)	M(3)	Loss-h	Loss-l	Delay-h	Delay-l
128-4-4	0.9	0.3	0.6	0.3	8	11	29	3.8e-6±3e-7	5.6e-8±7e-9	5.46±6e-2	3.671±9e-3
128-4-8	0.9	0.3	0.6	0.3	5	6	26	5.1e-6±2e-7	1.8e-7±2e-8	5.44±1e-2	3.364±1e-3
128-4-16	0.9	0.3	0.6	0.3	3	4	20	2.7e-5±2e-6	4.3e-6±6e-7	5.42±2e-2	3.290±1e-3

5.2.2 Results as a function of the input loads

With the same architecture the performance is represented as a function of the low traffic and the heavy traffic. If one input load increases, the cell loss probability and the cell delay may increase substancially. The saturation phase is shown on the figures (the load in the heavy traffic destinated output link is bigger than 1, losses increase, that is why the delay does not increase dramatically). In Fig. 13 for a load of 0.6 for heavy traffic and 0.33 for low traffic, the cell loss probability is about 10^{-5} for the low traffic and 10^{-4} for the heavy traffic. With the same load for the heavy traffic, for the low traffic of 0.4, the cell loss probability is about 10^{-4} for the low traffic and 10^{-2} for the heavy traffic.

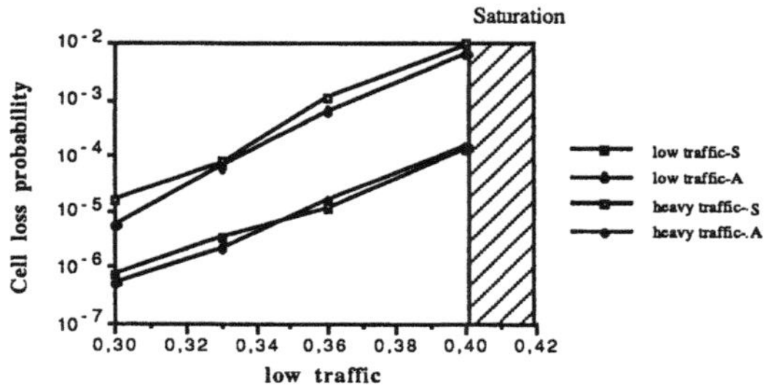

Figure 13. Cell loss probability as a function of the low traffic - SSSD case, p_h=0.6

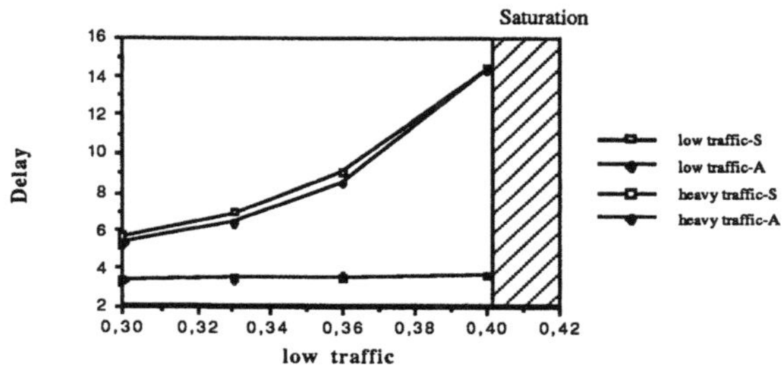

Figure 14. Cell delay as a function of the low traffic for SSSD case, C(128, 4, 8), p_h=0.6

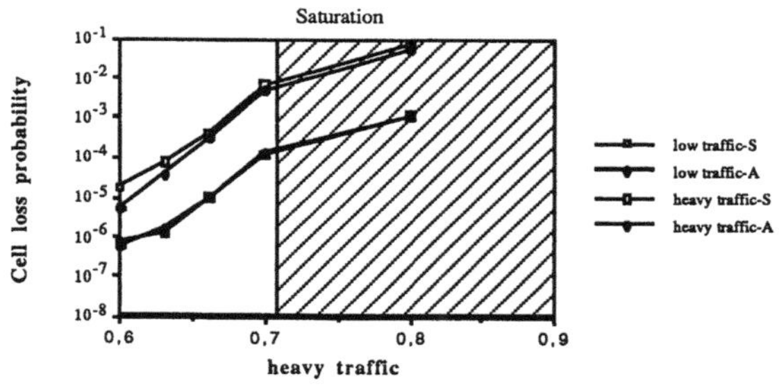

Figure 15. Cell loss probability as a function of the heavy traffic for SSSD case, p_l=0.3

The low traffic cell delay does not depend on the low traffic load (Fig. 14). The heavy traffic cell delay increases with the low load. The analytical model gives very good results.
We have the same conclusions for the cell loss probability and the cell delay as a function of the heavy traffic with load of 0.3 for the low traffic. In this case, for the 0.64 heavy traffic the cell loss probability for the heavy traffic is about 10^{-6} and for the low traffic 10^{-4}. For the heavy traffic these values are 10^{-3} and 10^{-1} (Fig. 15).

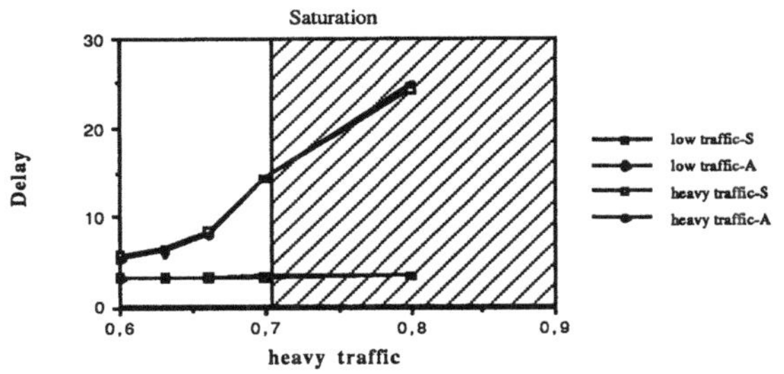

Figure 16. Cell delay as a function of the heavy traffic for SSSD case, C(128, 4, 8), p_l=0.3

The heavy traffic cell delay does not depend on the heavy traffic load. The low traffic delay increases with the heavy traffic load (in the saturation phase) Fig. 16. The analytical model results are very near the simulation results. Those results are given in order to show that the simulation results are not very different from the analytical ones, it partially validates the approximations, and emphasize the influence of the parameters.

5.3 Results MHS Case

The global input load is equal 0.8. We varied the heavy load and the low load keeping the mean input load constant. We studied the cell loss probability and the cell delay as a function of the memory size of the first stage.

The analytical model gives very good results. The influence of the memory size change on the cell delay is negligible. The cell delay of the heavy traffic is around 6.5 slots (for p_h=0.86) and 12 slots (for p_h=0.95) and the cell delay of the low traffic is around 5.2 slots (for p_l=0.78) and 5 slots (for p_l=0.75). The cell loss probability as a function of the memory size of the first stage shows an optimal value for a memory size equal to 9. From the cell loss probability as a function of the memory size in the first stage, it appears that the analytical model is better for the small memory size in the first stage, because if the memory size is smaller then the cell loss is larger in the first stage where the analytical model is exact. The cell loss for the heavy traffic is larger than for the low traffic. For larger memory sizes of the first stage, the cell loss for the heavy traffic is more important that the cell loss of the low traffic. The difference between the low traffic and the heavy traffic increases with the heavy load.

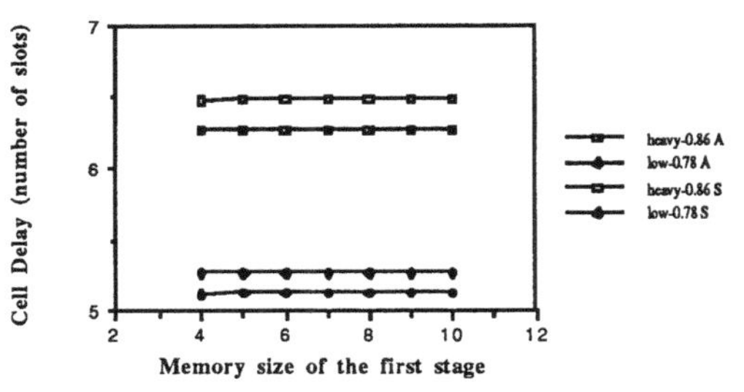

Figure 17. Cell delay as a function of the memory size of the first stage for MHS case C(128,4,8), p_h=0.86, p_l=0.78

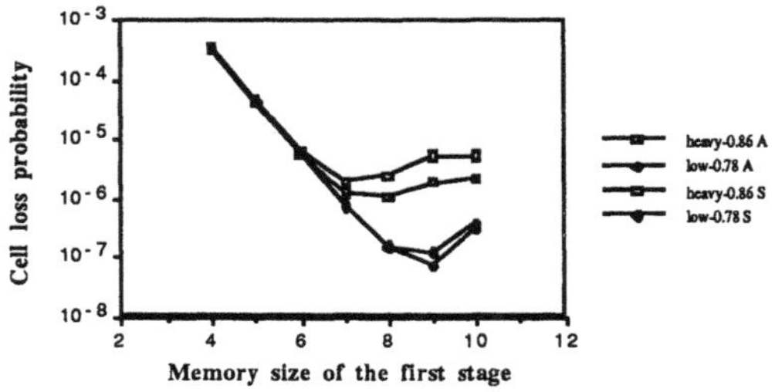

Figure 18. Cell loss probability as a function of the memory size of the first stage for MHS case C(128,4,8), p_h=0.86, p_l=0.78

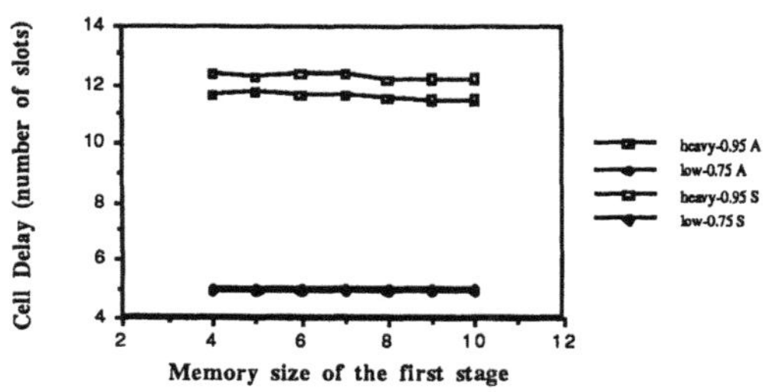

Figure 19. Cell delay as a function of the memory size of the first stage for MHS case C(128,4,8), p_h=0.95, p_l=0.75

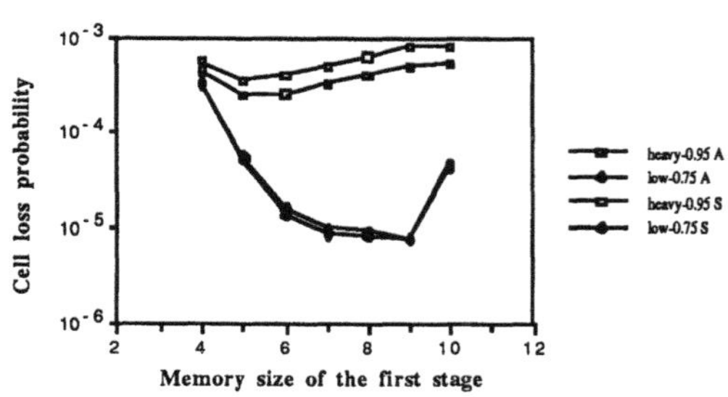

Figure 20. Cell loss probability as a function of the memory size of the first stage MHS case, C(128,4,8), p_h=0.95, p_l=0.75

Table 2 shows the best memory sizes for a global memory of 72x128, for several values of input and output links loads ($\Theta_h=p_h$ and $\Theta_l=p_l$) for heavy and low traffics. It presents the values of the loss probabilities and of total delays for heavy and low traffic. The confidence intervals are presented for these 4 performance parameters.

Table 2 : MHS traffic performance results for several loads and the best memory repartition.

Archit.	Θ_h	Θ_l	M(1)	M(2)	M(3)	Loss-h	Loss-l	Delay-h	Delay-l
128-4-8	0.95	0.75	6	7	46	4.1e-5±5e-5	1.6e-5±2e-6	12.27±6e-2	4.997±3e-3
128-4-8	0.92	0.76	6	8	44	4.1e-5±1e-6	7.5e-6±9e-7	9.03±3e-2	5.081±3e-3
128-4-8	0.89	0.77	7	8	42	1.0e-5±2e-6	9.8e-7±1.5e-7	7.410±8e-3	5.166±2e-3
128-4-8	0.86	0.78	8	9	38	2.5e-6±3e-7	1.5e-7±2e-8	6.479±3e-3	5.263±1e-3

6. CONCLUSION

A Clos switching network with non uniform traffic has been solved. The cell loss probability and the cell delay were computed. Those criteria were presented as functions of the buffer placement on different stages and also as functions of load input traffic. The values of the performance parameters have been computed for two traffic types: an heavy traffic on some inputs and a low traffic on the other input ports. It has been shown that the solution can be obtained in those cases, it could be done in similar cases.
The results show that the buffer placement in the switch architecture is very important.
The analytical model is good in the case when the parameters variation domain is acceptable for the simulations (reasonable time of simulation).
Those results were obtained for Bernoulli inputs. A work is in progress about the case of bursty sources sending IBP traffics. We are studying the output process of the switch, that is the input process of the next switch.

REFERENCES

1. CCITT, Draft Recommendation I.121 "On the broadband aspect of ISDN", COM XVIII-R55(C), Seoul, February 1988.
2. C. Clos "A study of non-blocking Switching Networks", Bell System Tech., T.32, 406-424, 1953.
3. A.-L. Beylot, M. Becker "Performance analysis of an ATM switch based on a three-stage Clos interconnection network", in proceedings of the IFIP conference NETWORKS'92, Trivandrum, India, 28-29 October 1992, pp.283-303.
4. A.-L. Beylot, M. Becker "Performance analysis of an ATM Clos switch with non symmetric switching elements and with output buffers", ITC seminar November 15-19, 1993, Bangalore India.
5. M. Karol, M. Hluchyj and S. Morgan, "Input vs. Output Queueing on a Space-Division Packet Switch", IEEE Trans. on com., Vol.35, N°.12, Dec. 1987.
6. M. D'Ambrosio and R. Melen "Performances analysis of ATM switching architecture : A review", COST 224, Final seminar, Paris, October 1991.
7. D. Chen and J. Mark, "A Buffer Management Scheme for the SCOQ Switch Under Nonuniform Traffic Loading", INFOCOM 1992.
8. H.S Kim and A.L. Garcia "Performance of Buffered Banyan Networks under Non Uniform Traffic Patterns", Proceedings Infocom 88 pp 344,353

9. S. Bassi, M. Dècina, A.Pattavina, "Performance analysis of the ATM shuffleout switching architecture under non-uniform traffic patterns", INFOCOM 1992.
10. H. Yamashita, H. Perros, S. Hong "Performance analysis of a shared buffer ATM switch under bursty arrivals", ITC'13, Copenhagen, June 1991.
11. T. Morris and H. Perros, "Performance modeling of a multi-buffered Banyan switch under bursty traffic", INFOCOM 1992.
12. D. Anick, D. Mitra, M. Sondhi "Stochastic theory of data-handling system with multiple source", Bell System Technical Journal, pp 1871-1894, October 1982.
13. B. Maglaris, D. Anastassiou, P. Sen, G. Karlsson, J. Robbins, "Performance models of statistical multiplexing for packet video communications", IEEE Trans. on Comm., Vol 36, n°7, pp. 834-844, 1988.
14. M. Becker, A.-L. Beylot, Z. Choueiri "A Markov Model of Sources for ATM Traffic : Superposition of On-Off Voice, Data and Video Sources", Proceedings of ISCIS VII conference, November 1992, Antalya, Turkey.
15. W. Willinger, W. Leland, D. Wilson, "Measurements of Ethernet Traffic", SigCom'93.
16. A. Monterosso, A. Pattavina "Performance analysis of multistage interconnection networks with shared-buffered switching elements for ATM switching", INFOCOM'92.
17. H. Kim, M.G. Widjaja, A. Leon-Garcia "Performance analysis of output buffered Banyan Networks with arbitrary buffer sizes", IEEE Infocom'91, April 1991.
18. C. Fayet, A.-L. Beylot, A. Jacques "BSS : Bus structure switching, architecture and performance analysis", proceedings of the IFIP workshop on modelling and performance evaluation of ATM technology, La Martinique, January 25-27 1993.
19. L. Kleinrock "Queuing systems", vol 1 : Theory, J. Wiley and son 1975.

Broadband Communications, II (C-24)
S. Tohmé and A. Casaca (Editors)
Elsevier Science B.V. (North-Holland)

Performance Evaluation of the ATM Ring-based Switching Architecture

Thomas Meuser, Jan Wolsky

Department of Computer Science (Info IV), Aachen University of Technology
Ahornstrasse 55, D-52074 Aachen, email: isatom@informatik.rwth-aachen.de

Abstract

The ATM Ring (ATMR) concept has been proposed as a distributed switching architecture based on a shared-medium ring network. It includes one of the cell mode protocols which are currently under discussion for access control in Gbit/s LANs, as well as for a GFC protocol at the UNI in ATM systems. The efficiency of the simple cell access mechanism is improved by slot reuse owing to destination release. Thereby the bandwidth is decentrally allocated by transmission quotas and controlled by a cyclic reset scheme. Our paper discusses the capability of the basic asynchronous service of ATMR to meet the requirements of universal broadband services. The performance evaluation is based on simulation results which have been obtained in heterogeneous traffic scenarios. It is shown that the mechanisms of ATMR are generally suitable to provide flexible and efficient bandwidth utilisation with a guaranteed service for each station. Bounded access delays and delay jitter however, are quite sensitive to the chosen window size, indicating the well known tradeoff between high throughput and strong real time requirements. Therefore, modifications and enhanced mechanisms have to be further discussed in order to provide end-to-end transport for a wide spectrum of applications with an acceptable quality of service.

Keyword Codes: C.2.1, C.2.5, C.4
Keywords: ATM (Asynchronous Transfer Mode) networks, Local networks, Performance evaluation

1. ATM COMPATIPLE SHARED-MEDIUM NETWORKS

The emerging ATM technology and progress in the design of optical networks, which today operate with data rates into the range of multiple Gbit/s, are stimulating the development of new communication protocols [1,2]. ATM has been adopted by CCITT as a universal transfer mode for the future B-ISDN in the public area. Strong efforts of the private network industry in also deploying ATM in local environments and some simplifications in installation are responsible for the occurrence of the ATM technology first in customer premises networks, which have been dominated by classical Local Area Networks (LANs) up to now [3]. A switch-based network that uses ATM trunks for interconnection of the hosts, might be the best choice for the design of a local infrastructure which supports ATM services. However there are also approaches for new shared-medium, multi-access LAN architectures operating in cell mode. The discussion about the pros and cons of both concepts is still going on.

Switches are better suited if networks grow in size and aggregate bandwidth, while keeping the access bit rate constant. Connection oriented communication is realised more efficiently by point-to-point access to switches. Multiple switches can be mesh-connected to form a large network supporting many hosts. This provides high flexibility in selecting different routes for establishing a connection, and faulty subsystems can also be isolated. On the other hand, in customer premises networks many applications will require efficient support of connectionless services as well as the emulation of traditional LAN-like services as provided by a shared-medium architecture. Multiple access to a single user-network interface (UNI) is a cost effective solution to offer B-ISDN services to hosts in private networks. Lower performance hosts share a common interface and do not need a dedicated (high speed) link into the switch. The problems of bandwidth allocation seem to be the same in both concepts, but the medium access control of shared-media networks compensates poor bandwidth utilisation of peak rate allocation. Furthermore it incorporates congestion control which is necessary if statistical multiplexing is used. For a number of economic and practical reasons, there may be a coexistence of shared-medium and switch-based architectures. Consequently, significant efforts should be spent on the problem of interworking both architectures to provide universal multimedia services independently of the underlying transmission technology.

For data transmission and admission control in shared-medium LANs, a slotted ring architecture applying spatial slot reuse is favoured as a transmission system [4,5]. This approach allows a significant gain in performance, providing network throughput far beyond the medium data rate. That basic mechanism, however, is unfair by giving preference to stations positioned upstream in the ring circulation direction. Dependent on the actual traffic distribution, without an additional fairness control, some stations could be handicapped or even completely starved. This starvation problem has been addressed in different protocols for access control like ATMR [6,7], CRMA-II [8,9], Distributor [5], and MetaRing [10]. These protocols are dedicated to networks operating as Gbit/s LANs, as well as to sharing a multiple access configuration of various users to a common UNI in ATM systems autonomously. All candidates operate at least on the dual counter rotating ring topology, which offers some advantages for efficiency and distributed operation. Although the control mechanisms used are quite different, they all employ spatial slot reuse with shortest path routing, providing fairness, immediate access to the medium under low load, and bounded access delays.

There are two basic strategies to designing the admission control. One way is dynamic unfairness detection with explicit backpressure. An alternative solution is the negotiation of a transmission quota for each station, which will be renewed periodically. The MAC protocol of ATMR follows the second approach. It provides fairness control with a distributed window mechanism, and a cyclic reset scheme with many similarities to the Orwell protocol [11]. The ATMR protocol has been proposed to provide end-to-end transport for a broad spectrum of services flexibly and efficiently via asynchronous transmission. It is assigned for the asynchronous admission control in a multi-service environment, and its basic mechanisms are also currently under discussion for a GFC scheme [12]. Through its topology of an access node system architecture it is assimilated to a local ATM system with distributed switching capabilities. The asynchronous service of ATMR should combine an efficient media access control with the characteristics of the ATM technology to achieve a direct connection of ATM networks and shared-medium LANs. This interworking requires that the virtual ATM connections are extended from the UNI to each host in the local environment, and that bursty cell arrivals can be served efficiently.

The problems in designing an appropriate MAC protocol are currently under discussion in the FDDI Follow-On LAN (FFOL) project [13] as well as in the framework of CCITT SG XVIII, which is working on a GFC scheme [14]. Today's broadband networks require a high grade of flexibility. They have to provide services for delay-sensitive traffic as well as for data with high loss-sensitivity. Thus the most important requirements for media access include [5]:

- high bandwidth utilisation and fairness between stations,
- scalability concerning geographical size, transfer rate and number of attachments,
- guaranteed throughput and bounded access delays,
- support for a wide variety of integrated services, e.g. data, graphics, video, audio,
- easy interworking with public networks, ATM based systems, and existing LANs and MANs.

Several performance studies of ATMR have been published, e.g. in [15,16,17], but all of them are restricted to networks with a uniform traffic distribution. The main purpose of this paper is to give a general performance evaluation of the ATMR protocol in inhomogeneous scenarios aswell. After a short review of the basic concepts of ATMR, our simulation model will be introduced. The load model has been built emulating a heterogeneous environment which includes constant and variable bit rate traffic sources. The performance analysis is based on event driven simulations [18], and it is done with particular regard to the above mentioned requirements. The final conclusions discuss the meaning of ATMR for the development of customer premises networks in the future.

2. BASIC CONCEPTS IN ATMR

The ATMR architecture has been developed by the NTT Communication Switching Laboratories (Tokyo) where an 156 Mbit/s prototype was tested in 1989. Meanwhile an ATM Ring has been designed with a data rate of 620 Mbit/s. Both implementations use the BiCMOS technology to realise an access control LSI.

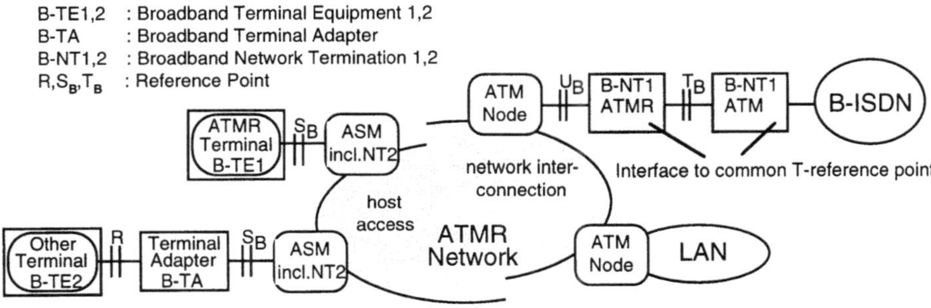

Figure 1. ATMR network architecture

The ATMRing represents a distributed switching architecture based on a shared-medium network. Through its distributed access control, ATMR provides network access for all stations, regardless of their location on the ring. It allows simultaneous and mostly unrestricted usage of network resources, and provides traffic regulation only if it is necessary, in the case of high and overloaded network conditions. ATMR has adopted the ATM slot format and the asynchronous transmission technique. This is one base for compatibility to B-ISDN and an easy interconnection to ATM´s UNI.

In customer premises terminal equipment that follows one of the B-ISDN reference-points S_B or R could be connected to the ATMR network directly, or via an additional terminal adapter (see figure 1). Any other functional grouping beyond the U_B reference-point defined for B-ISDN is possible. Thus the interconnections of LANs can also be realised by ATMR. On the other hand, ATMR can easily be connected to public B-ISDN with a high grade of interoperability. The advantage of that direct access from and to B-ISDN is, that all conversion functionality, which is sufficient for resizing and readdressing of non-ATM-compatible protocol data units, can be omitted.

Following the ATM Adaption Layer (AAL), which will be adopted without appreciable changes, connectionless and connection oriented services with variable and constant bitrate are supported by ATMR [7]. The three most common connection types *point-to-point*, *multicast* and *broadcast* communication can be maintained by the corresponding layers of the protocol.

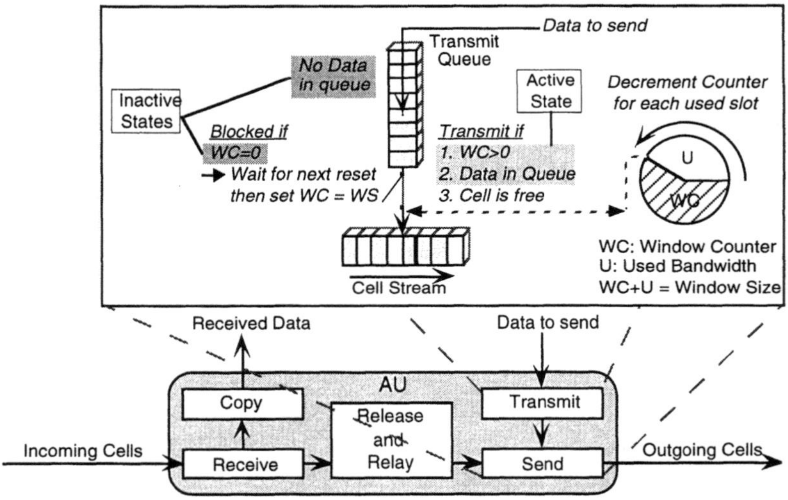

Figure 2. Access Control of ATMR

The basic access control protocol of ATMR operates in consecutive time periods, called *reset periods*, in which a transmission quota bounds the load each station is allowed to send (see figure 2). At the beginning of a reset period, each station sets an internal counter (*window counter*) to an initial value, the *window size*. This value determines the number of slots which are allowed to be used in this period. The window counter is decremented each time the access unit (AU) occupies a cell. A node can use each passing free slot until its window counter expires. Then the AU is forced into the inactive state. In this state, it is restrained from sending any data until a reset is received, which will start the next reset cycle. If a station has no more data to send, the AU will also pass over to the inactive state, but it may become active again without receiving a reset on arriving data at the transmit queue. By counting the window counter down to zero, each station is guaranteed a certain fraction of the available bandwidth within one reset period.

The reset mechanism itself is based on a monitoring system, in which each AU controls which station is the last active sender. If the last active AU detects the inactivity of all other AUs, it generates a reset by setting a flag in the ACF/GFC-field of next cell´s header, as soon as stops sending itself. On its way around the ring, this slot informs all stations about the start of the next reset period and causes them to reset their window counters. Monitoring of inactivity is performed as each active AU overwrites a busyaddress field in the header of each ATMR cell with its own address. So any AU which receives a slot with its own address in this field can be sure, that all other AUs are inactive at the moment. Due to the propagation delay on the ring, it could occur that more than one station detects the inactivity of the other nodes. Thus, they all initiate the reset-procedure and generate a reset. In that case the reset cell only runs up to the next station, which has also generated a reset, and there it is removed. Otherwise, the reset is not taken from the ring before that station which has generated it. This reset mechanism realises a distributed fairness control, which removes the unfairness of the pure slotted ring mechanism concerning preference or starvation of single stations.

There are some basic characteristics of the ATMR protocol which should be noted before discussing the simulation results. It is clear, that achieving fairness causes some overhead, which leads to decreasing network throughput. The only way to regulate the traffic flow in a cell mode transmission system is to force favoured stations to pass free slots unused to the handicapped stations. Applying the ATMR protocol, stations could be blocked mainly at the end of a reset period if they have already used their transmission quotas. Additional overhead occurs due to the round trip delay of the reset-signal. Thus access control may become inefficient when the frequency of resets increases in the case of window sizes being too small. Furthermore, in that case the protocol will react on any 'short-term' unfairness, which occurs if one station gets the opportunity to send much data temporarily.

On the other hand, the guarantee of bounded access delays depends on the chosen window size and the traffic pattern on the network. We can imagine a worst case in which a station gets the fortuity to consume its whole quota at the beginning of one reset period. In the next reset period it gets access to the medium at the very end. Thus the maximum access time is almost two times the reset period. The duration of the reset period can only be bounded by the total window size of all AUs, plus one round-trip delay as worst time to detect the inactive state of the ring. This time demonstrates the influence of the chosen window size on delay-sensitive data.

In this paper we do not evaluate the priority scheme of ATMR in detail. Therefore we only give a short description of the mechanism. To serve both delay-sensitive and bandwidth-consuming data, two reset priority classes have been installed. The first priority class is controlled by the reset which is activated by expired window counters as described above. A higher priority class is realised by an additional transfer completion reset (TCR). In each station a second transmit queue is implemented with its own counter for delay-sensitive data. A guaranteed bandwidth can be preallocated for this type of data which is served by the highest priority. Each time an AU detects an insufficient service for this data during the actual reset-period, a TCR is generated. This TCR restricts sending of each station to high priority transfer only. The next ordinary reset, which is generated if the counters of all high priority queues have been expired, resets the ATMRing to normal operation mode.

3. SIMULATION MODEL

Generally the basic network scenario consists of 20 equally spaced stations, connected by a counter rotating dual ring topology. The data rate of each ring is set to 1 Gbit/s and a ring length of 20 km is assumed. The slot size is equivalent to the format of an ATM cell, i.e. 53 Bytes with a payload of 48 Bytes. The ring latency is increased by an additional latency of one slot time at each station.

For our experiments we have defined two different types of traffic models. Using saturated traffic scenarios the basic behaviour of ATMR can be studied in an abstract system, showing its shortcomings that cannot be removed by any amount of technological breakthrough. Thereby, each station operates as saturated sender and a new packet is generated each time the current transmission has been completed. The packet length varies according to a truncated hyperexponential (H2) distribution with a coefficient of variation CV=2. Thereby, the mean and the maximum length is set to 700 Bytes and 4500 Bytes, respectively. The second model, called bursty scenario, includes variable as well as constant bit rate traffic, at which the data rate can be chosen to any value. The variable bit rate (**VBR**) source is characterised by alternating silence and busy periods as described in [19]. During fixed length busy periods, isochronous arrivals of fixed length packets occur. The packets´ length depends on the average load, which has to be generated by the VBR source. The duration of the idle periods is assumed to be exponentially distributed. Constant bit rate (**CBR**) traffic is generated at a station according to model an uncompressed video data stream. The information of each picture, which has to be transmitted at a constant rate, is represented by a fixed number of fixed length packets. That means, in the bursty scenario each isochronous data stream is modelled by a train model.

The investigated scenarios with saturated senders differ in the distribution of the destination addresses as shown in figure 3:

In scenario 1 (**S1**) all stations send to all destinations uniformly except to themselves, selecting the ring for transmission by the shortest path algorithm. If the distance is equal on both rings, stations with an odd number send on the counterclockwise (CCW) ring. In the other case the clockwise (CW) ring is used.

In scenario 2 (**S2**) station 2 sends to station 9 only, and station 16 is the single destination for station 1. All other stations send to all destinations uniformly except to themselves. In this traffic scenario there are two stations on each ring which get more opportunities to transmit than the others. E.g. on the CCW ring, station 2 benefits from the fact that station 1 sends only on the other ring, and station 9 is chosen as destination with a higher probability.

In scenario 3 (**S3**) station 19 is never chosen as destination. Thus, on each ring there is one station which would starve without any traffic regulation.

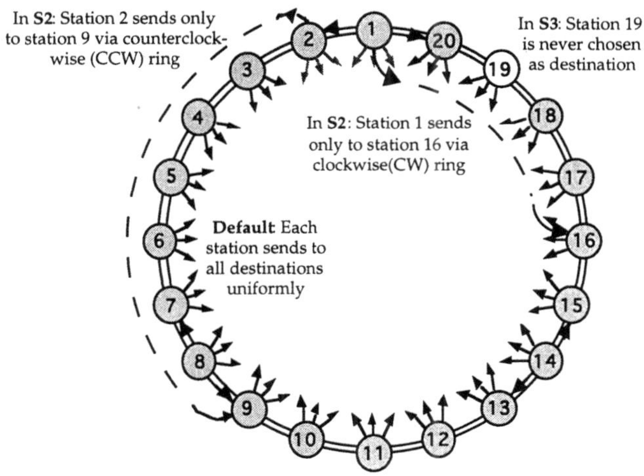

Figure 3. Investigated traffic scenarios

In the bursty scenario, the CBR stations generate a background load on the network by sending uniformly distributed to all CBR stations. Only four stations on the ring are set up as VBR stations. Between each pair of them point-to-point communication takes place by addressing all packets to a dedicated destination.

The following metrics are used in the performance evaluation: *Station throughput* is defined by the ratio of used slots to slots passed at a station during a simulation run. *Network throughput* is the sum of all stations throughput. The *end-to-end delay* is measured by adding the waiting time, which is equivalent to the time between queuing a packet into the transmit buffer and the start of its transmission, the transmit time for the packet and the propagation delay. Thereby, the transmit time also includes the waiting times for all slots which are necessary for the complete transmission of the packet.

The protocol analysis is based on simulations using the ATLAS tool (**A**nalysis **T**ool for **L**ocal **A**rea Network **S**imulation) which has been developed at our institute [20]. It provides powerful features for the definition and evaluation of queuing networks. It is based on the discrete event driven simulation technique. The computation of 99% confidence level is based on the batch means algorithm. In order to achieve statistical significance, the duration of the simulations has been chosen according to reach intervals of confidence of at most 5% of the mean values.

4. DISCUSSION OF THE SIMULATION RESULTS

An inherent feature of the ATMR approach is its capability of immediate medium access for all stations in the case of low to medium load conditions. Protocol´s operation only becomes important for maintaining fair access with tight delay bounds during periods of sustained heavy load or overload. Thus in our experiments, first the overall network performance provided by the ATMR protocol is analysed in saturated traffic scenarios. Thereby efficiency and fairness related to throughput as well as the guarantee of bounded end-to-end delays come into focus. The second part of this section includes a discussion of ATMR´s capability to serve bursty VBR traffic.

4.1. Overall network performance in homogeneous traffic scenarios

The duration of the control cycles, i.e. the time between two resets, is of important meaning for the performance of ATMR. A finer granularity of network control leads to more traffic regulation, since the protocol reacts on any 'short-term' unfairness, which occurs even in the case of a uniform traffic distribution. This unfairness is compensated for in longer control intervals and thus it is not detected by the protocol anymore. On the other hand, the coarser the granularity of network control the longer a station, which does not receive any free slot without the control, has to wait for medium access. In the ATMR protocol the duration of the control cycles is greatly influenced by the chosen window size. A larger transmission quota automatically leads to a lower frequency of the renewing processes. The network throughput of the ATMRing in relation to different window sizes is shown in figure 4. In each scenario the protocol´s efficiency can be improved with increased transmission quotas. It should be noted that a window size of only a few thousand slots is sufficient to approximate the optimal throughput-bounds reachable in theory of 760% (S1) and 700% (S2, S3) respectively. The results also indicate that the obtained throughput depends on the actual traffic distribution. The more the traffic flow becomes homogeneous, the less traffic regulation has to take place. Thus the throughput in the scenario with a uniform traffic distribution is higher than those in the heterogeneous scenarios. The differences give an idea of the protocol´s cost to compensate a given unfairness by an access control based on transmission quotas.

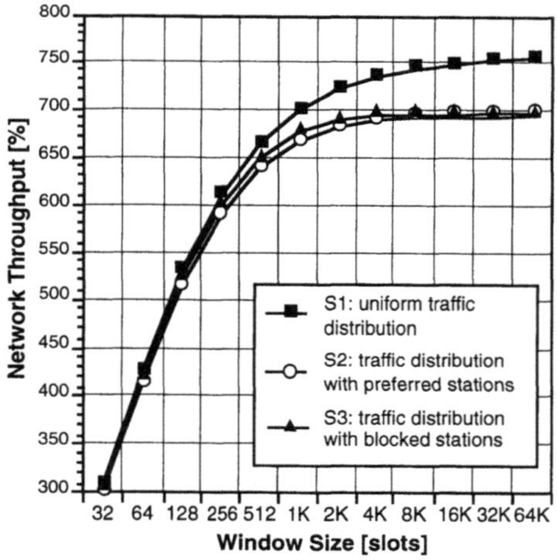

Figure 4. Total network throughput

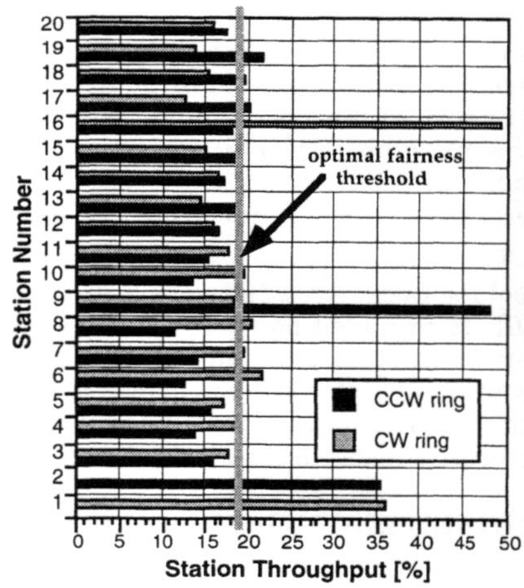

Figure 5. Station throughput in S2

The uncontrolled bandwidth-sharing, which results from the unhomogeneous traffic flow modelled in scenario S2, is shown in figure 5. Without any traffic control, on each ring, two stations would be given preference. It should be noted, that stations 1 and 2 use only the one ring, which is selected to reach station 16 and station 9, respectively, on the shortest path. Thus the indicated optimal fairness threshold will not be achieved by these two stations on their unused rings. On the CCW ring station 2 benefits from the fact that station 1 only sends on the other ring. The same holds for station 1 on the CW ring. Station 9 as well as station 16 are preferred since they are chosen as destination with a higher probability. Unlike this preference of single stations, in scenario S3 on each ring one station would completely starve in the case of missing traffic control by ATMR. Despite this dominant unfairness, in each scenario used in our experiments, an additional small inhomogeneity occurs as indicated in figure 5 by the alternating throughput pattern for stations with an odd and an even number, respectively. This is caused by the routing convention which is used in the case of equal distances between source and destination on each ring as described in section 3.

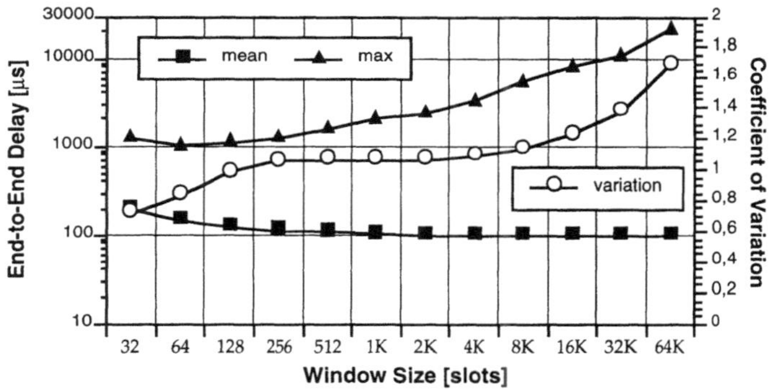

Figure 6. Delay versus window size in a homogeneous traffic environment

In the homogeneous traffic scenario (S1) we can restrict our discussion of the delay-characteristics on global results. As shown in figure 6, the mean values of the end-to-end delays decrease with an increased ATMR window size. The only variable part of a packet´s end-to-end delay is the waiting time for a sufficient number of free slots, which is necessary to transfer the regarded packet, and all packets which were already waiting in the queue when it arrived. As discussed above, the higher transmission quotas result in an improved throughput for each station. Thus more free slots are available during a certain period of time leading to reduced mean values. The traffic regulation, however, is weakened by higher window sizes as indicated by the maximum delay values. Although these values are obtained in worst cases only, they indicate a worse service for delay-sensitive applications. It should be noted that logarithmic scaling is used in figure 6. In our performance evaluations, delay jitter, which has more practical meaning than the mean and the maximum values, is indicated by the coefficient of delay variation. Real-time requirements of various user applications, especially compressed audio and video communications, include small jitter in the end-to-end delays to avoid information loss and extended buffer space at the destination. Furthermore, for multimedia applications, synchronisation of various data streams is necessary. As indicated by the results of the coefficient of variation in figure 6, the jitter increases with growing transmission quotas.

Now we want to examine the possibility of improving ATMR´s transmission behaviour by the implementation of priorities. As described in section 2, a high priority reset scheme can be realised by the transfer completion reset (TCR). We compare both operation modes by results obtained in scenario S1. If the TCR mechanism is enabled, we choose its parameters in order to provide a maximum delay of 1 ms. Thereby a separate queue has not been installed for the different priority classes. Thus during a simulation run, all stations operate either with low priority (TCR disabled) or with high priority (TCR enabled). Figure 7 shows the impact of the TCR on the delay characteristics. While the maximum values grow in proportion to the window size in the case of conventional access control, there is a clear restraint of these delays if the TCR is applied. Also the 99 percentiles could be decreased by the TCR mechanism, especially for mean window sizes. This effect is caused by the fact that a node which is blocked due to its lapsed window counter until the next reset, can gain access to the medium by the generation of an earlier TCR. This high priority reset smoothes the instable behaviour whilst it considers the delay threshold of the data in the transmit queues. It should be noted that in our experiments the results could not be improved more only due to the reason that all stations operate at the high priority and a TCR does not refrain any station from sending any data.

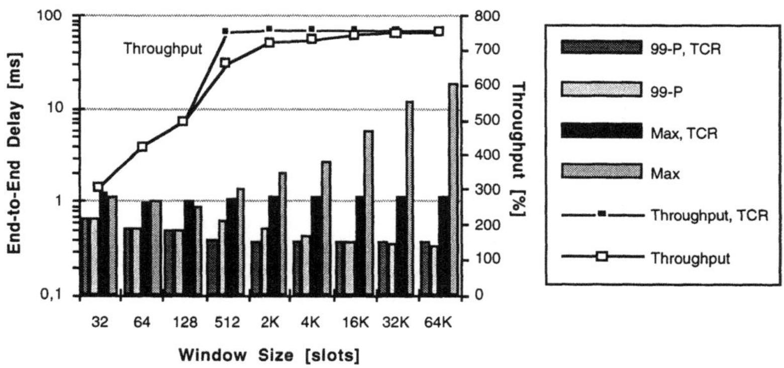

Figure 7. Influence of TCR on the protocol´s behaviour

The installation of the TCR also improves the bandwidth utilisation. Network throughput already approaches the theoretical maximum of 760% in the case of fairly small transmission quotas. The TCR always anticipates the ordinary reset. Thus higher window sizes do not cause any more changes to the throughput or delays. At least concerning the throughput, the TCR mechanism generates the resets in an optimised manner and prevents blocking of many stations as obtained at the end of a normal reset-period. The installation of both mechanisms, including separate queues (one controlled by the ordinary reset and the other by an additional TCR) provides a two priority class operation which supports delay sensitive data side by side with conventional user traffic. The throughput/delay tradeoff, however, can not be removed in general. A choice of smaller time parameters for the TCR leads to higher frequency of resets which results once more in additional protocol overhead.

4.2. Overall network performance in heterogeneous traffic scenarios

Now we change to a discussion of the protocol's behaviour obtained in heterogeneous traffic scenarios. As stated above, in scenario S2, ATMR is able to share the available bandwidth equally among all stations, although two stations are preferred on each ring. Fairness, however, is not restricted only to equal throughput. Also an acceptable delay characteristic should be guaranteed for each station. As shown in figure 8 ATMR provides - on average - the same delays for packets of both the 'normal' and the 'preferred' stations as a result of the fair bandwidth sharing. A strong access control by small window sizes also achieves equal values for the 99 percentiles of the delays. Contrary to the uniformity of the mean values with increasing transmission quotas, the percentiles of 'normal' and 'preferred' stations differ greatly. The less ATMR regulates the access to the medium the more the preferred stations operate in a 'send-and-blocked' manner. At the beginning of each control cycle these stations get many opportunities to access free slots, resulting in fast expiration of their transmission quotas. Then these stations have to wait longer until all other stations have also used their quota, as indicated by the values of the 99 percentiles. It should be noted that the 99 percentiles have more practical meaning than the mean values. Thus a growing window size could lead to some problems in serving delay-sensitive data.

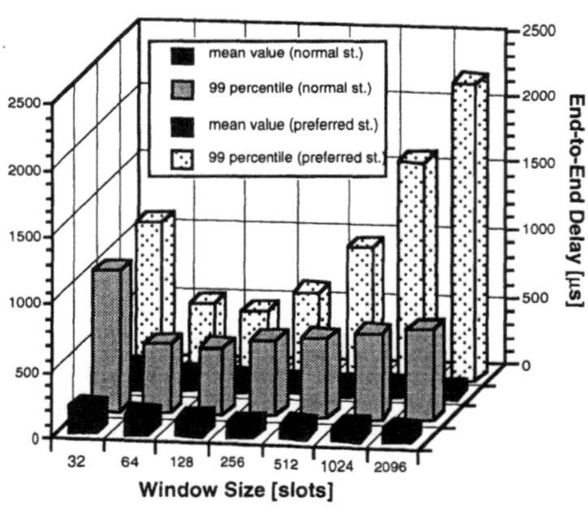

Figure 8. Delay characteristic

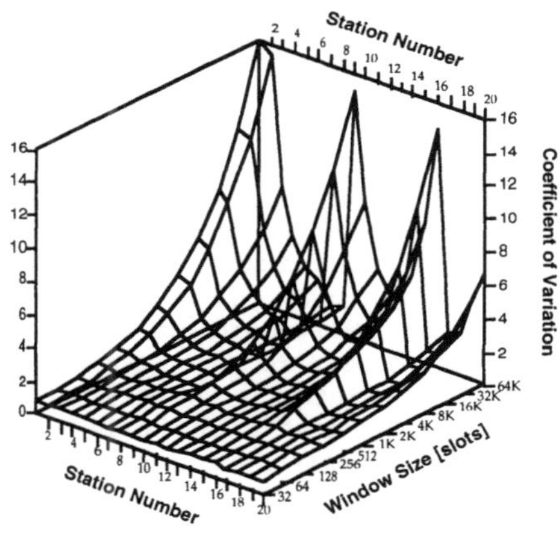

Figure 9. Delay jitter vs. window size

Figure 9 shows the same effect now illustrated by growing delay jitter. In the case of small window sizes the well bounded delays also reduce the variation of delays for all stations. Weaker traffic control leads to the above discussed 'send-and-blocked' behaviour of the preferred stations. The results, however, also show that not only the preferred stations suffer from this worse delay characteristic. The nearer a station is positioned to a preferred one, the more it is influenced by the irregular transmission behaviour of the preferred station. It gets more free slots if the preferred station has consumed its quota. On the other hand it gets more busy slots at the beginning of the reset period. Thus the delay jitter increases at these stations aswell. It should be noted that in figure 9 the values of both rings are superposed. In fact only the stations downstream of the preferred stations are affected on each ring.

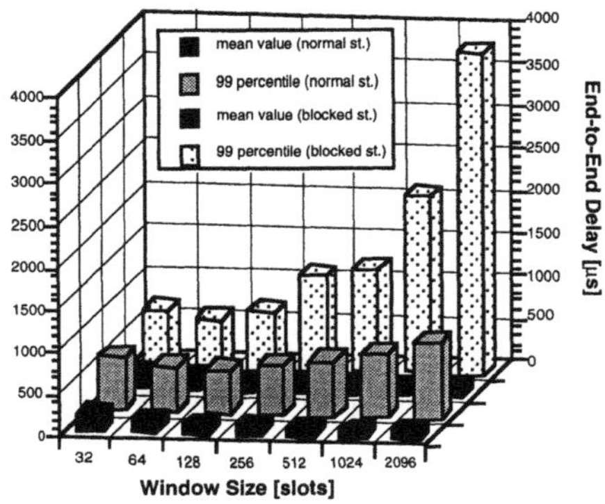

Figure 10. Delay characteristic

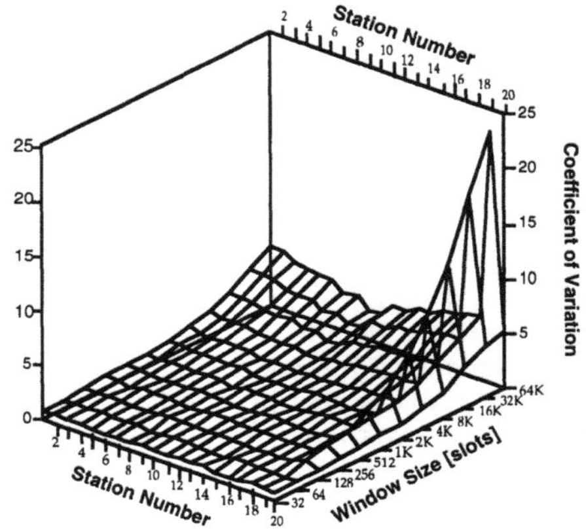

Figure 11: Delay jitter vs. window size

Although scenario S3 models a load characteristic which is inverse to that of S2, similar results can be obtained as shown in figure 10 and figure 11, respectively. The transmission behaviour, nevertheless, is quite different. While in scenario S2 the 'send-and-**blocked**' behaviour of the preferred stations dominates the network, now the handicapped station is forced into a '**blocked**-and-send' mode. Since no cells are ever addressed to the blocked station 19, in the saturated scenario it gets no opportunity to access any free slots at the beginning of each reset cycle. Thus it cannot utilise its transmit quota not before most of the other nodes become inactive, due to expired transmission counters. The 99 percentiles increase more strongly for blocked stations, whereas the mean values remain similar for both types of station again. Compared to the percentiles for the preferred stations in the former discussed scenario S2, the 99 percentiles of end-to-end delays of the blocked station increase more with rising window sizes. This is caused by the fact that the handicapped station is almost the only active user at the end of a reset period, whereas in scenario S2 the preferred stations indeed get more bandwidth at the beginning of the cycle, but they have still to compete with the normal stations. Also the delay jitter at station 19 is more strongly increased (see figure 11). But now this worse behaviour does not influence the neighbouring stations since the blocked station is dominated by the network. Contrary to that, in scenario S2 the preferred stations actively prevail the traffic flow.

In the last experiment including saturated senders, we increase the length of the underlying ring topology in scenario S2. Thereby the window size in the ATMR protocol remains fixed at 2096 slots per cycle. It should be clear that the higher the ring latency, the longer the duration, until the last active station detects the end of the actual control cycle, and the longer the duration of the reset procedure itself.

As shown by the results of figure 12, the high efficiency of bandwidth utilisation can nearly be preserved up to a ring length of 50 km. If the network extends to more than 100 km, however, the efficiency of the quota mechanism of ATMR collapses. This worse behaviour could be derived from the increased propagation delays of the reset signals due to an extending ring latency. When the reset signal arrives at one station, most of the other nodes have already used their quota. Now the station becomes active and can utilise its transmission quota very quickly. Thus at each station there is a time-gap between becoming blocked and the moment of renewing the quota. During this time all free slots have to pass unused. The number of simultaneous active stations decreases with higher ring lengths. In a worst case, only one station is active at a given time and the ATMR protocol is reduced to a simple token mechanism. A possible solution to overcome this problem is to increase every station's quota. But as discussed above, a higher quota leads to less restricted delay bounds, especially for handicapped stations.

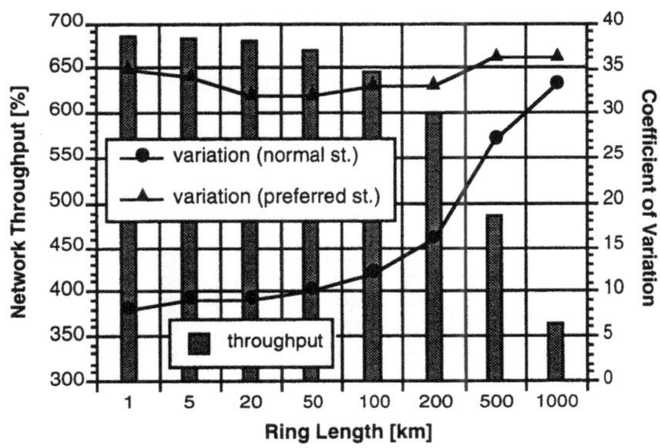

Figure 12. The protocol's behaviour under increasing ring length

To evaluate the protocol's capability for achieving tight delay bounds in the case of growing network sizes, we restrict to the metric of the access delays. This value is given by the time between two consecutive starts of packet transmissions. By increasing ring lengths, the metric of end-to-end delay would not be quite expressive enough because it is dominated by the propagation delay between source and destination. Thus the characteristics of the protocol would be covered. The protocol's behaviour as indicated above by the throughput results is also confirmed by the delay characteristics. As shown before, in scenario S2 the preferred stations operate in a 'send-and-blocked' mode. This is indicated by the high values for the coefficients of variation in figure 12. With increasing ring length, however, the jitter in access delays for normal stations approximates to the values of the preferred stations. This indicates that in extended networks the jitter caused by the higher ring latency dominates the jitter which results from the traffic distribution. Now, both types of station suffer from the delay between two consecutive resets.

4.3. The protocol's performance in serving bursty traffic

In the following, the capability of ATMR to serve variable bit rate traffic with high burstiness shall be examined. The degree of burstiness is measured by the ratio of the duration of busy and silence periods. In our experiments the burstiness is varied between 1:1 and 1:23. Again the network scenario consists of twenty stations where four VBR-stations, which generate variable bitrate traffic, are located nearby on a small ring segment (station 0, 1, 4 and 6). All other stations generate traffic with a constant bitrate (CBR). It should be noted that in the bursty scenario the stations do not operate as saturated sender anymore. The mean data rate at the VBR stations and the CBR stations is set to 10 Mbit/s and 40 Mbit/s, respectively. This results in an overall network load of 680 Mbit/s. To achieve a mean data rate of 10 Mbit/s at the VBR stations during each burst the traffic is generated with a rate of 20 Mbit/s (in the case of 1:1 burstiness) and 240 Mbit/s (in the case of 1:23 burstiness), respectively. In order to bound the delays, the TCR mechanism has been enabled.

Figure 13 summarises the results for the stations which are located on that ring segment where the VBR stations are positioned. In the case of the 1:1 burstiness the transmission behaviour of the CBR stations is not affected by the bursty traffic characteristic. The coefficients of delay variation show quite low values due to the immediate access under the low load conditions. The same is true for the VBR stations concerning the mean values. It is clear that the maximum values as well as the jitter slightly increase due to the changing delays for packets of one burst.

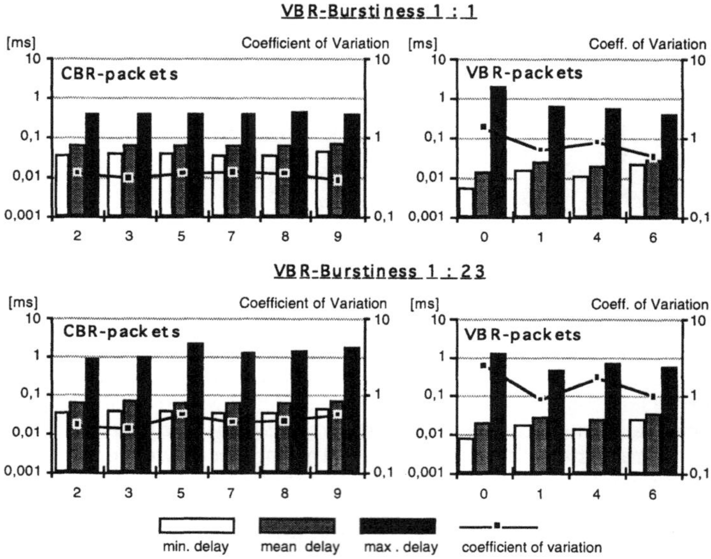

Figure 13. Delay characteristics under bursty traffic conditions

If we increase the burstiness at the VBR stations to a ratio of 1:23, it is clear that the CBR stations are influenced by the changing conditions of the common medium. The mean values of the end-to-end delays are only sligthly increased, indicating a stable transmission behaviour. The high burstiness only temporarily overloads the network. Then packets of the CBR stations experience higher delays leading to the higher coefficients of variation. The maximum delays, however, can still be bound to a few milliseconds by the TCR. Although a greater degree of burstiness is modelled by the VBR-stations, the mean data rate is not increased. The mean delays at these stations remain unchanged indicating that the protocol reacts relatively quickly to the changed network conditions. The maximum values can be bounded by the TCR in the same way as in the case of 1:1 burstiness. The jitter, however, is greatly increased as indicated by the coefficient of variation, especially for station 0 and station 4. The differences between single VBR stations can be explained by the number of CBR stations located in an upstream direction.

The more CBR stations located upstream, the less is the bandwidth which is available immediately when a new burst arrives. Thus the transfer of one burst may exceed one reset period.

Finally we evaluate the service provided by ATMR for a VBR station which is located downstream of a heavy user. In such a scenario the VBR station is strongly influenced by the transmission behaviour of the heavy user. The overall network load is increased from 1 Gbit/s up to 7 Gbit/s, whereas the data rate of the regarded VBR station is increased from 2 Mbit/s up to 14 Mbit/s in steps of 2 Mbit/s.

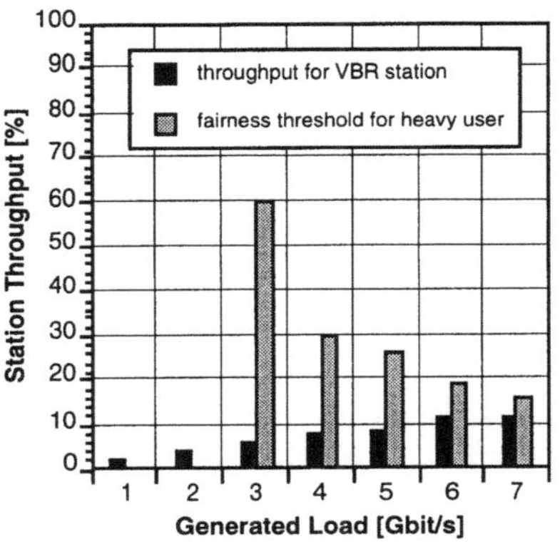

Figure 14. Throughput at VBR station

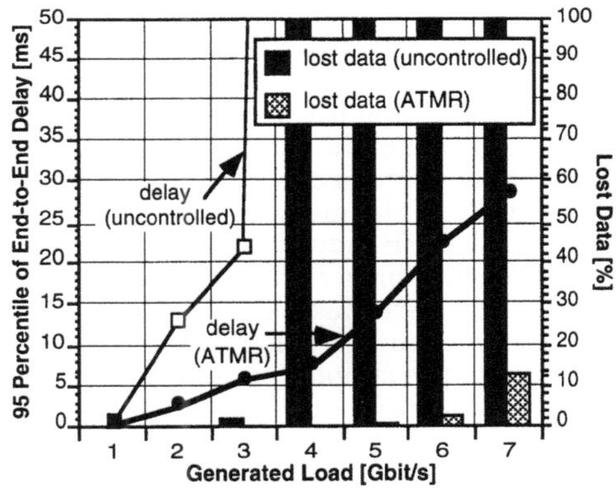

Figure 15: Delays for bursty traffic

A generated load of 1 Gbit/s and 2 Gbit/s, respectively, can be regarded as a low load condition in which no fairness threshold is necessary (see figure 14). The heavy user does not generate so much data that it has to be blocked by the protocol in order to prevent data loss at the VBR station. The amount of lost data at the VBR station in an uncontrolled network, as shown in figure 15, indicates that in our model the regarded station would be blocked in the case of a load of more than 3 Gbit/s. In order to provide a sufficient service for this handicapped station under these higher load conditions, the ATMR protocol compensates the inhomogeneous traffic distribution in the well known way. The more the overall network load increases, the smaller is the fairness threshold which bounds the bandwidth of the heavy user. The access control, however, can not achieve a total fairness among the heavy user and the VBR station. Under overloaded network conditions data is lost at the VBR station (see figure 15) although its throughput has not reached the calculated fairness threshold (see figure 14). This is due to the fact that the ATMR protocol operates 'memoryless'. If the VBR station has not utilised its transmission quota in one reset period since no burst has occurred, it can not save this bandwidth for the next cycle in which an arriving burst could exceed its available transmission capacity.

In figure 15 the delay characteristic of a VBR station is indicated by the 95 percentile of the end-to-end delays. Even under low load conditions the ATMR protocol reduces the delays compared to the values obtained in an uncontrolled network. If the generated load is low, ATMR provides immediate medium access and can bound the delays in an acceptable way. If the network becomes overloaded no results for packet delays can be obtained in the uncontrolled network since the VBR station is completely blocked. In the ATM Ring delays increase due to the increasing number of packets per burst. As long as all packets of one burst can be transmitted before the next burst arrives, the delay is caused by the waiting times of the last packets of each burst. Thus the delay bound depends on the given data arrival process. If a station´s buffer capacity and the bandwidth provided by ATMR are not sufficient to serve the increasing data rate, the delay bounds are more influenced by the overall network load and therefore they increase more significantly.

5. FUTURE EMPLOYMENT OF ATMR

This paper has discussed the performance of the ATM-based high-speed medium access protocol, called ATMR, with particular regard to the requirements of future communication applications. Data transmission is done in cell mode and based on the slotted ring architecture. The principle of spatial slot reuse, controlled by transmission quotas and a cyclic reset scheme, pushes the throughput of the ATMRing to a multiple of the medium´s data rate. Immediate access is provided under low load conditions. Bandwidth competition under high load or overload is solved efficiently by achieving a fair bandwidth for all stations. Choosing a higher window size allows better performance but leads to worse delay characteristics, especially for jitter and guaranteed delay bounds. It is shown, that a priority reset mechanism can improve the support of delay-sensitive services. Several strategies for tuning the window size and the installation of priority resets for satisfying QoS in a multiservices environment have already been proposed [7,16]. Nevertheless, ATMR will not change to a hybrid mode to support isochronous in addition to the asynchronous services. All data transmission remains purely based on an asynchronous transfer mode using ATM cell format only. This mode in operation as well as its scalability in data rate, the support of virtual connections, and the connection to a common interface (UNI), make ATMR compatible to the ATM technology.

Further study is needed to see how this protocol can support other configurations. The individual optimisation of the ATMR protocol according to the requirements of specific services has to be discussed in more detail in the future. This work includes investigations of both the parameter tuning according to changing network states, and the integration of above mentioned delay-sensitive services.

Although ATM was originally developed as the target solution for B-ISDN in the public area, there is a growing interest in local ATM systems. Besides its connectivity to wide area ATM networks, an ATM LAN supports multiple guaranteed classes of service. These features are an inherent part of the ATM technology due to its concept of virtual end-to-end connections. By exploiting some simplifications in installation, operation and management of a customer premises ATM system, and by meeting the requirements of LAN specific applications, a simple and cheap switch-based ATM system can be considered as a realistic alternative to current LAN architectures.

A direct comparison of shared-medium and switch-based ATM LANs seems to be quite difficult. One problem is to find a fair network configuration, e.g. which transmission and switching capacity should be selected, or in which geographical area the nodes should be located. Another problem is to find a broad range of traffic scenarios to discover all advantages and disadvantages of both architectures. Last but not least the interpretation of the results will depend on the preferences of the examiner. At the moment the development of ATM LANs benefits from significant commercial support in it´s standardisation, as well as from the vision of a universal network based on ATM. Nevertheless, new LAN protocols operating on a shared medium can also meet the requirements of future user applications and network architectures. Therefore strong effort should be spent to provide interoperability of both architectures, and to design common interfaces which are efficient and cheap. The ATMRing is an approach which represents a distributed system of small ATM switching modules, interconnected by a shared medium. Thus it integrates both concepts and can be considered as one possible architecture for future customer premises networks.

REFERENCES

[1] Kleinrock, L.: *The Latency/Bandwidth Tradeoff in Gigabit Networks*, IEEE Communications Magazine, vol. 30, no. 4, 1992.

[2] Kung,H.T.: *Gigabit Local Area Networks: A Systems Perspective*, IEEE Communications Magazine, vol. 30, no. 4, 1992.

[3] IEEE: *Focus on ATM LANs*, IEEE Network Magazine, vol. 7, no. 2, 1993.

[4] Fink, R.L.; Ross, F.E.: *Following the Fiber Distributed Data Interface*, IEEE Network Magazine, vol. 6, no. 2, 1992.

[5] Gvozdanovic, R.: *A Review of High Performance Protocols for the FDDI Follow on LAN*, Proc. of EFOC/LAN´92, Paris 1992.

[6] Honda, T.; Imai, K.; Kasahara, H.; Ito, T.: *ATMR: Ring architecture for Broadband Networks*, Proc. of Globecom´90, San Diego 1990.

[7] Kasahara, H.; Imai, K.; Morita, N.; Ito, T.: *Distributed ATM ring-based switching architecture for MAN and B-ISDN access networks*, Proc. of first IFIP Conference on Broadband Communications, Estoril 1992.

[8] van As, H.R. et al.: *CRMA-II: A Gbit/s MAC-Protocol for Ring and Bus Networks with Immediate Access Capability*, Proc. of EFOC/LAN´91, London 1991.

[9] Lemppenau, W.; van As, H.R.; Schindler, H.R.: *A 2.4 Gbit/s ATM Implementation of the CRMA-II dual-ring LAN and MAN,* Proc. of EFOC&N, The Hague 1993.

[10] Cidon, I.; Ofek, Y.: *Metaring - A Full-duplex Ring with Fairness and Spatial Reuse*, Proc. of IEEE INFOCOM, San Francisco 1990.

[11] Falconer, R.M.; Adams, J.L.: *Orwell: A Protocol for an Integrated Service Local Network,* Advances in Local Area Networks, Kümmerle et al. (ed.), IEEE Press, 1987.

[12] Okada, T.; Ohnishi, H.; Morita, N.: *Traffic Control in Asynchronous Transfer Mode*, IEEE Communications Magazine, vol. 29, no. 9, 1991.

[13] Horvath, S. et al.: *The FDDI Follow-On LAN Working Paper*, Rev. 23, 1992.

[14] Hou, X.; Niemegeers, I.G.: *The Concept of Fairness and a GFC Protocol*, Proc. of IEEE Conference on Local Computer Networks, Minneapolis 1992.

[15] Ohnishi, H. et al.: *ATM Ring Protocol and Performance,* Proc. of International Conference Communication, IEEE ICC´89, 1989.

[16] Ito, T. et al.: *Performance Analysis of a High-Speed Ring Network in a Multi-Services Environment,* Proc. of 11th Intern. Conference Computer Communication, Genoa 1992.

[17] Bach, C.; Richter, S.: *Performance Evaluation of an ATM-based high-speed slotted ring network*, Proc. of EUROPTO´93, Berlin 1993.

[18] Wolsky, J.: *Einsatzmöglichkeiten von ATM-Prinzipien in Gbit/s-Netzwerken,* Diploma Thesis at Aachen University of Technology, Aachen 1993 (in German).

[19] Martini, P.; Werschmann, G.: *Connection Oriented Data Service in DQDB*, Proc. of ITG/GI-Conference - Communication in Distributed Systems, Munich 1993.

[20] Davids, P.: *ATLAS - Reference Manual*, Department of Computer Science, Aachen University of Technology, Version 5.0, Nov. 1991 (in German).

TRAFFIC

Broadband Communications, II (C-24)
S. Tohmé and A. Casaca (Editors)
Elsevier Science B.V. (North-Holland)

PERFORMANCE MODELLING AND TRAFFIC ENGINEERING FOR BROADBAND COMMUNICATION NETWORKS

Paul J. Kühn

University of Stuttgart
Institute of Communication Switching and Data Technics (IND)
Seidenstraße 36, D-70174 Stuttgart
Tel. :+49-711-121 2479, Fax: +49-711-121 2477, email: Kuehn @ ind.uni-stuttg.d400.de

Performance Modelling and Traffic Engineering provide methods and tools to analyze the network performance, to size network resources and to evaluate the efficiency of protocols, in particular, when the network load is governed by stochastic traffic flows. They form the only feasable technique during the phase of standardization and design. The contribution addresses the key modelling, analysis and engineering aspects for broadband communication networks based on shared media and switched architectures.

1. INTRODUCTION

New services such as high speed data communication, video on demand or multimedia reveal quite different traffic characteristics, especially high speed data with variable bitrates. To support these services adequately, the new broadband communication networks have to be flexible to accept these types of traffic and to guarantee a specified quality of service (QOS) and network performance, respectively. Performance modelling and traffic engineering may provide the necessary methods and tools by which network resources could be dimensioned and by which grade of service parameters could be estimated based on modelling techniques.

The contribution addresses at first existing and future services. Typical attributes which characterize these services are their dynamic bitrate requirements in terms of average and peak values, the correlation of peak periods, the tolerable loss and delay figures, connection frequency and holding times.

The third part introduces some fundamental networking concepts such as multiplexing, switching, transfer modes connection and path concepts from which performance and network management requirements are motivated.

In the fourth and fifth part, current and future broadband communication network architectures are addressed. They appear mainly in two categories: High Speed Local and Metropolitan Area Networks (HSLAN, MAN) at one hand and Broadband-ISDN on the basis of the Asynchronous Transfer Mode (ATM) at the other hand. Both network types, however, have been designed for different purposes; therfore they show quite different concepts with respect to connection types, transfer modes, protocols and traffic control. Especially, the interworking of both network types will be of prime importance. The main architectural structures are shortly addressed.

The sixth part focuses on the important concept of the interconnection of LAN/MAN at one hand and the B-ISDN at the other.

The QOS requirements of the various services have to be defined from which network performance requirements could be derived such that an acceptable satisfaction of the user results. However, resource sizing and management have to be done with great care to keep network costs sufficiently low by a high utilization of the resources. This problem turns out to be rather critical due to the various requirements of different services. The way out can only be found by intelligent traffic control mechanisms such as usage parameter control (UPC), connection acceptance control (CAC), the control of traffic losses and delay variations, routing control and resource management. Although many useful approaches are already known, the lack, however, is a realistic verification, as most of the new services are not yet known. Therefore, the standardization process is still in progress and more research is needed on these topics.

The next part addresses the related methods for the traffic characterization, modelling of traffic and system components, analytical and simulative performance evaluation methods, and traffic engineering techniques and practises. This field has a long history in communications. Although principal results of the past may be useful to explain certain phenomena, a solid analysis and application requires new methods to capture the new aspects. Some key results will be addressed to show the approach and the capabilities of these methods.

Finally, modelling and traffic engineering methods depend largely on their quality. Therefore, verification of the assumptions and results are necessary from which refined methods result which are reliable for practical applications. First steps into these directions are in progress based on first ATM switching system demonstrators and terminal equipment with new coding facilities. To perform the related tests, a new set of traffic generator and analyzer equipment is necessary which has been developped during the last years. In particular the contribution addresses experiments which are in progress with the RACE project EXPLOIT and its demonstrator setup at the Basel location.

2. EXISTING AND FUTURE SERVICES

2.1 Existing Services

Current applications are digital telephony and various other telematic services based on 64 kbps channels such as ISDN-telephony and audio conferences, ISDN-videotex, ISDN-telefax. Initial videoconference services have been taken up based on a circuit switched broadband network operating at 140 Mpbs; this service will, however, be replaced later on by B-ISDN videoconferencing with possibly much lower, but variable bitrates. On the other hand, data communication services based on packet switching are widespread through connectionless datagram transfer in LANs with transmission rates up to 16 Mbps. In WANs the standardized X.25 based packet service dominates providing connectionoriented or connectionless packet transfer from a few kbps up to 64 kbps. In some countries a 2 Mbps packet service has been taken up, too. The current situation originates from various conditions such as parallel developments in telephone and data networks, the lack of adequate technologies, little or no cooperation of standardization bodies, existing terminal equipment, and the like. This has led to a rather colourful coexistence of networks differing in speed, switching and protocols. Interoperation is only possible through expensive interworking equipment.

2.2 Future Services

Future networking aims at a much higher degree of service integration within one universal communication network with worldwide standardized interfaces and protocols. This starts at first with a standardization of Service Types and Service Attributes. Figure 1 gives an overview of the two major Service Types and classes together with their definition and application examples as they have been defined by CCITT (now renamed as ITU-TS). These service types have to be complemented by Service Attributes which characterize the service in more detail and which allow for the estimation of network resource requirements. Typical Service Attribute parameters are:

- Constant / Variable Bitrate (CBR / VBR)
- Average and Peak Bitrate
- Duration and Correlation of Peak Bitrates
- Connection Holding Time
- Tolerable Bit Error Rate (BER)
- Tolerable End-to-End Information Delay
- Tolerable Connection Setup / Takedown Delay.

Parameters of these Service Attributes are highly dependent on the applied source coding technique, signal processing techniques, or user behaviour. Current state of the art reveals a wide spectrum of parameters, as, e.g.,

CBR	voice:	64 kbps	VBR	voice:	12-24 kbps
CBR	video:	140 Mbps	VBR	video:	1- 5 Mbps.

The standardization of related parameters is one of the urgent questions for future networking. Because of the technological dependence of these parameters and the involved risk of a premature standardization, however, network architects tend more and more to a network solution which is independent of the individual service characteristic, i.e., to switched communication networks supporting services of arbitrary constant or variable bitrates. This would allow that enhanced or up to now unknown services can be integrated later on without changes of the networking technology. Necessary changes would primarily affect the (private) terminal equipment and soft-controlled parameters inside the network, such as for connection acceptance control.

Besides the Service Attribute parameters mentioned before, future services may also differ to the current ones by their procedural characteristics. Typical examples are:

- Multimedia Communication
- Universal Personal Communication (UPC)
- Multicast Communication.

In a Multimedia communication, several communication types may be used simultaneously with dynamic changes during a Multimedia call. A user may apply video and data communication in addition to voice with a rather flexible control. The UPC is a further extension of the digital telephone, within which a user carries a personal ID, through which any type of communication could be established at any time and at any point. This service requires Intelligent Network (IN) services, i.e., network data base access and fast and reliable signalling procedures. Multicast applications differ from Distributive Services; contrary to broadcast of programs or other services, any user could make use of group communication for an effective and timely distribution of information among a defined group of recipients. Special forms of mobile networks, in particular satellite networks, seem especially adequate for such types of applications.

2.3. Networking Requirements

In conclusion of the short illustration of existing and future services, some requirements for future networks can be derived:

- Arbitrary transmission speed up to several hundreds of Mbps
- Capability of integration of all services
- Fast and efficient information transfer for constant and variable bitrate traffic
- Low information loss and delay, possibly user-defined or service-dependent
- Fast and efficient network control, i.e. connection acceptance, connection management, and overload control
- Flexible and dynamic network configuration
- Flexible terminal addressing independent of user location
- Evolutionary development with the integration of existing services
- Efficient network management support.

In the next chapter, major developments in networking concepts will be addressed.

3. NETWORKING CONCEPTS

3.1 Multiplexing, Switching, and Transfer Modes

Communication architectures are defined in the framework of the layered OSI Reference Model. The transmission medium, the physical layer (1), and the media access layer (2a) provide a fast and efficient information transfer; their functions are technology-dependent and will be understood as networking functions in a closer sense. Functions of the logical link (2b), network (3) and transport layers (4) refer to the transport functions, whereas functions of the session (5), presentation (6) and application layers (7) are more related to information processing functions; we will consider them as networking functions in the wider sense. The following discussion will concentrate more on the technology- dependent networking functions.

Multiplexing defines the way how various information streams are carried through one media. There are different multiplexing principles such as frequency, wavelength, code and time division multiplexing. For modern optical communication networks, wavelength and time division are the most important ones. Wavelength multiplexing provides several distinct channels on one media and can be considered like separate media. In time division, two basic principles can be distinguished, the Synchronous Time Division (STD) and the Asynchronous Time Division (ATD), c.f. Figure 2. In STD, time is devided into frames. Each frame is devided into slots. Equally numbered slots of consecutive frames form a STD channel, e.g. with 64 kbps transmission rate. A physical channel is identified by the position of the slot within the frame.

In ATD, time is again devided; each slot carries an information unit of variable length (also called a frame or a packet) or of constant length (called a cell). Packets or cells may appear totally asynchronously. They carry a Virtual Channel Identifier (VCI) within their header; units with identical labels (VCIs) form a virtual channel. There are several cell synchronization concepts. First, cells are self-delineating by usage of the Header Error Control. Cells may also be put into the payload of SDH frames; alternatively, synchronization cells could be used as shown in Figure 2.

Connections are switched physical or virtual channels between terminal equipment across the networks, i.e., a connection is the concatenation of physical or virtual channels of consecutive links. Physical connections use the circuit switching principle, whereas virtual connections are based on the packet switching principle. Thus, a physical connection provides always a constant transmission rate according to the underlying slot capacity;

virtual connections provide a variable transmission capacity dependent on the frequency of packet or cell transmissions.

With these notions, three main transfer modes may be defined, Synchronous, Packet, and Asynchronous Transfer Mode (STM, PTM, ATM):

STM - Synchronous Transfer Mode

- Multiplexing of Physical Connections based on STD
- Circuit Switching Principle
- Constant Bitrate Traffic Streams
- Integration of Traffics with Different but Constant Bitrates

PTM - Packet Transfer Mode

- Multiplexing of Virtual Connections based on ATD
- Packet Switching Principle with Variable Block Length (Packets)
- Bursty Traffic Streams
- Flow Control, Error Recovery
- Special Case: Connectionsless (Datagram) Transfer

ATM - Asynchronous Transfer Mode

- Multiplexing of Virtual Connections based on ATD
- Packetized Switching of Fixed Sized Cells
- Bursty Traffic Streams
- Lightweight Protocols: No Flow Control
- Integration of Traffics with Arbitrary Cell Rates

3.2 Connection and Path Concepts

As already mentioned two basic switching concepts exist, circuit (CS) and packet switching (PS). CS is always connection-oriented (CO), i.e. an end-to-end physical connection has to be established first. PS can either be CO, using the virtual connection principle, or it can be connection-less (CL), where packets are routed independently link by link from the origin to the destination. Within the CO virtual connection concept, various additional functions can be supported such as fast switching without routing, sequence control, data flow control, and error detection and correction.

In typical LANs, a CL distributed PS principle is used only, whereas MANs feature hybrid switching, i.e. a CO, CS service (isochronous service) and a CL, PS service like in LANs. The future B-ISDN supports principally only a CO, PS service (ATM); however, various concepts exist for the support of CL services, see Chapter 6.

A Virtual Path (VP) is a switched path between Users or between Users and Network Specialized Resources transparently, indicated by Virtual Path Identifiers (VPI). Within these "pipes", several virtual connections may be established.

3.3 Performance and Network Management Requirements

The various services should be supported by the network with respect to their specific sensitivities as, e.g., information loss. At the other hand, the network operator is interested in a high utilization of the network components. Thus, networks have to meet various performance criteria as throughput, utilization, connection and call blocking, cell or packet loss, connection setup delay and information transfer delay. These quantities have to be defined precisely with respect to reference points, reference connections, reference configurations, and with respect to their mathematical meaning.

Distributed systems impose a rather severe problem for their management which has to be supported automatically. In OSI, five major objectives are defined and standardized for the management: configuration, fault detection and correction, performance, accounting and security. A proper management structure has been defined starting with the identification of Managed Objects, their representation within Management Data Bases, protocols for layer-specific management functions, interlayer plane management, systems and network management. For instance, the automatic overload detection and avoidance is a typical network management function which cannot be handled manually anymore.

4. HIGH SPEED LOCAL AND METROPOLITAN AREA NETWORKS

4.1 Configurations

During the last 15 years, Local Area Networks (LAN) have been extensively introduced for local computer communication. They are mainly based on a bus or ring type medium topology operating with Carrier Sense Multiple Access with Collision Detection (CSMA/CD), or Token-Passing Media Access Control (MAC), respectively. These MAC-schemes are only useful for transmission rates up to about 20 Mbps over short distances up to a few kilometers [1,2,3]. These types of LANs operate on single packet (datagram) transfer, where packets are of variable lengths ranging from a few bytes up to several kbytes. On top of this basic MAC-services, virtual connections may or may not be used.

Currently, High Speed LANs (HSLAN) and Metropolitan Area Networks (MAN) are introduced, whose extension may reach up to 50...100 km, and whose transmission rates are in the order of 100 Mbps. Figure 3 illustrates a typical networking scenario of FDDI (Fiber Distributed Data Interface); this HSLAN operates as backbone network for the interconnection of LANs, powerful Mainframes, CAD Workstations, File Servers, or Interworking Units with public WANs. FDDI is based on a dual-ring topology with an extension up to 200 km, up to 1000 terminal connections, 100 Mbps transmission rate, and a timed-token-based MAC allowing for 3 kinds of packet traffic classes. In a later version, FDDI-II, a circuit-switched isochronous traffic class has been added by which isochronous connections can easily be accommodated, e.g., for PBX-interconnection purposes. The bandwidth management for the packet traffic types is based on the timed-token-access scheme; in this scheme, the time a station is allowed to hold the token, i.e. to send, is timer-controlled. This results in controlled token circulation times such that real-time conditions can be met. For further details we refer to the ANSI-Standard [4].

A competitive candidate to FDDI-II is the new IEEE 802.6 standard DQDB (Distributed Queue, Dual Bus), cf. [5]. This MAN is based on a dual bus topology and a reservation type protocol for packet MAC, see Figure 4. DQDB allows for two traffic classes, isochronous circuit switched and asynchronous packet switched. Whereas the isochronous connections require - as for FDDI-II - a connection setup procedure, the asynchronous packet access follows through a distributed access mechanism based on reservation requests and distributed counters in each station forming a spacially separated FIFO queue.

New developments direct to transmission rates of several hundred Mbps up to 1.2 or 2.4 Gbps [6]. These MANs use modified MAC schemes. An example of this development ("FDDI Follow On") is the CRMA (Cyclic Reservation Multi Access), cf. [6].

4.2 Protocol Architecture

Communication systems are functionally defined by their layered protocol architectures. The basis for the architectural structure is the ISO Protocol Reference Model (RM) for Open Systems Interconnection (OSI). In the RM, a layered approach has been defined. Entities of one layer communicate with each other by exchange of Protocol Data Units (PDU) according to a specific peer-to-peer protocol. A layer (N) provides a specific service to the adjacent layer (N+1); the service of a layer is invoked / provided through Abstract Service Primitives (ASP) and by the exchange of Service Data Units (SDU). On the "surface" of an entity, Service Access Points (SAP) are defined; between two SAPs connections may be established supporting the communication between these entities.

The distributed access and switching capabilities of HSLAN/MAN networks require a modified protocol layering which is reflected by the intermediate MAC Layer between the Physical Layer (or the Physical Media Dependent Layer, PMD) and the Data Link Layer (Logical Link Control, LLC), see Figure 5. In Figure 5, the principal functions of these layers are also indicated. The MAC and LLC layers are also referred to as layers 2a and 2b, respectively.

5. BROADBAND - ISDN

5.1 Configuration and Services

A Reference Configuration defines all network components and their reference points in the path between two Terminal Equipments (TE). For the Narrowband-ISDN, the network access comprises the access line which is terminated by the Line and Network Terminators LT and NT1, respectively. An optional NT2 provides PBX services. The access protocol (D-Channel Protocol) is defined between the TE and the Exchange Termination (ET). Between the Local Switch (LS) Functions (Exchanges) various CS channels with different bitrates, permanently switched channels, packet service and the signalling network functions (CCIS) are defined.

In the B-ISDN, this Reference Configuration is complemented by Broadband Capabilities based on ATM, which also takes care of packet switching services, see Figure 6.

The B-ISDN is based on a fully optical fiber transmission network operating at about 150 Mbits/s and 600 Mbits/s at the User Network Interface (UNI) and internally of the network, respectively. It provides a very fast and efficient connectionoriented bearer service, i.e., setup / clearing of virtual connections and packet (cell) transfer between the terminal service access points of the ATM bearer service. The Asynchronous Transfer Mode is a virtual connection based fast packet switching technique using small fixed sized packets, called cells. Cells are consisting of 53 octets, where 5 octets are reserved for the cell header and 48 octets are used for user information with respect to the basic ATM Layer of the network. Each header contains, among others, virtual channel and virtual path identifiers (VCI, VPI), payload and priority indicators, and one octet for a 1 bit header error forward correction and for the self-delineation of cell boundaries. All higher services, both for the user and the control context, are multiplexed into this ATM bearer service; the corresponding functionalities are transformed by the ATM Adaptation Layer (AAL).

The ATM bearer service is able to integrate any connectionoriented service such as circuit switched type services with constant bitrates as well as services with highly variable bitrates as in case of computer, video and packetized voice communication. Necessary functions for end-to-end synchronization, segmentation of stream-type data or connection handling for connectionless type services have to be provided by the AAL such that at the ATM layer all functions can be performed fast and effective by hardware support.

5.2 Protocol Architecture

The basic structure of the protocol block of the lower layers is shown in Figure 7, consisting of the Physical, ATM, and ATM Adaptation Layer on top of which both Higher Layer stacks are built upon for the user and for the control contexts, respectively. The multi-plane concept of ISDN is also visible. A more detailed view of the principal three lower layers provides Figure 8.

The lower layers of the B-ISDN Protocol Model are divided into the Physical Layer (Ph), with sublayers PMD (Physical Media Dependent) and TC (Transmission Convergence), ATM Layer, and the ATM Adaptation Layer (AAL), which again is subdivided into a Segmentation and Reassembly Sublayer (SAR) and a Convergence Sublayer (CS). This layered approach defines all necessary support functions as a basis for user and control information transfer.

The PMD provides mainly bit timing functions for the underlying Physical Medium, including line coding, insertion/extraction of timing information.

In the TC sublayer a transmission frame (e.g., according to the Synchronous Digital Hierarchy, SDH) may be used to transport ATM cells within its payload. It is also possible, as another option, that ATM operates on a completely unframed channel. Cell Delineation is provided by each cell header error correction mechanism (HEC), by which the header of each cell is protected against errors (1 error correction). If no information transfer is requested, empty cells are inserted.

The ATM sublayer is the most important one for ATM networks. Cells of 53 octets length are the basic transport units, consisting of octets payload and 5 octets control (header), see Figure 9. Within the header field, VCI and VPI information is carried which allows to identify to which connection a cell belongs to, and through which the cell is switched from source to destination along the before established virtual connection. At each switching point, a VPI/VCI translation is performed. Cells of different connections are multiplexed according to the ATD principle.

Also, payload type and priority indications are part of the short header field defining the type of the cell (user/network) or the priority status, respectively. ATM virtual connections are set up by the use of the signalling subnetwork. The signalling system is also integrated into the ATM-transport network.

The SAR sublayer provides the segmentation of larger messages into ATM cells and vice versa. SAR - PDUs are formed and transported as the payload of an ATM cell, as illustrated in one case of the AAL classes in Figure 9. Currently, four basic classes of AAL services are defined, differing with respect to Time Relation, Bitrate and Connection Mode. The AAL CS sublayer provides all the necessary adaptation functions between the higher user and control layers and the ATM layer.

For the control, an enhanced signalling system No.7 will be used for the transfer of control information for the connection management and other administrative functions. Signalling channels will be formed at the ATM layer which are negotiated by a metasignalling procedure.

The User Information is classified into the four classes within which CS-type video, VBR video, CO and CL data transfer are provided.

5.3 Connection and Path Concepts in B-ISDN

As mentioned before, B-ISDN operates on connection-oriented basis using the Virtual Connection (VC) and Virtual Path (VP) principles. Figure 10 illustrates at first the hierarchy consisting of VCs which are embedded within VPs; both are multiplexed into the physical Transmission Path (TP).

Figure 11 demonstrates the use of VPs as a transparent pipe between Users or between Users and Network Specialized Resources. Virtual Paths are established by a Network Management function.

6. INTERCONNECTION of LAN/MAN AND B-ISDN

6.1 Data Communication Scenarios

In the short term, most applications will be expected from Private Local Area Networks (LAN) or Metropolitan Area Networks (MAN). Since these networks operate on quite different protocols, Interworking Units (IWU) have to be inserted between the networks for the necessary protocol conversion. In the long term, a more ISDN-like structure may evolve. Both principal scenarios are illustrated in Figure 12.

The interconnection of MAN and B-ISDN imposes, however, a number of problems due to the principally different functional principles of both as contrasted in the following:

MAN	**B-ISDN**
Shared Medium Access	Centralized Connection Acceptance
Distributed Switching	Centralized Switching
CL VBR Data Transfer	CO CBR / VBR Data Transfer
CO CBR Data Transfer	
No Usage Parameter Control	Usage Parameter Control
Data Buffering within TE	No Data Buffering in TE

These differences have to be compensated by the Interworking Unit. The standardization development, however, directs partially to an adjustment of both functional principles. This can be seen in the IEEE 802.6 development, where the slot size and structure coincides, largely with the cell size and format of B-ISDN.

6.2 MAN - MAN Interconnection Options

There are several options how MAN could be interconnected. These options are illustrated in the series of Figures 13 - 17. Figure 13 shows a structure of direct interconnection using leased broadband channels. A much more flexible scheme is presented in Figure 14, where a MAN itself is used as a switching or backbone system for the interconnection of regional MANs ("MAN Switching System", MSS). Figure 15 shows a solution where the Virtual Path concept of B-ISDN is applied and Figure 16 where for each MAN-to-MAN interconnection a Virtual Connection is established. Figure 17, finally, shows the option where a so-called Connectionless Service Function (CLSF) is applied. The CLSFs are network-internal service functions which receive data segments (datagrams) from MANs through VPs using ATM cells. At the CLSF the routing function is applied to the first cell which contains the destination address. Then, from the CLSF all successive cells of the segment are forwarded either to the destination or to another CLSF. This concept can also be augmented by fast reservation protocols for a proper resource management of the ATM network.

6.3 IWU Architecture

Figure 18 presents a principal protocol architecture stack of the IWU between a MAN and the ATM network. Typical characteristics of the IWU are

- high real-time throughput
- address translation
- filtering and routing
- connection management
- buffering
- segmentation and reassembly
- usage monitoring
- overload control.

7. TRAFFIC CONTROL AND ENGINEERING

7.1 Traffic Control

ATM has been recommended by CCITT as the basis for the future B-ISDN. It is applicable for both local and wide area networks [7,8].

This promising integration technique, however, rises a number of new problems due to the higher degree of resource sharing compared to the conventional Synchronous Transfer Mode STM. Part of these questions are highly dependent on the user or traffic source characteristic and have, therefore, to be considered on the basis of stochastic traffic flows. Packetized traffic sources reveal not only VBR traffic flows, even more important are the statistical correlations among its discretized information packages (cells). The coincidence of peak bitrate or buffer requests of many connections sharing the same transmission, switching, or buffer resources may cause excessive delays or even information losses which must be kept within strict limits to meet the overall quality of service (QOS) figures in an B-ISDN.

To guarantee a properly defined QOS to the user, the network has to control the traffic flow in the network accordingly. This control is not easy due to the different time scales of dynamics or due to the limited effectivity of hardware-implementable algorithms. Therefore, a whole set of control schemes will be applied whose actions are related to the various levels of activity: connection, burst and cell level.

Figure 19 illustrates the principal view of a multi-level traffic source distinguishing 4 levels for the connection, dialog, bursts and cells. These levels typically show dynamics within the minute, second, millisecond and microsecond order, respectively. Additionally, high correlations are observable not only between but also within each of these levels.

At connection level, the connection acceptance control (CAC) decides whether a new connection could be accepted under the requested traffic and QOS parameters. Traffic parameters are typically peak and average bitrates and some measure of the frequency and length of peak periods. QOS parameters are related to the cell loss and cell delay, in particular the cell delay variation (jitter). Contrary to the simpler case of STM-based networks, in ATM-based networks a new connection cannot be accepted only on the basis of the available network resources; its effect on the QOS parameters of all other already existing connections sharing the same resources with the new connection has also to be taken into account. Routing and global network management for congestion control are other control actions operating on the connection level.

At burst and cell level, a source policing control (Usage Parameter Control, UPC) monitors the user data flow of each connection with respect to its negotiated traffic parameters. If a violation is detected an adequate action has to be taken such as immediate dropping or marking of excessive cells to become candidates for losses in case of congestion. The best known source policing function is the Leaky Bucket algorithm, a pure up and down counter; other algorithms operate on cell counts within sliding or jumping time windows. Cells which have been detected outside a counting range may be dropped or may be marked as candidates for dropping in case of congestion. The problem is not completely solved yet due to the difficulty to police peak and average rate at the same time.

Buffering is an inherent ATM control mechanism to resolve short-term shared resource blocking. Buffers have to be dimensioned to a low buffer overflow probability, typically 10^{-8} and less. Specific buffer schemes may be used to adapt the traffic loss to specific requirements of particular services by spacial priorities by which low priority cells are pushed out of the common buffer or by partial buffer reservation for high priority cells.

A final set of problems results from the design of high capacity switches and global network operation. Fast packet switches for ATM traffic must be designed in a highly parallel structure to meet the throughput requirements. Design and operation of switch elements, buffering as well as internal routing strategies are responsible for the performance of ATM switches.

The global network performance, in particular the end-to-end performance observable by the user, depends in a high degree on all network component performance figures and on globally operating strategies such as routing of connections, connection acceptance algorithms, bandwidth management, and network management. Of particular interest is the cell delay variation (CDV) of an end-to-end virtual channel connection; the CDV must be kept reasonably small as buffers have to be applied for its compensation for the class 1 (circuit emulation) service class where end-to-end synchronization has to be maintained.

7.2 Traffic Modelling, Analysis and Engineering

Traffic Engineering of high speed communication network comprizes a set of particular problems such as

- Modelling of the statistical characteristics of traffic sources and services
- Identifying traffic volumes and traffic matrices (origination-destination pairs)
- Performance Analysis of the key resource sharing problems
- Dimensioning of network resources
- Tuning of control parameters
- Traffic and Performance Monitoring
- Network Planning according to load and QOS requirements including resource / cost optimization.

This opens a rich field of research and engineering issues.

In the following, a number of key examples will be given which illustrate the nature of problems.

7.2.1 Traffic Source Modelling

Whereas CBR sources are modelled simply by the connection holding time distribution and transmission bitrate, VBR sources require additional details, reflecting the bursty nature of VBR traffic. Typical details are

- peak-to-mean bitrate ratio ("burstiness factor")
- lengths and distributions of burst and silence periods
- correlation effects between bursts, especially periodic peaks resulting out of synchronization mechanisms.

Figure 20 illustrates three examples of CBR/VBR source traffic.

Various mathematical models have been suggested and applied such as

- Superposition of CBR streams with random phase shifts (N*D)
- Poisson (Markov) Process (M)
- Discrete time Bernoulli Process (BP)
- Markov-Modulated Poisson Process (MMPP)
- Markov-Modulated Deterministic Process (MMDP)
- Markov-Modulated Bernoulli Process (MMBP)
- Generally Modulated Deterministic Process (GMDP)
- Markovian Arrival Process (MAP)
- Discrete Time Markovian Arrival Process (DMAP)
- Batch Arrival Processes
- Fluid Flow Approximation.

Two models are illustrated in Figures 21 and 22, where a number of the beforementioned processes are special cases of. For further details we refer to the literature, e.g. [9-12]. Most of these models try to capture at least two levels of the dynamics, such as the cell and the burst level. More than two levels are difficult to model; its value is doubtful also since performance evaluation must be limited to a certain level of detail, either from a practical point of view or simply for reasons of feasability.

The mathematical description of traffic source models is another problem. Most of the beforementioned models reveal a number of parameters which have to be quantified either from application experience or from statistical analyses of measurements.

Models may additionally be subdivided into

- source traffic generator models and
- source traffic descriptive models.

Generator models allow a direct implementation, e.g. within a simulation program or within test equipment. These models also allow a more or less direct modelling for analytic treatment. Descriptive models provide statistical parameters such as the number N (t) of events during a time interval of length t and its statistical measures such as mean, variance etc.

It has also been observed that in high speed communication systems randomization effects may apply; this effect is based on a statistical theorem, that the superposition of n individual, arbitrary point processes results in a Poisson Process as n tends to infinity and when all individual processes contribute with vanishing arrival rates ("thinning requirement"). Then, network components, such as a finite capacity buffer, can be modelled much easier by a resulting M- or BP-Process. However, correlation effects have to be taken serious: it may need hundred and more processes to justify this assumption.

7.2.2 Finite Capacity Buffer Analysis

Figure 23 shows the mostly applicable model for a multiplexing device in cell-based networks, the ΣG/D/1/S model. In this model a superposition of n individual arrival streams forms the arrival process. There is one server with deterministic service time (D, referring to the constant cell transmission time) and a finite buffer of capacity S. The important QOS parameters are

- the cell loss probability
- the cell delay variation,

possibly individually for each arrival stream.

The analysis of such models depends strongly on the arrival process assumptions and vary from Embedded Markov Chain (M/D/1, MMPP/D/1, BP/D/1, MMB /D/1) to Fluid Flow Approximation.

The basic results of a finite buffer single server model with bursty arrivals are demonstrated in Figure 24. The exact solution reveals a characteristic bend. For smaller buffer sizes, the random (M) characteristic is sufficiently accurate to capture the loss effects; for larger buffer sizes, the burst behaviour causes a much slower decay. Both effects are sufficiently accurate approximated by the M/D/1/S model for small capacities and by the Fluid Flow Approximation for larger capacities. In fact, an overall approximation can be derived from the simple superposition of both individual results [13].

As a conclusion of this result we may conjecture that it is hardly possible to buffer whole bursts: besides the long delays, extremely large buffer capacities would be required. Therefore, traffic control has to operate quick enough, that loss rates can be maintained as small as 10^{-9} with reasonably small buffer sizes.

Similarly, delay variations can be obtained. Figure 25 illustrates an example in form of the complementary delay distribution function, where similar approximations are possible [14].

7.2.3 End-to-End Cell Delays

An end-to-end path of a virtual channel connection may be modelled by a serial configuration of finite buffer models with intruding "background" traffic, cf. Figure 26. The individual end-to-end cell delays of a particular reference connection can be approximated by a recursive procedure, whereby the correlation between delays in adjacent stages are taken into account [14]. Figure 27 illustrates typical results of the delay distribution which have also been in good agreement with simulations.

7.2.4 Shared Media Multiplexers

Shared Media Multiplexer models result from Local Area Networks (LAN) and Metropolitan Area Networks (MAN). For isochronous traffic (circuit switching), the media can be modelled as a multi-channel loss system for n classes of sources with individual holding times and bitrate requirements, cf. Figure 28. For this model, various classes of bandwidth assignment strategies may be applied:

bandwidth management:	complete sharing complete partitioning class limitation sum limitation reservation
positional strategies:	arbitrary adjacent equally spaced
hunting strategies:	sequential random gap oriented.

For packet traffic, most LAN/MAN models result into a variant of a polling system, cf. Figure 29: a single server (channel) serves g queues in turn. The control strategy reflects the particular Media Access Control mechanism. There is a great variety of mechanisms:

token passing:	exhaustive service gated service limited-N service limited-T service priority service
slotted ring:	idle slot indication credit mechanisms
reservation:	distributed queue, dual bus (DQDB) cyclic reservation multi access (CRMA).

These strategies cover also very actual schemes as used in FDDI and IEEE 802.6 standards.

7.2.5 Usage Parameter Control (UPC)

UPC algorithms are applied by the network operator, individually for a Virtual Channel Connection (VCC) or a Virtual Path Connection (VPC). They aim at monitoring of the user traffic flow with respect to the negotiated traffic parameters, e.g., the average/peak cell rates. Actions may be marking or dropping of cells, violation indication to the call control or even traffic enforcement by shapers to smooth traffic peaks.

The most prominent algorithm is the "Leaky Bucket" algorithm, cf. Figure 30. Variants of this algorithm have been standardized in the ATM Forum, so, e.g., to police the sustained bitrate and the peak bitrate. From the performance point of view it is important to know how well these algorithms are working and, in particular, how the parameters, such as the bucket size, have to be dimensioned.

7.2.6 Loss Control

Traffic losses can be controlled primarily by the size of buffers (which is a static means by system design) or dynamically by a restrictive Connection Acceptance Control (CAC). In addition to that, more sophisticated buffer schemes may be applied in connection with the CLP marking: loss-sensitive traffic may have a spacial (buffer) priority or it may have access to extended buffers compared to less loss-sensitive traffic; for details we refer to [15].

7.2.7 Connection Acceptance Control (CAC)

CAC algorithms operate on a connection level at the connection set up time. In STM-based networks, on each consecutive link from source to destination the requested bandwidth must be available and assignable. Particular CAC rules have to be applied to protect higher bitrate traffic and to give a good QOS to it such as the individual traffic class connection or call loss probabilities.

CAC is more complicated for ATD-based networks. Besides connection loss or connection rejection probabilities, individual cell loss probabilities have to be regarded. For this, the individual source characteristics (burst behaviour) have to be considered, as a new connection may also affect the QOS of already established other connections sharing common resources such as buffers, transmission links or virtual paths. New connections may be accepted as long as the sum of all average bitrates is below the link or path capacity; in practice, the sum bit rate must be kept considerable below the link/path capacity since the bursty behaviour may cause instantaneous peaks during which excessive losses of cells may be experienced. Thererfore, the cell loss limit has to be regarded and the acceptance of new connections has to be reduced accordingly. As the coincidence of peaks may appear quite rarely, it may be possible that the sum of all peak bitrates of the accepted connections exceeds the link capacity but all QOS requirements are still met; in that case a "multiplexing gain" may be achieved.

For the evaluation of CAC algorithms various schemes have been suggested, such as [16]

- convolution algorithm
- equivalent bandwidth algorithm
- multi-phase algorithms

by which individual service state distributions are convolved to estimate the link state or by which an equivalent bandwidth is defined to follow a similar CAC procedure as for STM. These algorithms may be too costly to perform in real time or they are too complicated. Therefore, multi-phase algorithms may be applied.

In the first phase of B-ISDN, only peak rate CAC algorithms will be applied. Figures 31 and 32 illustrate the problem of CAC.

7.2.8 Routing

Routing algorithms define the procedure by which a path is searched and defined for network wide connections. Routing algorithms are based on the network topology, the current state of the loading and on fast signalling techniques to perform the control. There is a rich theory and experience on routing. B-ISDN adds more criteria to it as the route selection may also be based on service protection or on packing methods (aggregation or separation of connections of same type).

8. TRAFFIC VALIDATION

Although the addressed traffic engineering items are by no means exhaustive, the dimension of the problem may be anticipated. For a solid traffic engineering, methods and assumptions have to be validated by measurements. Many new applications or services do not yet exist, nor do the ATM based networks. Therefore, validation of traffic assumptions and analysis/engineering procedures have to be validated.

One method of validation is simulation. In a computer simulation the simplifying assumptions of analytic techniques may be validated. What simulation is not able to provide are real source traffic models or the detection of extremely rare events. For the first problem only measurements on real sources and systems can help whereas for the second new simulation techniques may be applied such as highly parallel simulation to generate much more events or rare event simulation by which transforms are applied to concentrate on rare events.

Various RACE projects provide - for the first time - ATM testbeds on which measurements and tests can be performed (RACE Projects EXPLOIT, TRIBUNE and others). It requires besides the testbed test equipment. Figure 33 shows a principal test configuration by which a "System Under Test" is fed by both specific traffic generated artificially by an ATM Traffic Generator and by Background Traffic to provide a realistic loading. The test cells of the artificially generated traffic may carry additional information like connection Id, Sequence Number, and Time Stamp to be processed at the receiving ATM Traffic Analyzer.

ITU-TS has defined in its document I.356 a layered model for performance measurements, cf. Figure 34. In this model, the reference points are identified between which the layer-specific parameters have to be measured which characterize the Network Performance.

The objectives of traffic measurements are:

- cell loss ratio
- cell misinsertion rate
- cell sequence integrity
- cell transfer delay
- cell delay variation

Besides these objective values, subjective tests can be made, e.g., tests on the transmission quality of VBR video-transmission under certain Network Performance parameters. Other tests are addressed to the verification of UPC and CAC algorithms, to the validation of theoretical traffic assumptions and analysis results, to the validation of resource management algorithms, or to the validation of the efficiency of packet transfer mechanisms in case of LAN interconnection through B-ISDN using indirect methods (VCC; VPC), direct methods (CLSF) or inband Fast Reservation Protocols.

CONCLUSION

Starting with a summary on new services and network types principal problems have been discussed which are related to performance modelling and traffic engineering for such services and networks. Key elements for this are the VBC source traffic characterization and the consideration of the various traffic control procedures. There is an impressive volume of work already in the literature; despite that, most of the critical issues are not finally solved due to the uncertainty about real ATM traffic sources, user behaviour and adequate traffic control algorithms. It has been pointed out that validation of traffic models along with the initial testbed experiments is of prime interest equally for researchers, manufacturers, network operators and finally also for the users.

REFERENCES

1. W. Stallings, Local Networks. Macmillan Publ. Company, New York, 1987.
2. D. Hutchison, Local Area Networks. Addison-Wesley Publ. Company, Workingham, 1988.
3. K. Kümmerle, J.O. Limb, F.A. Tobagi (Ed.), Advances in Local Area Networks. IEEE Pres, IEEE, New York, 1987.
4. Fibre Distributed Data Interface (FDDI), ISO 9314-1, -2, -3. ANSI Association, New York.
5. Distributed Queue Dual Bus (DQDB) Subnetwork of a Metropolitan Area Network (MAN). IEEE Standard 802.6.
6. H.R. Müller et al., New Access Schemes for Gbit/s LANs and MANs, Proceedings "High-Capacity Local and Metropolitan Area Networks", 1990. Springer-Verlag, Berlin, 1991, pp. 41-57.
7. M. de Prycker, Asynchronous Transfer Mode, Solution for B-ISDN. E. Horwood Ltd., Chichester, 1991.
8. R. Händel, M.N. Huber, Integrated Broadband Networks, An Introduction to ATM-Based Networks. Addison-Wesley Publ. Ltd., 1991.
9. J. Roberts (Ed.), COST 224 Performance Evaluation and Design of Multiservice Networks. Final Report, Oct. 1991. Commission of the European Communities, ECSC-EEC-EAEC, Brussels/Luxembourg, 1992.
10. V.A. Bolotin, J.G. Kappel, P.J. Kuehn (Eds.), Teletraffic Analysis of ATM Systems. IEEE Selected Areas in Communcations (JSAC), April 1991, Vol. 9, No. 3.
11. A. Jensen, V.B. Iversen, C.D. Pack (Eds.), Proceedings of the 13th International Teletraffic Congress (ITC), Copenhagen, 1991. North Holland Publ.
12. J. Y. Hui, Switching and Traffic Theory for Integrated Broadband Networks. Kluwer Academic Publishers, Boston, 1991.
13. H. Kröner, T. Theimer, U. Briem, Queueing Models for ATM Systems - A Comparison. Proc. 7th ITC Specialist Seminar, Morristown, N.J., October 1990, paper 9.1.
14. H. Kröner, M. Eberspächer, T. Theimer, P.J. Kühn, U. Briem, Approximate Analysis of the End-to-End Delay in ATM Networks. Proc. IEEE INFOCOM '92, Florence, paper 7 C.2.
15. H. Kröner, G. Hébuterne, P. Boyer, A. Gravey, Priority Management in ATM Switching Nodes. IEEE JSAC Vol. 9, No. 3 (April 1991), pp. 418-427.
16. H. Kröner, Evaluation of Connection Acceptance Algorithms for ATM Networks. Dissertation (forthcoming), University of Stuttgart, 1994.

INTERACTIVE SERVICES

Service	Definition	Examples
Conversational Services	Bidirectional (dialogue) communication between TE's without intermediate buffering of user information	- video-telephone - video-conferencing - high-speed interactive data communication - file transfer
Messaging Services	User-to-user communication with intermediate buffering of user information	- message handling - voice/video mail - document mail - high resolution images
Retrieval Services	Retrieval of information stored in public domain on demand of individual users	- Hifi audio retrieval - video programs - mixed mode document retrieval - data retrieval

DISTRIBUTION SERVICES

Service	Definition	Examples
Broadcast Services without individual presentation control	Continuous information flow distributed from a central source accessed by an unlimited number of users without control on the start and order of presentation	- television programs - audio programs
Broadcast Services with individual presentation control	Individual access to broadcast services with control on the start and presentation of the information by the user	- interactive videotex - videography

Figure 1. Service Types for B-ISDN

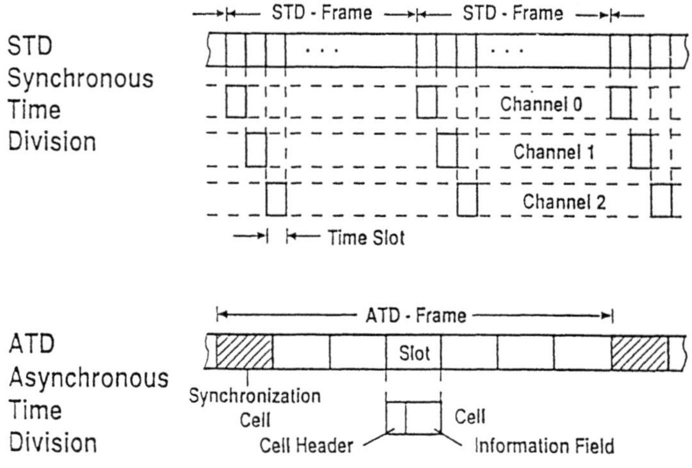

Figure 2. Multiplexing Principles

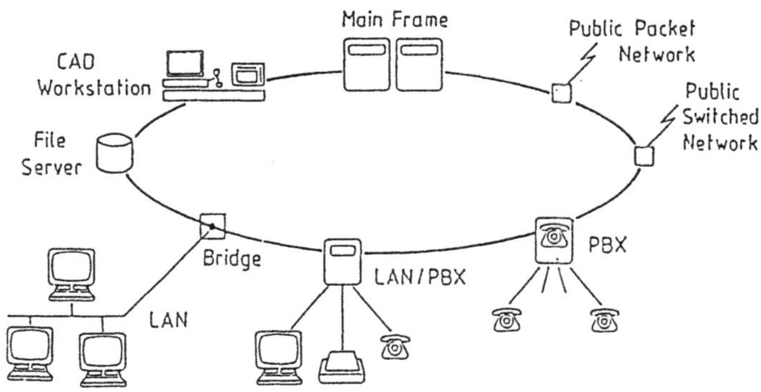

Figure 3. FDDI Backbone Application Example

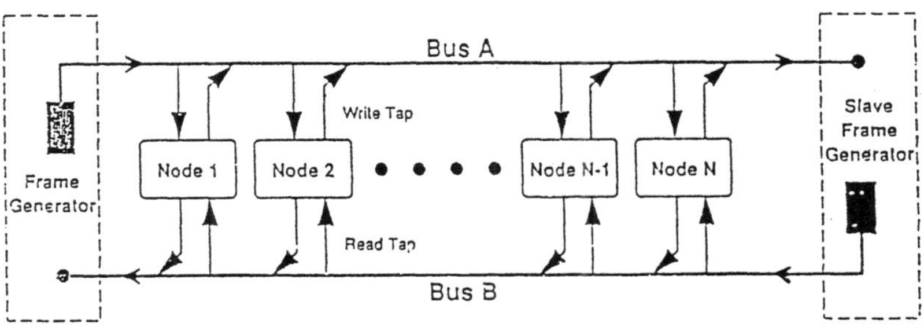

Figure 4. DQDB MAN Principal Structure

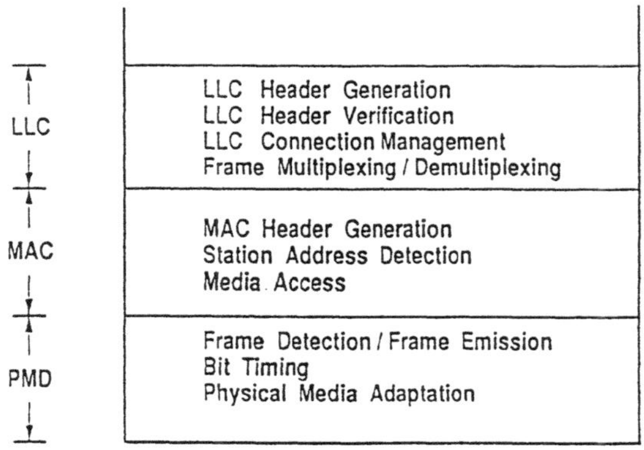

Figure 5. LAN /MAN Lower Layer Protocol Functions

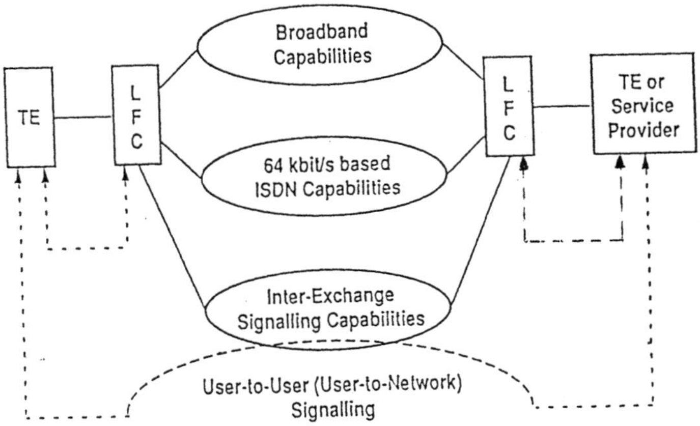

Figure 6. **B-ISDN Reference Configuration**
TE Terminal Equipment
LFC Local Switching Functional Capabilities

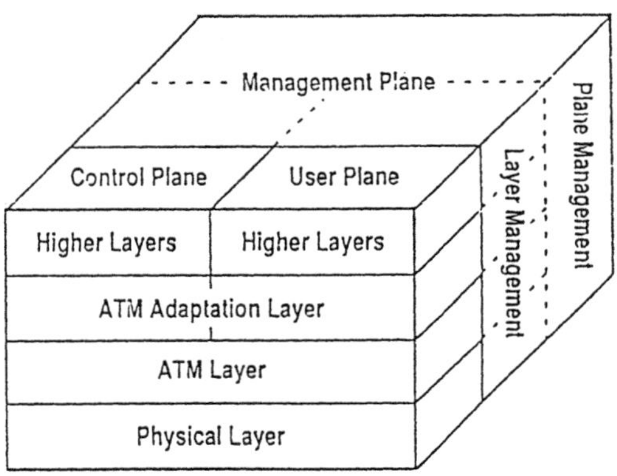

Figure 7. **B-ISDN Protocol Block**

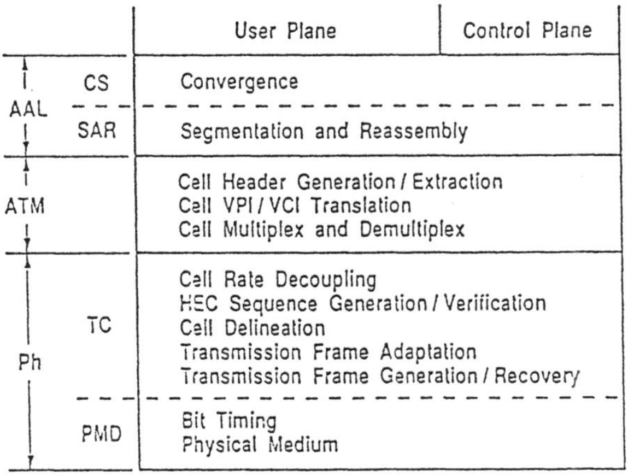

Figure 8. B-ISDN Lower Layer Protocol Functions

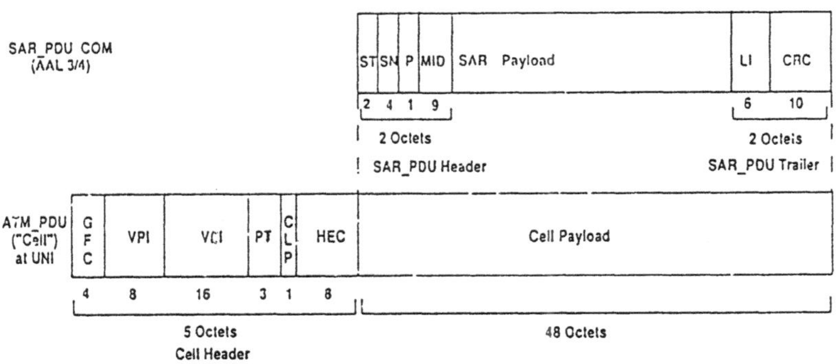

Figure 9 ATM Cell Structure

GFC	Generic Flow Control	ST / SN	Segment Type / Sequence Number
PT	Payload Type	P	Priority
CLP	Cell Loss Priority	MID	Multiplexing Identifier
HEC	Header Error Control	LI	Length Indicator
		CRC	Cyclical Redundancy Check

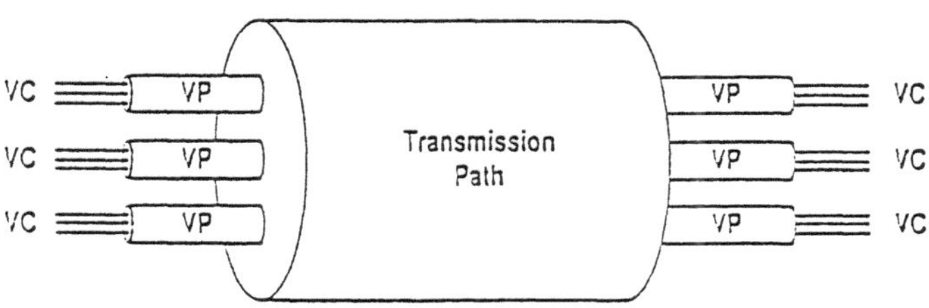

Figure 10. Transmission Path, Virtual Path and Virtual Connection

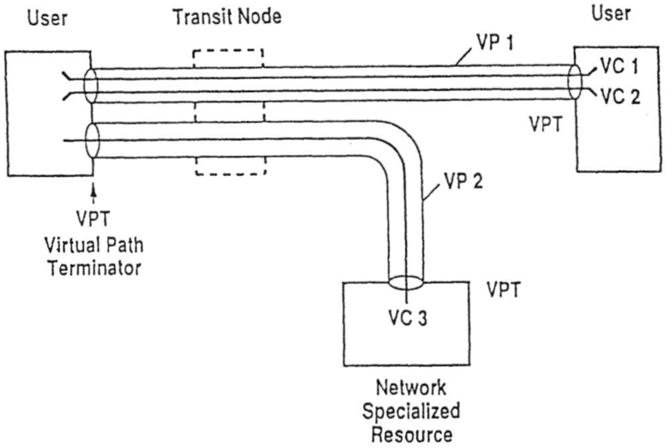

Figure 11. Application of Virtual Path Concept

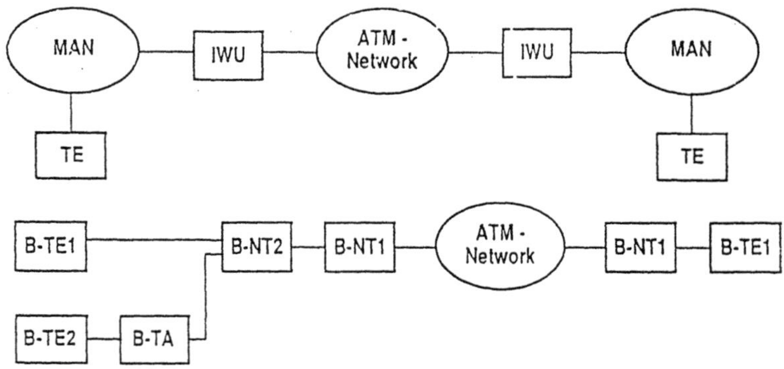

Figure 12. Data Communication Scenarios
Upper Part: Initial Situation
Lower Part: Mature Situation

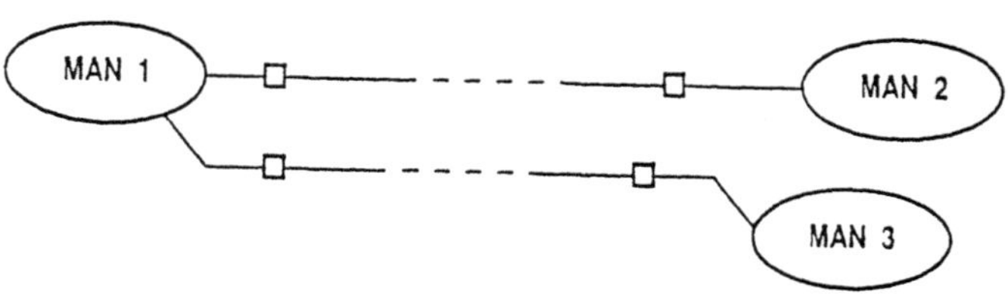

Figure 13. Direct Interconnection of MAN's by Leased Channels

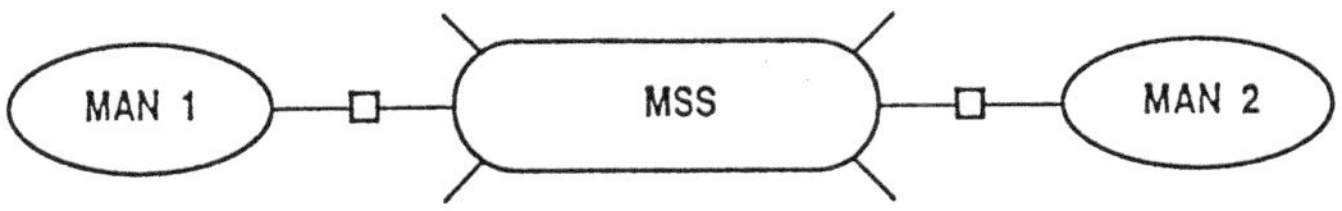

Figure 14. MAN Interconnection by MAN Switching System

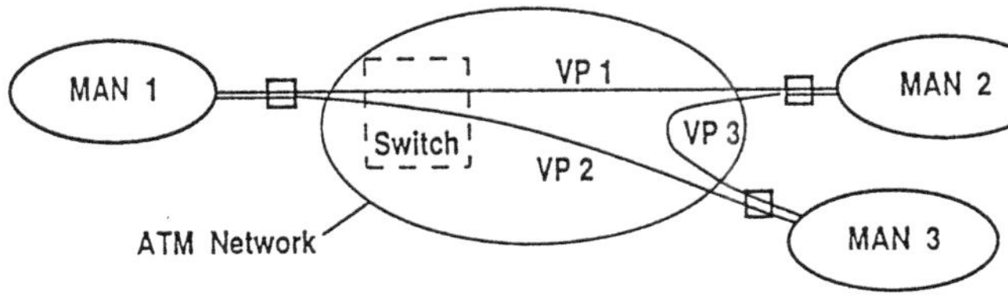

Figure 15. MAN Interconnection by Virtual Path Concept

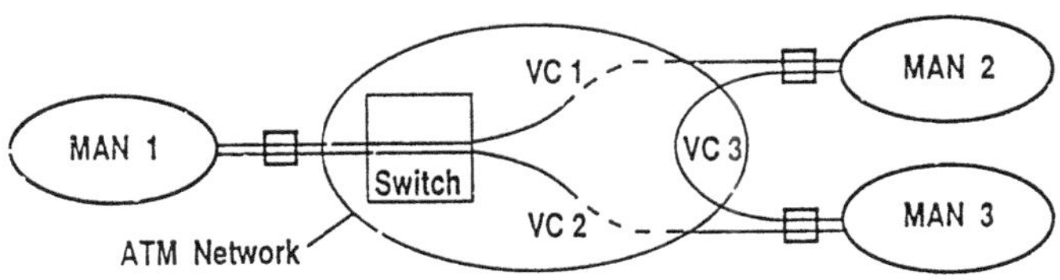

Figure 16. MAN Interconnection by Virtual Connection Concept

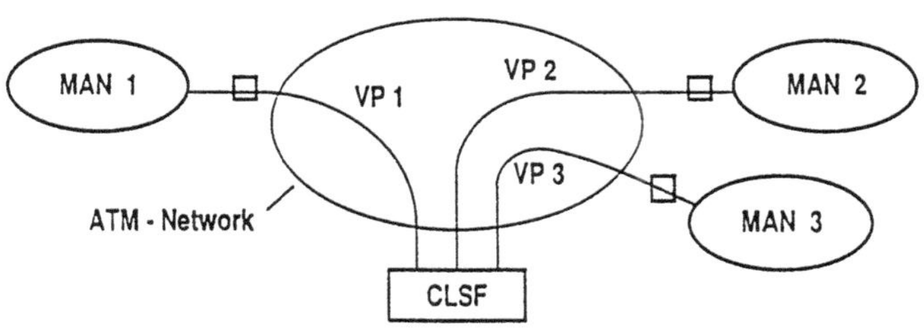

Figure 17. MAN Interconnection by Connectionless Server
CLSF Connectionless Service Function

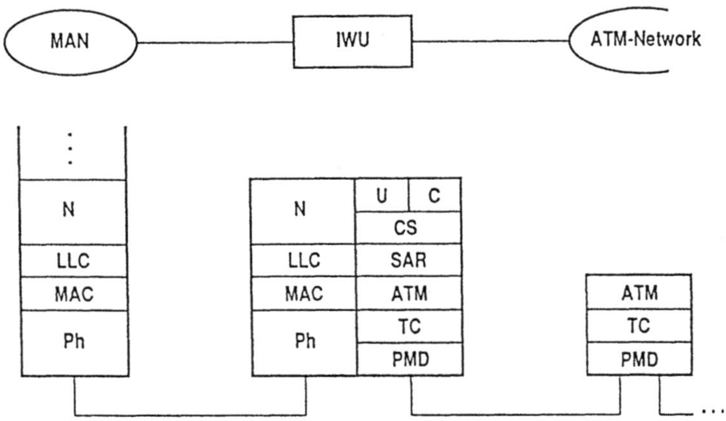

Figure 18. MAN-ATM IWU Protocol Architecture

Ph	Physical	PMD	Physical Media Dependent
MAC	Media Access Control	TC	Transmission Congress
LLC	Logical Link Control	ATM	Asynchron. Transfer Mode
N	Network	SAR	Segmentation And Reassembly
		CS	Convergence Sublayer
		U, C	User, Control (Context)

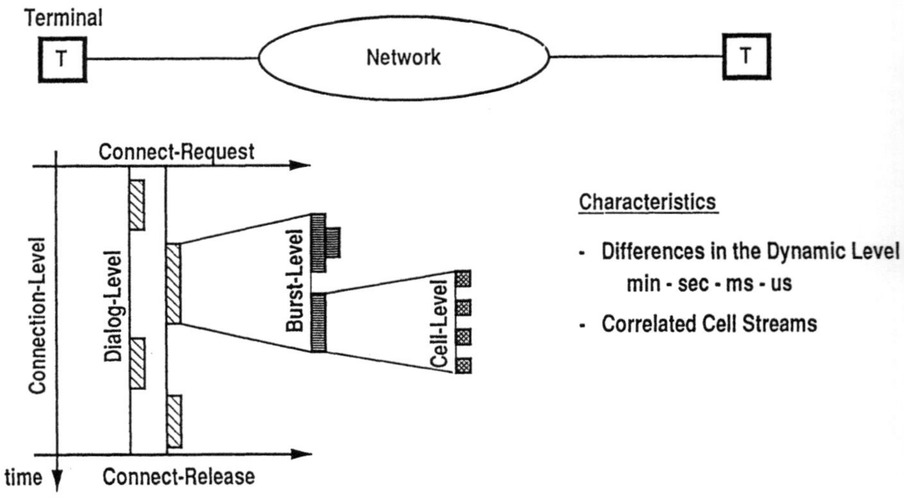

Figure 19. Multi-Level Traffic Source for VBR Traffic

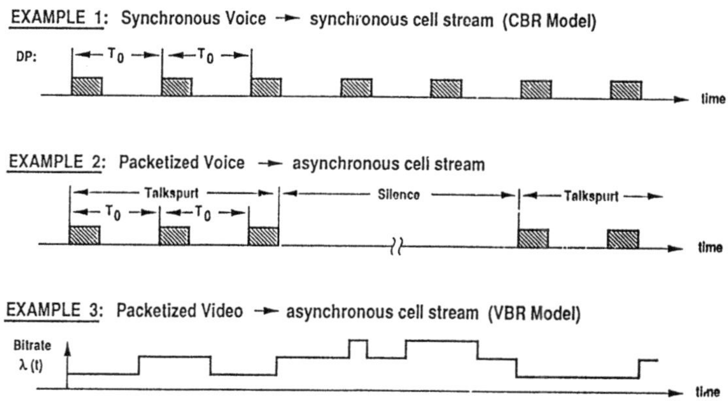

Figure 20. Examples for CBR and VBR Source Traffic

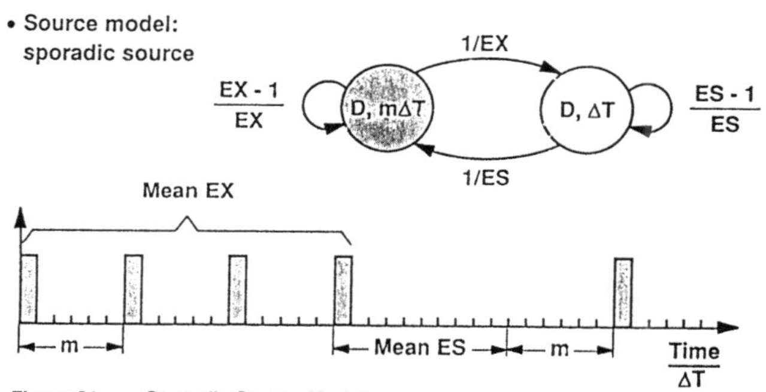

Figure 21. Sporadic Source Model

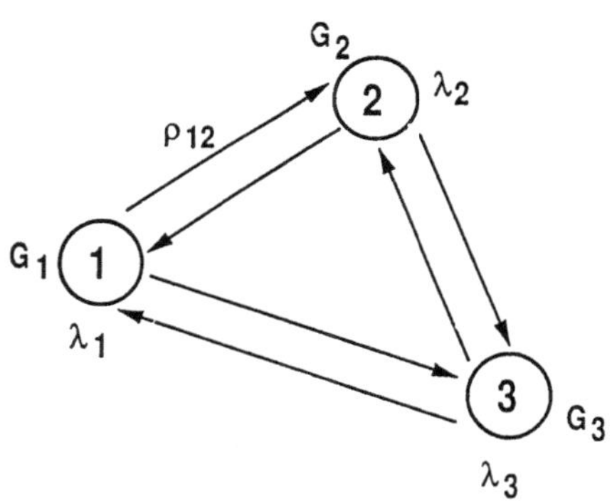

State 1 low cell rate λ_1, constant interarrival time

State 2 medium cell rate λ_2, constant interarrival time

State 3 peak cell rate λ_3, constant interarrival time

Generally distributed state durations (G_1, G_2, G_3)

Arbitrary state transition probabilities (ρ_{ij})

Special Cases

MMDP Markov Modulated Deterministic Process

MMPP Markov Modulated Poisson Process

GEO Geometrically Distributed Cell Burst Length
Geometrically Distributed Silence Length

Figure 22. 3-state GMDP Source Model

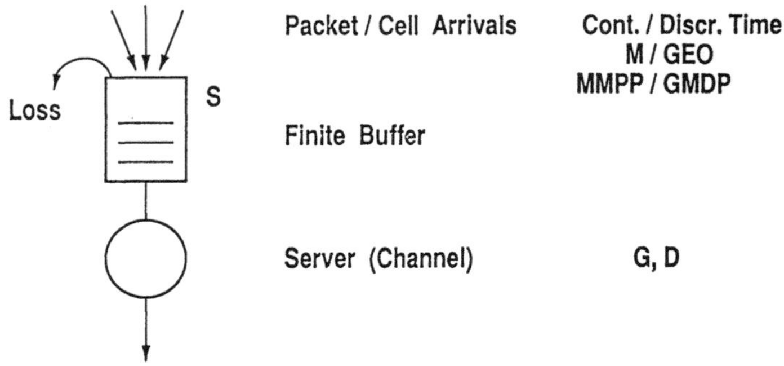

Figure 23. Finite Capacity Single Server Model with Multiple Input Streams

- 10 sources
- Offered load: 50%
- Burstiness: 5

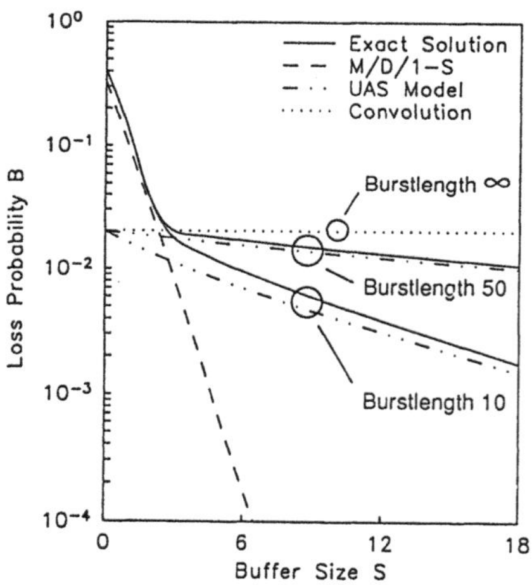

Figure 24. Cell Loss Probability of a Finite Capacity Single Server Model

- Cell interarrival time : 10 ΔT
- Average passive time : 2000 ΔT
- Average active time : 500 ΔT
- Number of sources : 30
- Buffer size : 50

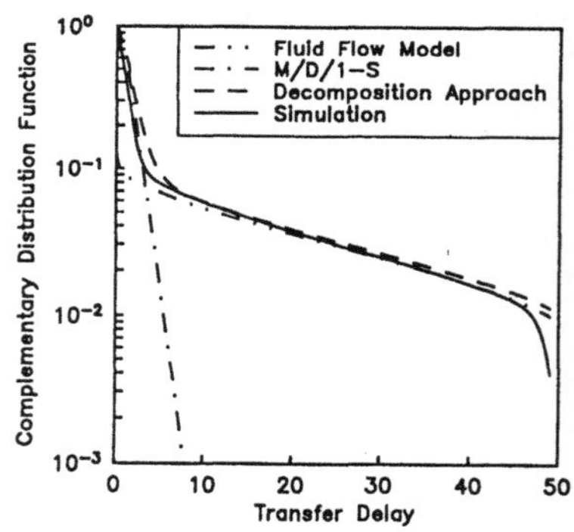

Figure 25. Cell Delay Distribution of a Finite Capacity Single Server Model

UAS Model ≙ Fluid Flow Model

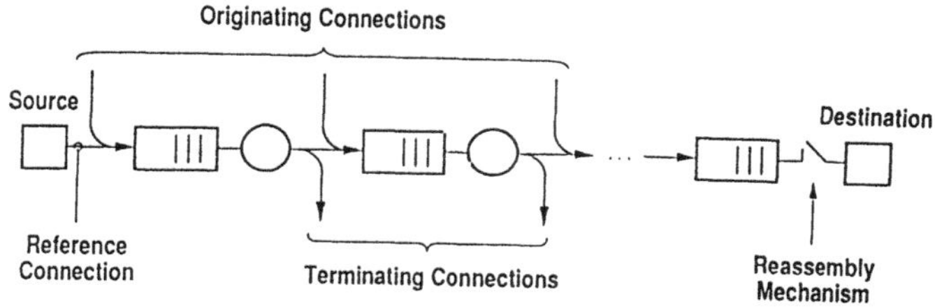

Figure 26. End-to-End Reference Connection Model

- **Same source parameters as in the previous examples**
- **12 connections at each node**
- **Na = Nd = 4 after each node**

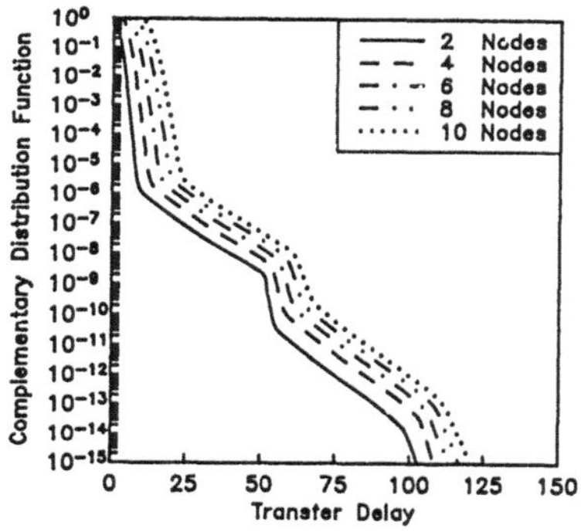

Figure 27. Cell Delay Distribution of a Serial Finite Capacity Delay System with 2 to 10 Nodes

Na Number of Arriving Connections
Nd Number of Departing Connections

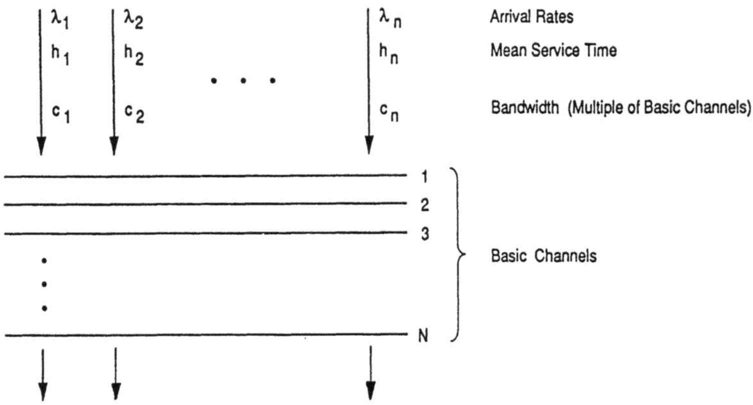

Figure 28. Traffic Model for a Shared Media Multiplexer for n Classes of Isochronous Traffic Sources

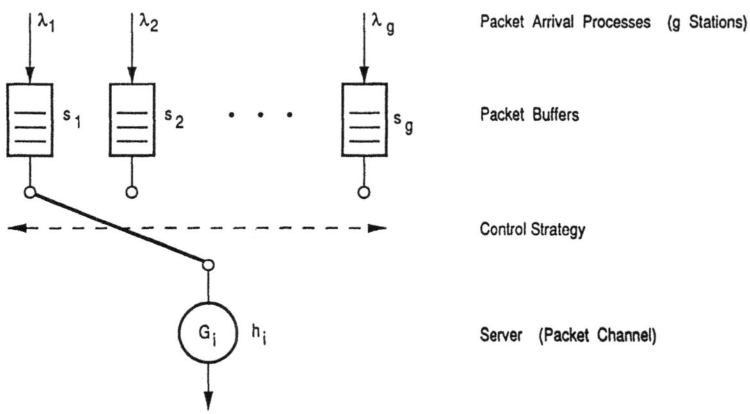

Figure 29. Traffic Model of a Shared Media Multiplexer for g Stations with Arbitrary Packet Arrival Processes

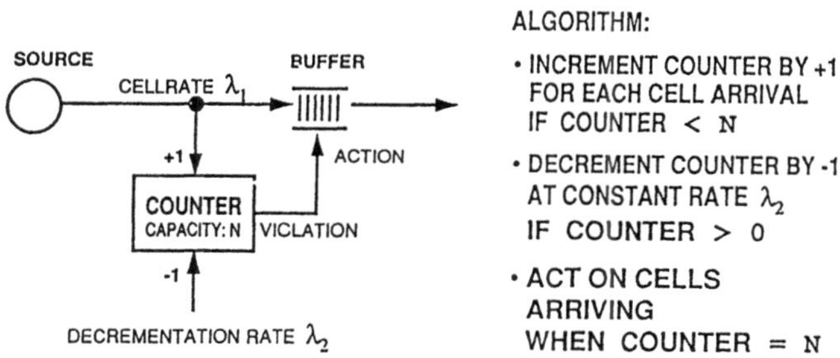

Figure 30. Principle of the "Leaky Bucket" Source Policing Algorithm for UPC

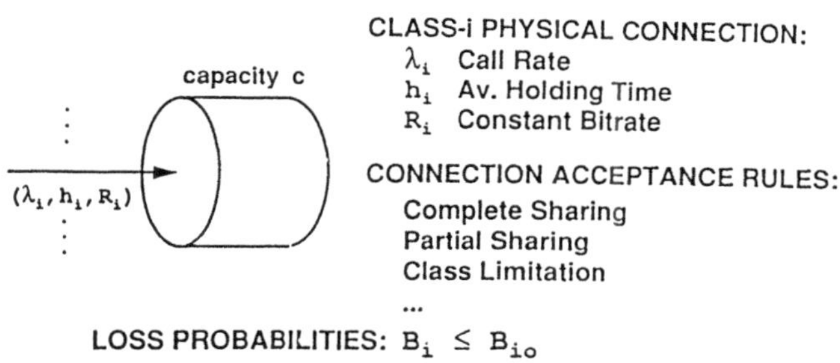

Figure 31. Principle of Connection Acceptance Control for STM-Based Networks

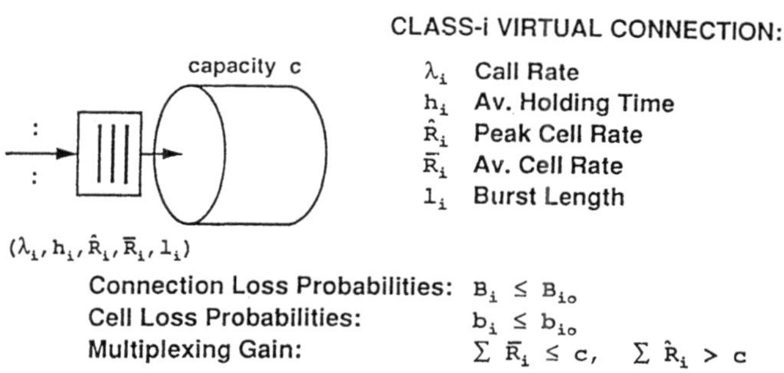

Figure 32. Principle of Connection Acceptance Control for ATM-Based Networks

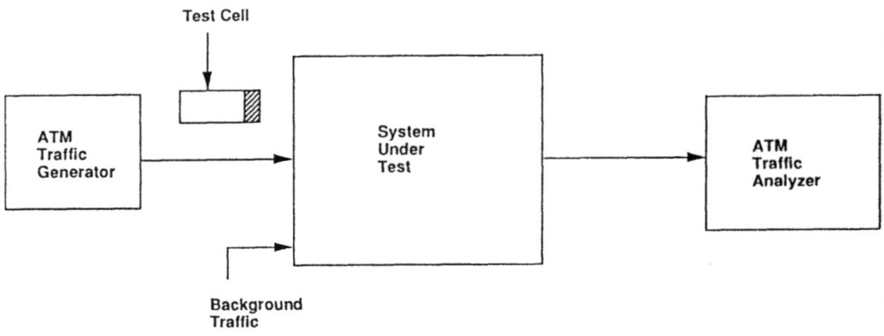

Figure 33. Test Configuration for ATM Traffic Measurements (Principal Configuration)

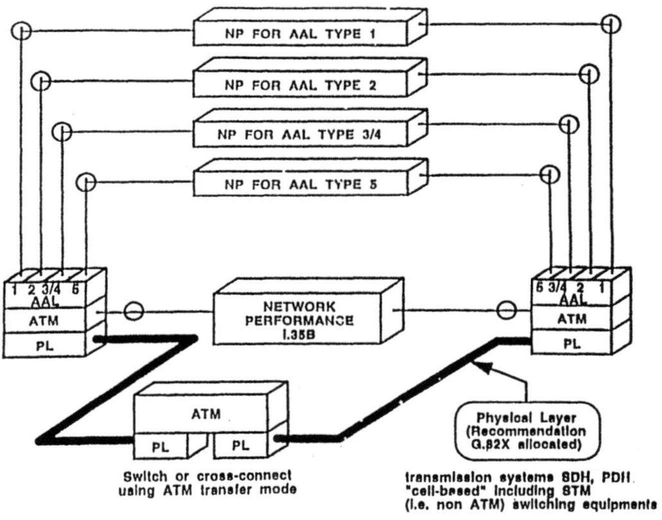

Figure 34. Layered Reference Model for Performance Testing in B-ISDN According to ITU-TS Recommendation I.356

Broadband Communications, II (C-24)
S. Tohmé and A. Casaca (Editors)
Elsevier Science B.V. (North-Holland)

A Comparison of Models for VBR Video Traffic Sources in B-ISDN

Oliver Rose[1] and **Michael R. Frater**[2]

[1] *Institute of Computer Science*
University of Würzburg, Am Hubland, 97074 Würzburg, Germany
Email: rose@informatik.uni-wuerzburg.de

[2] *Dept. Electrical Engineering, University College, University of New South Wales*
Australian Defence Force Academy, Canberra ACT 2600, Australia
Email: mrf@truk.ee.adfa.oz.au

Abstract

Variable bit rate (VBR) video is expected to be a major source of traffic for the Broadband Integrated Services Digital Network (B-ISDN). To dimension the network, there is a need to predict the network performance. Traffic models are often used as an aid in this process, and many have been proposed as suitable for the modeling of VBR video sources. However, further work is required in the area of verifying the accuracy of models, and little comparison between models is available. In this paper, several of these models are presented and their properties discussed, including their ability to predict accurately different aspects of network performance, with an emphasis on cell loss statistics. Finally, some time is spent reviewing a number of open questions in modeling video streams and their superposition, and the analysis of queueing systems with this kind of input.

1 Introduction

Variable bit rate (VBR) video is expected to be a major source of traffic for the Broadband Integrated Services Digital Network (B-ISDN). For any multi-media application based on video data from remote sources, such as information systems or production supervision systems, the high quality transmission of video sequences via a telecommunication network is a core function.

A video sequence consists of a series of frames, each containing a two-dimensional array of pixels. The number of frames per second and the number of lines per frame and pixels per line vary from country to country. For each pixel, both luminance and chrominance information is stored. If a television signal is digitally coded, a bandwidth of approximately 160 Mbps is required to transmit it. A variety of compression algorithms is used to reduce the data rate. Modern techniques can achieve acceptable quality for consumer applications at rates of less than 4 Mbps.

One of the properties of video sequences is that some sequences are harder to code efficiently than others. In most currently used coders, each frame is coded using approximately the same number of bits. This means that the sequence can be transmitted efficiently over a constant bandwidth channel. Such coders are often referred to as constant bit rate (CBR) coders. The drawback of this approach is that some frames are coded at a higher quality than others.

One solution to this problem is to allow the number of bits per video frame to vary in a manner that tends to keep the video quality constant. Coders that behave in this way are known as variable bit rate (VBR) coders. Such coders do not make efficient use of channels with a constant bandwidth, since a channel bandwidth equal to the peak rate of the coder must be available. However, if a variable bandwidth channel is available, as is proposed for the B-ISDN, a smaller bandwidth can be allocated. It is hoped that it will be possible to allocate a bandwidth close to the mean rate of the source. If the ratio of the peak to mean rates is large, then a large saving can be made in the required bandwidth compared to a constant bandwidth channel. Hence, VBR coders can be attractive if they are able to provide essentially the same quality as CBR coders, but at a lower average rate.

In order to study the behavior of this type of source, traffic models are used. Numerous traffic models that are based on independence assumptions, e.g. Poisson streams, are not suitable for modeling video traffic (see [23, 18, 20, 16, 12]) because traffic generated by video coders is highly correlated. To overcome this problem, several models have been proposed in literature during the recent years.

There are two major types of traffic model:

1. models that give a description of the sequence of the numbers of bits for each video frame generated by a video coder;

2. models describing the cell process of a video coder used as a traffic source of a cell based telecommunication network such as an ATM network.

In this paper, a variety of models of each category, both for frames and cells, is presented and discussed. The modeling of layered coders, i.e. coders transmitting cells with priorities, will not be discussed here (cf. [5] and [17]). An overview of models for frame sequences is given in Section 2. Section 3 deals with models for cell sequences. A range of general issues related to the modeling of VBR video traffic is discussed in Section 4, including the statement of a number of open questions.

2 Models for frame sequences

In this section, models for the process that generates the sequence of the number of bits for each video frame are described. With each model, some indication is given of the usefulness of the model for predicting network performance.

In order to apply these models to traffic studies, it is necessary to model separately the process that breaks up the data into a form where it can be passed on to the network. In the case of B-ISDN, this process would involve placing the data into cells to be transmitted on the network. There are many ways in which these cells could be transmitted into the network. For example:

1. as soon as cells become available from the coder, they could be transmitted at the maximum rate of the input link;
2. cells could be transmitted with a constant inter-arrival time during each video frame.

A number of traffic studies have been carried out using the first approach (e.g. [8, 3].) Other studies have used the second (e.g. [2].) In [8], the predictions of cell loss probability obtained by simulation using several models are compared to those obtained by simulation using a set of real video-conference data. In [3, 2], a similar approach is used to compare the models with data obtained from coding a movie and television programs. Reference will be made to these traffic studies in connection with their results for individual models.

In the remainder of this section, models for the number of bits per video frame will be discussed. Autoregressive processes are discussed first, followed by the more general class of Markovian processes. The following notation will be used. Let X_n be random variable describing the amount of data carried by frame n of the sequence. If the model is used to describe a superposition of several video streams X_n denotes the sum of the frames sizes of the individual streams.

2.1 Autoregressive processes

In [13], [15], and [21] a first-order autoregressive process (ARP) with Gaussian innovations is used to model the bit-rate of a single source. A first-order ARP, also known as an AR(1) process, obeys the update equation

$$X_{n+1} = a + bX_n + \omega_n,$$

where a and b are constants, and $\{\omega_n\}$ is a so called innovation process, in most cases a Gaussian noise process[1]. The parameters are set so that the mean, the variance and the short-term autocorrelation of the experimental data are matched. A disadvantage of this simple model is the fact that it can only match short-term correlations. To overcome this phenomenon the authors of [13] propose to use higher-order ARP's. However, it has been found (see e.g. [8]) that even higher order ARP's do not provide a good match between the autocorrelation functions of real data and those for the model. Another approach that has been suggested is to use several first-order ARP's, with the selection of a certain ARP controlled by a Markov chain [15].

It has also been observed that the use of autoregressive processes in traffic models does not lead to an accurate estimate of the cell loss probability for large buffer sizes [8, 3]. For small buffer sizes, especially where the cell inter-arrival time is constant within a video frame, the AR(1) model appears to provide a good estimate for the cell loss probability [2].

Another approach to overcome the inaccuracies of the simple models is presented in [19]. The authors combine three processes. The first process is an ARP to capture the short-term autocorrelation. The second ARP is for the long-term autocorrelation. A third process is introduced to account for the bit-rate peaks observed by the authors during scene changes in their experimental video data. To model these peaks, a Gaussian random variable controlled by a three-state Markov chain is used. Most of the time the Markov chain stays in state 0 and no extra bit-rate is added. If this state is left, it will take two frames to get back to state 0. During this period the bit-rate is increased.

A further tool is provided by the TES (Transform-Expand-Sample) modeling methodology proposed by Melamed et al[14]. The core of this methodology is a first-order ARP with innovations following a common, but arbitrary, distribution and the use of modulo-1 arithmetic. For example, Fig. 1 shows the representation of a step function innovation density.

In this way a sequence U_n covering the range $[0, 1]$ is produced recursively, which can be represented by a walk around a circle with unit length. Often, the sample paths of this sequence show "discontinuities", which are overcome by a simple transformation. The transformed sequence can be used to generate random sequences with an arbitrary marginal distribution by using the inversion of this distribution. The main problem of the TES modeling is to find the adequate distribution for the innovations and the parameter for the transformation. It is necessary to make a good choice, because the distribution and the parameter determine the autocorrelation function of the generated sequence.

2.2 Markov-chain controlled processes

In [13] a finite-state, continuous-time Markov process is proposed to model the bit-rate of the superposition of video sources. Each state of the Markov chain represents a certain

[1] An ARP could be described as a linear Gauss-Markov process with an equilibrium point away from the origin.

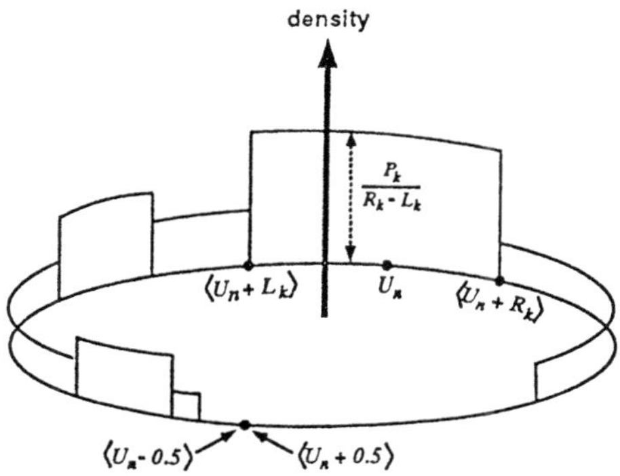

Figure 1: TES modeling scheme

bandwidth. If we have M states, labelled 0 to $M-1$, and a peak bit rate of p Mbps. being in state i means to have an output of $\frac{i}{M-1}p$ Mbps.

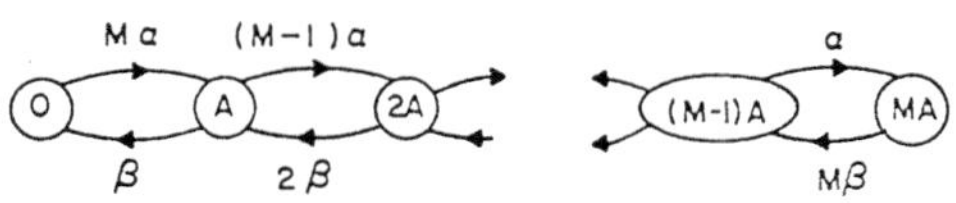

Figure 2: Markov chain of a one-dimensional birth-death model

Due to the characteristics of the statistical data, the authors decided to restrict themselves to a birth-death model, i.e., only state transitions to direct neighbor states are allowed (Fig. 2). They fixed the number of states, the peak bit-rate, and the state transition rates in order to fit the mean, the variance, the bell-shaped probability distribution of the frame sizes and the autocorrelation function of their experimental data. It is important to mention, that they assume that the autocorrelation function of the experimental data can be approximated by means of an exponential function.

Due to the fact that this model covered only small changes in the bit-rate it was extended in [22]. The one-dimensional state space was replaced by a two-dimensional one.

One dimension is used for small jumps in the aggregate bit-rate, the other dimension for larger jumps (Fig. 3). Thus correlations on two different time scales can be modeled. whereas no comparison to experimental data was provided.

Another approach is presented in [17], where a Markov chain is used to model the bit-rate of a single video source. The number of states is determined by the ratio of peak number of cells per frame to the standard deviation of the number of cells. The transition probabilities are set to the maximum likelihood estimates for these transitions calculated from experimental data. Issues concerning the correlation properties of the process generated by this model are not considered.

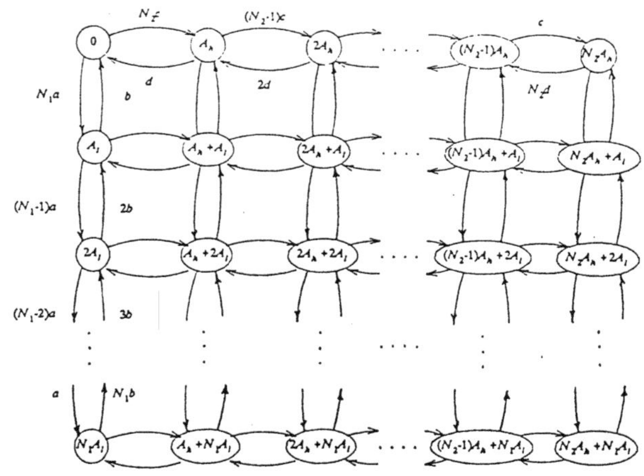

Figure 3: Two-dimensional Markov chain

A number of issues are important to consider in such a model that do not apply to previous examples. They include:

1. the computational cost of calculating the parameters of the transition matrix;
2. the accuracy with which these parameters can be estimated;
3. the numerical tractability of very large transition matrices.

One way in which these issues can be addressed is through state aggregation, i.e. assigning a state to a small range of rates rather than a state to each possible rate.

In the DAR(1) model [9], a finite-state Markov chain is used to generate the sequence of the number of bits associated with each video frame. The transition matrix of this Markov chain is given by

$$P = \rho I + (1 - \rho)Q \tag{1}$$

where ρ is the autocorrelation coefficient and I the identity matrix. In practice, it has been observed [8] that the negative binomial probabilities $(f_0, f_1, \ldots, f_K, f_K^c)$, defined by:

$$f_k = \binom{k+r-1}{k} p^r (1-p)^k \tag{2}$$

$$f_K^c = \sum_{k>K} f_k \tag{3}$$

are a good choice for the rows of Q.

The parameter K represents the maximum number of cells that can be generated by the coder for a single video frame; r and p are derived from the mean rate m of the data sequence, and its variance σ^2, via the relations:

$$m = \frac{r(1-p)}{p} \tag{4}$$

$$\sigma^2 = \frac{r(1-p)}{p^2} \tag{5}$$

This model matches the autocorrelation function of real data more closely than the AR(1) model, and provided the most accurate estimates of cell loss probability among the models investigated in [8]. However, for large buffer sizes, the use of this model results in a significant error in loss probability estimates.

The DAR(1) model can be seen as assuming that the bit rate of a coder is approximately constant within a scene, with sharp changes occurring at scene boundaries. The scene length distribution is exponential. In [3], this idea is taken further, in allowing for an arbitrary distribution of for the lengths of scenes. For a number of sequences examined, it was found that the use of a Pareto distribution for the scene lengths resulted in good predictions of network performance. One effect of this is that the sequence of the number of bits per video frame is no longer Markovian. It was also observed that this alteration to the DAR(1) model resulted in a significantly improved match to the autocorrelation function.

It has also been observed that the negative binomial distribution does not match well the tail behavior of the stationary distribution of the bit rate. In [4], it is proposed that a Pareto distribution should be used to model this tail. It is also verified that this distribution provides a good match to the tail observed in real data sets. So far, no studies are available that verify that this model results in better predictions of network performance than the DAR(1) model.

Two further approaches for modeling the sequence of numbers of bits per video frame are described in [1]. Both models are based on the discrete-time batch Markovian arrival process (D-BMAP). The first model is used for modeling a superposition of video sources, and can be seen as a discrete-time analogue of the birth-death model (model B) of [9]. The second model, which models only a single VBR video source, attempts to take into account higher level properties of the video sequence being coded. This is in contrast with other models that are based only on a statistical analysis of measurements of coder output, and take no advantage of knowledge of how this output stream is generated.

In this model the states of the Markov process are divided into disjoint sets, where each set of states, together with its transition probabilities, represents one class of video scenes with certain characteristics. Each set has a special starting state which is used to produce a large amount of data after the scene change. This was incorporated into the model because the experimental data have shown that large peaks in the bit-rate occur when scene changes take place. However, in the paper no parameter fitting procedure, and no comparison of the model and experimental data is provided.

A serious problem with the D-BMAP models arises due to the fact that the superposition of sources described with D-BMAPs is normally made by Kronecker products. Therefore, if we superpose M sources described with n states, the composite source is modeled with n^M states. As observed above, there are serious difficulties in storing very large matrices, in addition to the numerical problems associated with performing calculations on them. State-aggregation may provide a partial solution, and allow the number of states to be reduced to $O(n)$. However, this implies an assumption that the correlation structure of the superposition is of the same complexity as the one of a single source. Verifying such an assumption may present many difficulties in practice.

3 Models for cell sequences

The models of the above section have in common that the stochastic process that they describe is independent of the way in which the video sequence is transmitted via the communications network. The following models are more closely related to the mode of transmission, because they describe sequences of cells produced by a video coder and not longer frame sequences. To the best of our knowledge, there is no literature to verify the accuracy of estimates of network performance made using these models.

3.1 Markov-chain controlled processes

In [7] a two-state Markov modulated Poisson process (MMPP) is used to model the packet arrival stream of the superposition of voice sources. The following characteristics of the model and the experimental data are matched: mean arrival rate, the index of dispersion for counts (IDC) for a given time instant t_1, the long term IDC ($t \to \infty$), and the third moment of the number of arrivals in $(0, t_2)$. The authors point out that their model can also be used for other types of sources. The number of states might be increased to achieve a better fit to the correlation structure of the process. The use of the MMPP in modeling of video traffic is often criticized because it ignores the fundamental periodicity of this traffic associated with the frame rate (see e.g. [20]).

A very general starting point for the modeling of video sources is presented in [10]. The authors do not dedicate their concept called "unified stochastic model for the packet stream from periodic sources" directly to the modeling of video sources, but it could be used for this purpose.

It is assumed that the calls of different classes form a Poisson stream. The length of each individual call is following a discrete phase type distribution, where each class has its own representation (α, T). The set of transient states of each call is partitioned into two disjoint subsets of active (A) and silent (S) states (Fig. 4). Packets are only transmitted if an active state is entered. The sojourn time in every state is constant and set appropriate to the sampling rate of one packet. The authors present some examples and claim that their model is well suited for shaping the IDC according to a given process. The major

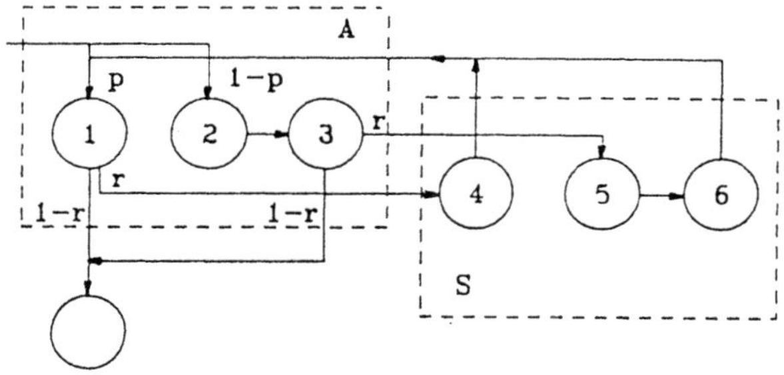

Figure 4: Example for the D-PH model

disadvantage of this model is its analytical intractability due to a large number of states, but it should be easy to implement it in simulation studies.

3.2 Autoregressive processes

In [6] the authors are modeling the periodic cell process of a video coder with an autoregressive moving average process. They present an algorithm that uses two Gaussian sequences to capture the autocorrelation structure of the experimental data. The output of this algorithm is transformed to match the mean and variance of the data. The authors report that the differences between measured and fitted data are acceptably small.

4 Discussion

In Sections 2 and 3, a large variety of models for VBR video sources has been presented, and their properties described. In this section, some general comments on the modeling of video traffic in cell-based networks are presented, along with some suggestions for future work in the modeling of this traffic.

In attempting to model VBR video traffic, it is important to define precisely what is being modeled, and to be clear on what aspects of network performance are to be predicted. It is only when this information is available that it makes sense to ask whether or not a model is useful. In the literature, there is a great diversity of justifications for the validity of different models. These range from particular properties of network performance (such as cell loss probability) to more esoteric quantities, such as the autocorrelation function of the sequence of the number is bits in each video frame. The bottom line is that the predictions of network performance have to be useful; this means that if you want to estimate cell loss but not delay, you may use a different model to the one you would use if you wanted to estimate delay and not cell loss. In this context, it does not matter directly whether the model is an accurate model of the traffic, only that those aspects of the traffic which impact on network performance be modeled well enough that the results are useful.

In evaluating the quality of a model, the sensitivity of the cell loss probability to such parameters as the mean arrival rate should be considered. It has been observed that a very small change in the mean arrival rate (say $\sim$1 %) can cause a very large change in the cell loss probability (perhaps as large as an order of magnitude.) Given that the mean rate is unlikely to be known to this accuracy in a real network, there is no point in attempting to obtain highly accurate estimates using a traffic model; in fact it can be argued that under these circumstances, large errors in the cell loss probability are quite tolerable [3].

The target application is also important in considering the effect of correlation between successive video frames. If a multiplexer with a small buffer is to be used, this correlation is not important [8, 3]. If a large buffer size is to be used, then this correlation becomes crucial to estimating network performance. In general, this correlation is not modeled well by any model that assumes that the autocorrelation function decays exponentially as the lag increases, as is the case with most models described here.

Where a number of video sources are feeding a switch, cross-correlation between sources must be considered as well as the autocorrelation of individual sources. In predicting network performance, cross-correlation is important because, if two video streams are highly correlated, the probability is high that both sources have peaks in their bit-rate at the same time. Cross-correlation between sources is a significant extra source of complexity in system modeling. The most common approach is to assume that all video sources are independent. From network dimensioning point of view, this assumption is dangerous, because it is a "best case" assumption. On the other hand side, the number of possible cross-correlation scenarios is infinite varying from completely synchronized identical streams to independent streams. In practice, it may be necessary to consider several correlation scenarios, at least the worst case, i.e. synchronized sources, and the best case, i.e. independent sources.

Another criterion that should be considered is the analytic tractability of models. While this is a lower priority than the accuracy of estimates of network performance, it is important to have available models for which parameters such as delay and cell loss probability can be calculated. Without this, it is difficult to understand the general characteristics of traffic, as opposed to being able to simulate a range of special cases.

Many modern video coding algorithms display a number of deterministic properties that are not taken into account by any of these models. For example, in MPEG [11] streams, there are three types of frames, each using a slightly different coding scheme:

"I" frames use only intra-frame coding, based on the discrete cosine transform and entropy coding;

"P" frames use a similar coding algorithm to "I" frames, but with the addition of motion compensation with respect to the previous "I" or "P" frame;

"B" frames are similar to "P" frames, except that the motion compensation can be with respect to the previous "I" or "P" frame, the next "I" or "P" frame, or an interpolation between them.

Typically, "I" frames require more bits than "P" frames. "B" frames have the lowest bandwidth requirement. The differences in coding result in different traffic characteristics for the different frame types.

The frames are arranged in a deterministic sequence, typically "IBBPBBPBBPBBI...". It is reasonable to expect that such a sequence would have very different behavior to one coded using all "I" frames. Further work is required in the modeling of this type of traffic. One major limitation is the lack of large sets of experimental data.

This question is related also to the fact that many different types of VBR video exist, using different coding algorithms, e.g. video-telephony, tele-conference, MPEG. There is very little data available on how these services differ in the demands that they place upon networks, and the effect of the coding algorithm on traffic characteristics.

In order to determine the accuracy of predictions of network performance provided by a model, it is necessary to make some comparison with results obtained using real data. This requires access to a large amount of data, especially where estimates of cell loss are to be made. The use of data obtained from 10 second video sequences is unlikely to provide any useful information on cell loss, where losses may only occur once per hour in a network. Such comparisons are made only rarely in the literature, but are vital in the verification of traffic models.

This requirement for long sequences of data is not easy to meet. Real-time coders are very expensive, and do not allow any variation to be made in the coding algorithm to determine the effect on the output traffic. On the other hand, software coders allow great flexibility, but are very slow.

Clearly, more work is required in the area of understanding the properties of video traffic before it is feasible to transmit variable bit rate services over networks of ATM switches. This requires both more knowledge of the relationship between coding algorithms and characteristics of traffic generated, and a better understanding of the impact of variable rate services on the network.

References

[1] C. Blondia and O. Casals. Statistical multiplexing of VBR sources: A matrix-analytic approach. *Performance Evaluation*, (16):5–20, 1992.

[2] M. R. Frater and O. Rose. Cell loss analysis of broadband switching systems carrying VBR video. Technical report, University of Würzburg, 97074 Würzburg, Germany, to be published 1993.

[3] M. R. Frater, P. Tan, and J. F. Arnold. Modelling of variable bit rate video traffic in the broadband ISDN. In *Proc. Australian Broadband Switching and Services Symposium*, Wollongong NSW, July 1993.

[4] M. W. Garrett. *Contributions toward real-time services on packet switched networks*. PhD thesis, Columbia University, 1993.

[5] M. Ghanbari. Two-layer codec of video signals for VBR networks. *IEEE Journal on Selected Areas in Communications*, 7(5):771–781, June 1989.

[6] R. Grünenfelder, J. P. Cosmas, S. Manthorpe, and A. Odinma-Okafor. Characterization of video codecs as autoregressive moving average processes and related queueing system performance. *IEEE Journal on Selected Areas in Communications*, 9(3):284–193, Apr. 1991.

[7] H. Heffes and D. M. Lucantoni. A Markov modulated characterization of packetized voice and data traffic and related statistical multiplexer performance. *IEEE Journal on Selected Areas in Communications*, SAC-4(6):856–868, Sept. 1986.

[8] D. P. Heyman, A. Tabatabai, and T. V. Lakshman. Statistical analysis and simulation study of video teleconference traffic in ATM networks. *IEEE Transactions on Circuits and Systems for Video Technology*, 2(1):49–59, Mar. 1992.

[9] P. A. Jacobs and P. A. W. Lewis. Time series generated by mixtures. *J. Time Series Analysis*, 4(1):19–36, 1983.

[10] G. Latouche and V. Ramaswami. A unified stochastic model for the packet stream from periodic sources. *Performance Evaluation*, (14):103–121, 1992.

[11] D. Le Gall. MPEG: A video compression standard for multimedia applications. *Communications of the ACM*, 34(4):46–58, Apr. 1991.

[12] M. Livny, B. Melamed, and A. K. Tsiolis. The impact of autocorrelation on queuing systems. *Management Science*, to appear 1993.

[13] B. Maglaris, D. Anastassiou, P. Sen, G. Karlsson, and J. D. Robbins. Performance models of statistical multiplexing in packet video communications. *IEEE Transactions on Communications*, 36(7):834–844, July 1988.

[14] B. Melamed and B. Sengupta. TES modeling of video traffic. *IEICE Transactions on Communications*, (12):1292–1300, Dec. 1992.

[15] M. Nomura, T. Fujii, and N. Ohta. Basic characteristics of variable rate video coding in ATM environment. *IEEE Journal on Selected Areas in Communications*, 7(5):752–760, June 1989.

[16] I. Norros, J. W. Roberts, A. Simonian, and J. T. Virtamo. The superposition of variable bit rate sources in an ATM multiplexer. *IEEE Journal on Selected Areas in Communications*, 9(3):378–387, Apr. 1991.

[17] P. Pancha and M. E. Zarki. Bandwidth requirements of variable bit rate MPEG sources in ATM networks. In *Proceedings of the Conference on Modelling and Performance Evaluation of ATM Technology. Martinique*, pages 5.2.1–25, Jan. 1993.

[18] S. qi Li and J. W. Mark. Performance of voice/data integration on a TDM system. *IEEE Transactions on Communications*, COM-33(12):1265–1273, Dec. 1985.

[19] G. Ramamurthy and B. Sengupta. Modelling and analysis of a variable bit rate video multiplexer. In *Proceedings of the Infocom '92*, pages 6C.1.1–11, 1992.

[20] V. Ramaswami. Traffic performance modeling for packet communication. whence, where and whither. In *Proceedings of the Third Australian Teletraffic Seminar 1988, Melbourne*, Nov. 1988.

[21] J. W. Roberts, J. Guibert, and A. Simonian. Network performance considerations in the design of a VBR codec. In *Proceedings of the ITC-13 Workshop on Queueing, Performance and Control in ATM*, pages 77–82, 1991.

[22] P. Sen, B. Maglaris, N.-E. Rikli, and D. Anastassiou. Models for packet switching of variable-bit-rate video sources. *IEEE Journal on Selected Areas in Communications*, 7(5):865–869, June 1989.

[23] K. Sriram and W. Whitt. Characterizing superposition arrival processes in packet multiplexers for voice and data. *IEEE Journal on Selected Areas in Communications*, SAC-4(6):833–846, Sept. 1986.

Broadband Communications, II (C-24)
S. Tohmé and A. Casaca (Editors)
Elsevier Science B.V. (North-Holland)

Indices of Dispersion and ATM Traffic Characterization

Fabrice Guillemin[a] and Alain Dupuis[b]

[a]Centre National d'Etudes des Télécommunications, 22300 Lannion, France. Tel : (33) 96.05.13.46, Fax : (33) 96.05.39.45, e-mail : Fabrice.Guillemin@lannion.cnet.fr

[b]Centre National d'Etudes des Télécommunications, 22300 Lannion, France, Tel : (33) 96.05.28.07, Fax : (33) 96.05.39.45, e-mail : Alain.Dupuis@lannion.cnet.fr

In this paper, we study the influence of the Index of Dispersion for Intervals sequence on the behavior of a finite-capacity, single-server queue, namely the $MMBP_2/D/1/K$ queue. It turns out that this traffic characteristic cannot be used in addition to the offered load for evaluating the impact of the input process on the queue. The phenomenon of *local overload* or *burst scale congestion* plays a predominant role.
Keywords : ATM, Performance evaluation.

1. INTRODUCTION

In the near future, the Asynchronous Transfer Mode (ATM) will be used for supporting Broadband-Integrated Services Digital Networks, which will be able to cope with a wide a range of bandwidth requirements as well as bit rate variations within a same connection (Variable Bit Rate traffic). So far, the CCITT (see CCITT Recommendation I.371) has focused on the peak cell rate. However, a major challenge for ATM networks will consist in performing statistical multiplexing. For this purpose, it will be essential to adequately control traffic and allocate resources. These tasks may be achieved by accurately characterizing traffic. The situation is however complicated by the fact that ATM traffic may be unpredictable (e.g., interactive services) and *"bursty"* (e.g., data services), some applications being yet unknown.

Within the debate on ATM traffic characterization, several parameters have been introduced [10,12] (e.g., peak/mean cell rate ratio, burst length, etc.). From a theoretical viewpoint, the most promising ones seem to be the so-called Indices of Dispersion [5,6], namely the Index of Dispersion for Intervals (IDI) and the Index of Dispersion for Counts (IDC). (Theoretical background on these indices can be found in [1].)

These measures have been considered for ATM traffic characterization because of their relationships with Central Limit type convergence theorems and thus, they prove very useful in diffusion approximations of queueing systems [2]. In particular, it is stated in [14] that *"the limiting value of the IDI sequence completely characterize the effect of a general stationary arrival process on the congestion of a multiserver queue in heavy traffic"*. Nevertheless, the relevance of this traffic characteristic has to be checked in the absence of heavy traffic limit conditions.

In this paper, we study the impact of an *irregular* arrival process, namely the so-called 2-stage Markov Modulated Bernoulli Process ($MMBP_2$, for short), on a single-server FIFO queue, namely the $MMBP_2/D/1/K$ queue. Some motivation for this model is given in Section 2, where we analyze how the IDI sequence impacts the rejection probability. From this study, it turns out that the IDI sequence cannot *completely* characterize the impact of the input process on the queue behavior. In truth, further information on the relationships between the time structure of the input process and the server characteristics is necessary.

2. MODELLING THE CELL ARRIVAL PROCESS AT AN OUTPUT BUFFE

2.1. Preliminaries

Consider the reference configuration depicted in Figure 1, involving an ATM switch with δ "slotted" incoming links and based upon an output queueing architecture : the slot duration on each input link is denoted by D. We are interested in modeling the arrival process seen by an output queue. During a cell transmission time D, δ cells may arrive at the switch. Because of the polling system located at the input stage, we assume that potential arrival times at an output queue are equidistant, namely each $\theta = D/\delta$ units of time a cell may arrive. θ will be taken as time unit in the following.

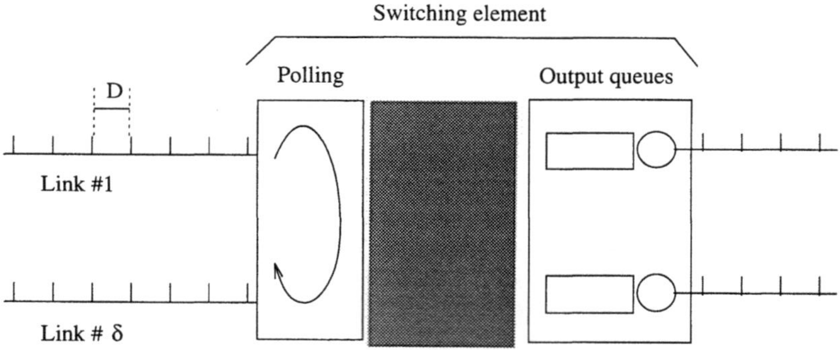

Figure 1. Reference configuration – switching element based on an output queueing architecture.

Dimensioning output queues is usually performed under the Balanced Traffic Assumption [3]. Specifically, the arrival process at an output queue is assumed to be Bernoulli (regular traffic). However, depending on the resource allocation and traffic control schemes adopted (especially in the Usage/Network Parameter Control), bursts of cells belonging to the same connection may arrive at the switch. In such a case, the cell arrival process at an output queue may be irregular (or "bursty"). To model this situation, we assume that the arrival process at a queue has two phases : a high instantaneous arrival rate phase, which corresponds to the arrival of a burst seizing the output queue considered, and another phase with a lower instantaneous arrival rate. We thus introduce an arrival process belonging to the family of Markov Modulated Bernoulli Processes.

2.2. The MMBP_2 arrival process

The two-stage Markov Modulated Bernoulli Process (MMBP_2) is a phase-type, discrete-time stochastic process [4], namely a Bernoulli process whose intensity is modulated by a Markov chain taking values in a two-phase space, say, $E = \{\psi_0, \psi_1\}$. Let φ_r denote the phase between the $(r-1)^{\mathrm{rst}}$ and the rth arrivals. When $\varphi_r = \psi_j$, the inter-arrival time I_r between these arrivals has a shifted geometric distribution, i.e.,

$$\text{for } k \geq 1,\ \Pr\{I_r = k/\varphi_r = \psi_j\} = p_j.(1-p_j)^{k-1} \tag{1}$$

where $p_j \in (0,1]$.

Just after an arrival, a new phase is drawn according to the transition matrix

$$P = \begin{pmatrix} 1-\omega_0 & \omega_0 \\ \omega_1 & 1-\omega_1 \end{pmatrix}. \tag{2}$$

Remark : Similar models have been considered in [7–9]. However, in the model considered here, arrivals trigger off phase changes whereas the phase evolves independently of arrivals in the other models.

2.3. Traffic characteristics

The stationary distribution of the phase process $\{\varphi_r\}_{r\geq 1}$ is $\eta = (\eta_0, \eta_1)$ where $\eta_0 = \frac{\omega_1}{\Omega}$ and $\eta_1 = \frac{\omega_0}{\Omega}$ with $\Omega = \omega_0 + \omega_1$. The mean arrival rate λ of the process is defined by [4]

$$\lambda = \frac{p_0 p_1 \Omega}{p_0\omega_0 + p_1\omega_1}. \tag{3}$$

The n-fold power matrix P^n being given by

$$P^n = \begin{pmatrix} \dfrac{\omega_1 + \omega_0(1-\Omega)^n}{\Omega} & \dfrac{\omega_0 - \omega_0(1-\Omega)^n}{\Omega} \\ \dfrac{\omega_1 - \omega_1(1-\Omega)^n}{\Omega} & \dfrac{\omega_0 + \omega_1(1-\Omega)^n}{\Omega} \end{pmatrix}, \tag{4}$$

we have for $\ell \leq k$,

$$E[I_k.I_\ell] = E[I_k.I_\ell/\varphi_\ell = 0].\eta_0 + E[I_k.I_\ell/\varphi_\ell = 1].\eta_1 = \frac{1}{\lambda^2} + \left(\frac{1}{p_0} - \frac{1}{p_1}\right)^2 \frac{\omega_0\omega_1}{\Omega^2}(1-\Omega)^{k-\ell}. \tag{5}$$

Hence,

$$\mathrm{Cov}(I_\ell, I_k) = \left(\frac{1}{p_0} - \frac{1}{p_1}\right)^2 \frac{\omega_0\omega_1}{\Omega^2}(1-\Omega)^{k-\ell} \text{ and } \mathrm{Var}[I_\ell] = \frac{2\omega_1}{\Omega p_0^2} + \frac{2\omega_0}{\Omega p_1^2} - \frac{1}{\lambda^2} - \frac{1}{\lambda}. \tag{6}$$

The IDI sequence $\{c_k^2\}_{k\geq 1}$ is then given by [14]

$$k \geq 1,\ c_k^2 = \frac{\mathrm{Var}[I_1]}{\mathrm{E}[I_1]^2} + \frac{2}{k.\mathrm{E}[I_1]^2}.\sum_{j=1}^{k-1}(k-j).\mathrm{Cov}(I_j, I_{j+1}) = \Gamma_a + \Gamma_i\left[1 - \frac{1-(1-\Omega)^k}{k.\Omega}\right] \tag{7}$$

where

$$\Gamma_a = \lambda^2 \left[\frac{2\omega_1}{\Omega p_0^2} + \frac{2\omega_0}{\Omega p_1^2} - \frac{1}{\lambda^2} - \frac{1}{\lambda}\right] \text{ and } \Gamma_i = 2\lambda^2 \left(\frac{1}{p_0} - \frac{1}{p_1}\right)^2 \frac{\omega_0\omega_1(1-\Omega)}{\Omega^3}. \quad (8)$$

Note that the limiting value $c_\infty^2 = \Gamma_a + \Gamma_i$.

Γ_a is the squared coefficient of variation of the inter-arrival time in the stationary regime (i.e, when the $MMBP_2$ arrival process is stationary) while Γ_i is a component which expresses the cumulated correlation between successive inter-arrival times.

2.4. The $MMBP_2/D/1/K$ queue

We consider a finite capacity, single-server queue with constant service times, at which customers arrive according to an $MMBP_2$ process. We adopt the following simultaneity principle : *arrivals see departures* [4]. Specifically, potential arrival times are the instants k^- and potential departure times from the queue are the instants k^+ for $k \geq 0$.

Let Φ_k and Q_k denote the phase of the $MMBP_2$ process and the queue length at time k^+, respectively. To make the process $\{Q_k, \Phi_k\}_{k\geq 1}$ Markovian, we introduce a supplementary variable as follows. Let d_k denote the position of the server in the current service. More precisely, $d_k = d$, $d \in \{1, \cdots, D\}$, if the service time of the cell being processed is already $d-1$ time units long at time k^+ (see [4] for more details).

Let $\{\tau(k,m,d)\}$ denote the stationary distribution of the Markov chain $\{Q_k, \Phi_k, d_k\}_{k\geq 1}$. $\{\tau(k,m,d)\}$ satisfies the following Chapman-Kolmogorov equations

$$\begin{aligned}
&\left[1_{\{k>0\}} + p_m 1_{\{k=0\}}\right] \tau(k,m,d) = \\
&(1-p_m).\left[\tau(k,m,d-1).1_{\{k>0\}}.1_{\{d>1\}} + \tau(k+1,m,D).1_{\{k<K\}}.1_{\{d=1\}}\right] \\
+ \;& \omega_m \sum_{m'=0,1} p_{m'} \left[\tau(0,m',0).1_{\{k=d=1\}} + \tau(k-1,m',d-1).1_{\{k>1\}}.1_{\{d>1\}}\right. \\
+ \;& \left.\tau(K,m',d-1).1_{\{k=K\}}.1_{\{d>1\}} + \left(\tau(k,m',D).1_{\{0<k<K\}} + \tau(K,m',D).1_{\{k=K-1\}}\right).1_{\{d=1\}}\right]
\end{aligned}$$

These equations are solved by using standard methods, namely the so-called MCV method [11].

Now, the probability distribution $\{\pi(k,m,d)\}$ at customer arrival epochs is obtained by using the conditional form of the "Bernoulli Arrivals See Time Averages" property [4], which yields

$$\pi(k,m,d) = \frac{p_m \tau(k,m,d)}{\lambda}. \quad (10)$$

The rejection probability in the $MMBP_2/D/1/K$ queue is then

$$\pi_{\text{loss}} = \frac{1}{\lambda} \sum_{m,d} p_m \tau(K,m,d). \quad (11)$$

3. ANALYSIS OF THE IMPACT OF THE INDEX OF DISPERSION FOR INTERVALS

To analyze the impact of the IDI sequence on the behavior of the $MMBP_2/D/1/K$ queue, the $MMBP_2$ arrival process is parametrized by using the mean arrival rate λ, the squared coefficient of variation Γ_a, the correlation term Γ_i, and the peak over mean ratio B. (The phase 0 is arbitrarily chosen as the most active, that, is $p_0 > p_1$.) After some algebra, we obtain

$$\begin{cases} p_0 = B.\lambda, \quad p_1 = \dfrac{B-1}{\frac{B}{2}\left(\frac{\Gamma_a+1}{\lambda}+1\right)-\frac{1}{\lambda}}, \\ \omega_0 = \dfrac{p_1(p_0-\lambda)}{\lambda(p_0-p_1)} \dfrac{1}{1+\dfrac{p_0p_1\Gamma_i}{2(p_0-\lambda)(\lambda-p_1)}}, \quad \omega_1 = \dfrac{p_0(\lambda-p_1)}{\lambda(p_0-p_1)} \dfrac{1}{1+\dfrac{p_0p_1\Gamma_i}{2(p_0-\lambda)(\lambda-p_1)}}. \end{cases} \tag{1}$$

In numerical applications, we have set the service time $D \triangleq \delta$ equal to 16 units of time (16 × 16 switch), the limiting value of the IDI sequence $c^2_\infty = \Gamma_a + \Gamma_i$ equal to 20, and the queue capacity K equal to 100 cells. The load of the MMBP_2/D/1/K queue is then $\rho = \lambda.D$.

We first analyze the behavior of the rejection probability π_{loss} with respect to the squared coefficient of variation Γ_a, for a given value of the load ρ and several values of the ratio B. Results are plotted in Figures 2 and 3 for $\rho = 0.5$ and $\rho = 0.8$, respectively. In these figures, Γ_a ranges from 0 to 10 but can actually range from 0 to more than $c^2_\infty = \Gamma_a + \Gamma_i$ (as long as $\omega_1 \leq 1$). The curves have been truncated because π_{loss} is almost constant beyond $\Gamma_a = 10$.

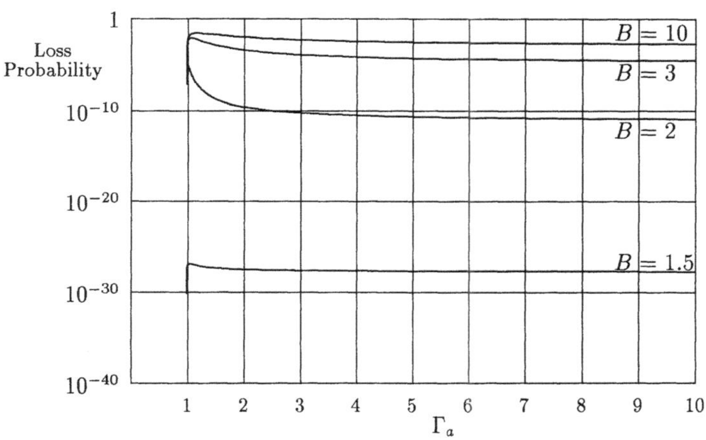

Figure 2. Loss probability vs. squared coefficient of variation Γ_a for a load $\rho = 0.5$ and different values of B – queue capacity $K = 100$ cells, service time $D = 16$, and limiting value of the IDI sequence $c^2_\infty = 20$.

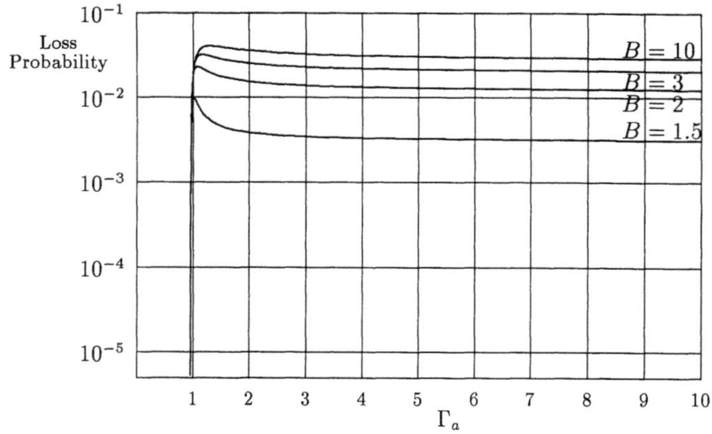

Figure 3. Loss probability vs. squared coefficient of variation Γ_a for a load $\rho = 0.8$ and different values of B – queue capacity $K = 100$ cells, service time $D = 16$, and limiting value of the IDI sequence $c_\infty^2 = 20$.

For the case $\rho = 0.5$, it clearly appears from Figure 2 that the rejection probability greatly depends on the ratio B, even if, for a given value of B, the rejection probability seems to be rather constant over the domain $\Gamma_a \geq 2$. The dependence on B is less significant for $\rho = 0.8$ as seen in Figure 3. Note moreover that for $B = 1.5$ and $\Gamma_a = 5$, the rejection probability ranges from 2.23×10^{-28} for $\rho = 0.5$ up to 3.25×10^{-3} for $\rho = 0.8$. These phenomena may be explained as follows.

In earlier studies, it was recognized that the value $\rho^\star = \frac{1}{B}$ is a *local overload* threshold [4] or equivalently, that beyond the value $\rho^\star$, the queue experiences *burst scale congestion* [13]. More precisely, for $\rho > \rho^\star$, the queue is overloaded during the phase ψ_0 since the instantaneous arrival rate p_0 is greater than the server rate equal to $\frac{1}{D}$, although the queue is in average not overloaded (the queue load ρ being less than one). The queue behavior then becomes very sensitive to the characteristics (overload duration, number of cells generated during an overload period, etc.) of the input process. In particular, the greater the value of B, the larger the magnitude of local overload, the larger the rejection probability.

In the situation considered in this paper, if $\rho = 0.5$, the queue is never overloaded for $B = 1.5$ and the rejection probability is then very small. For $B > 2$, local overload occurs and the rejection probability becomes very large. For example, when $\Gamma_a = 5$, it ranges from 2.23×10^{-28} for $B = 1.5$ to 5.78×10^{-5} for $B = 3$. For $\rho = 0.8$, the local overload threshold value for B is $B^\star \triangleq \frac{1}{\rho} = 1.25$. Hence, for $B = 1.5$, 2, 3, or 10, local overload occurs. The rejection probability does not depend very much on B, certainly because it has reached some asymptotic behavior.

Figure 4 displays the IDI sequence $\{c_k^2\}_{k\geq 1}$ for $\rho = 0.5$ and different values of Γ_a. Note that for given values of λ, Γ_a, and c_∞^2, the IDI sequence does not depend on B since $\Omega = (\lambda + \Gamma_a - 1)/(\lambda + c_\infty^2 - 1)$. However, rejection probabilities obtained for $B = 1.5$ and $B = 3$ are dramatically different. *Therefore, two input processes may have the same IDI sequence and their impacts on a queue may be completely different.*

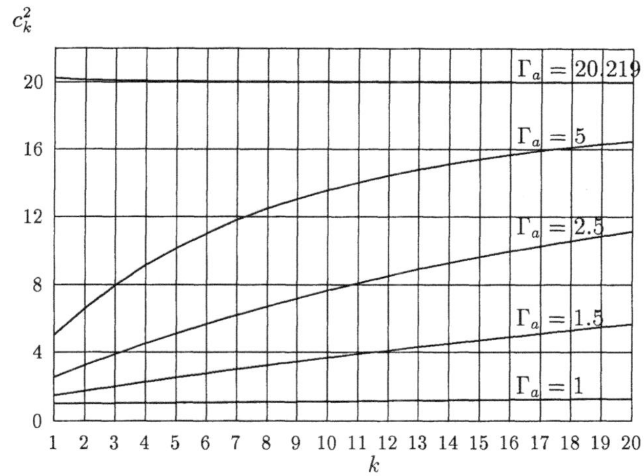

Figure 4. IDI sequence for a load $\rho = 0.5$ and different values of the squared coefficient of variation Γ_a – limiting value of the IDI sequence $c_\infty^2 = 20$.

Moreover, curves obtained for $\Gamma_a = 5$ and $\Gamma_a = 20.219$ are quite different but rejection probabilities are almost equal (2.56×10^{-28} and 1.98×10^{-28}, respectively). *It follows that even if two input processes have completely different IDI sequences, they may induce roughly the same rejection probability in a queue.*

4. Conclusion

In this paper, we have studied the influence of the Index-of-Dispersion for Intervals sequence on the behavior, especially the rejection probability, of a finite capacity queue, namely the $MMBP_2/D/1/K$ queue. It turns out that this traffic characteristic cannot be used in addition to the offered load for evaluating the impact of the input process on the queue. In fact, *local overload* or equivalently *burst scale congestion* is the predominant phenomenon to take into account when analyzing the behavior of the queue.

Results presented here confirm the fact that performance of a queue rapidly decreases when local overload occurs, as already observed in [4,13]. From a practical point of view, this indicates that resource allocation and traffic control schemes for ATM networks should protect network buffers from burst scale congestion.

REFERENCES

1. D.R. Cox and V. Isham. *Point processes.* Chapman and Hall, London, 1980.
2. K. Fendick, V. Saksena, and W. Whitt. Investigating dependence in packet queues with the Index of Dispersion for Work. *IEEE Trans. Commun.*, 39:64 –70, August 1991.
3. A. Gravey, G. Hébuterne, P. Boyer, and J.R. Louvion. Buffer sizing in an ATM switch for both ATM and STM traffics. *Int. Journal of Digital and Analog Cabled Systems*, 2:247–253, 1989.
4. F. Guillemin, J. Boyer, and A. Dupuis. Burstiness in broadband integrated networks. *Performance Evaluation*, 15(3):163–176, 1992.
5. R. Gusella. Characterizing the variability of arrival processes with indexes of dispersion. *IEEE J. Sel. Areas Commun.*, February 1991.
6. I.W. Habib and T.N. Saadawi. Multimedia traffic characteristics in broadband networks. *IEEE Communications Magazine*, pages 48–54, July 1991.
7. O. Hashida, Y. Takahashi, and S. Shimgawa. Switched Batch Bernoulli Process and the discrete-time SBBP/G/1 queue with application to statistical multiplexer performance. *IEEE JSAC, Special Issue on Teletraffic Analysis of Communications Systems*, pages 394–401, April 1991.
8. M. Hirano and N. Watanabe. Traffic characteristics and a congestion control scheme for ATM networks. *Int. Journal of Digital and Analog Communications Systems*, Vol. 3(2):211–217, March-April 1990.
9. J.Y. Le Boudec. An efficient solution method for Markov models of ATM links with loss priorities. *IEEE JSAC, Special Issue on Teletraffic Analysis of Communications Systems*, pages 408–417, April 1991.
10. W.E. Leland. LAN traffic behavior from milliseconds to days. In *ITC Specialist Seminar*, October 1990.
11. G. Pujolle and D. Potier. Réseaux de files d'attente à capacité limitée avec des applications aux systèmes informatiques. *RAIRO-Informatique*, Vol. 13 (2), 1979.
12. J. Roberts, editor. *Performance evaluation and design of multiservice networks.* European Economic Community, October 1991.
13. J. Roberts. Variable-bit-rate traffic control in B-ISDN. *IEEE Communications Magazine*, pages 50–56, September 1991.
14. K. Sriram and W. Whitt. Characterizing superposition arrival processes in packet multiplexers for voice and data. *IEEE J. Select. Areas Commun.*, Vol 4(6):833–846, 1986.

Broadband Communications, II (C-24)
S. Tohmé and A. Casaca (Editors)
Elsevier Science B.V. (North-Holland)

Analysis of Usage Parameter Control Algorithms for ATM Networks

Henning Clausen† & Peter Ryberg Jensen‡
hclausen@daimi.aau.dk & pryberg@daimi.aau.dk

Computer Science Department
Aarhus University
Ny Munkegade, Bldg. 540
DK-8000 Århus C
Denmark

Phone: +45 89 42 32 85
Fax: +45 86 13 57 25

Abstract

In the past years a lot of effort has been devoted to the standardisation of B-ISDN and Asynchronous Transfer Mode (ATM). In this paper we investigate four Usage Parameter Control (UPC) algorithms which control the flow of traffic sent from the sources to the network.

We model the four UPC algorithms by Coloured Petri Nets and conduct validation and performance analysis of the UPC algorithms. The validation includes reaction on an excess situation, speed-of-reaction and sensitivity to parameter changes. In the performance part cell delay is examined. Furthermore, the complexity of the implementation is considered.

The investigation show the Triggered Jumping Window algorithm to be the most suitable but also the Leaky Bucket algorithm is found to be appropriate.

Changes to the Sliding Window and Exponentially Weighted Moving Average algorithms are proposed which would improve these algorithms.

†Now at Meta Software Corporation, 125 Cambridge Park Drive, Cambridge MA 02140, USA.
(henning@metasoft.com)

‡Now at Mjølner Informatics, Gustav Wiedes vej 1, 8000 Aarhus C, Denmark.
(pryberg@mjolner.dk)

1 Introduction

The emerging demand for higher transfer speeds has lead to the standardisation of a broadband network called Broadband ISDN (B-ISDN). The B-ISDN network uses Asynchronous Transfer Mode (ATM) to transmit data.

During the last few years a lot of discussion has been devoted to the choice of algorithm for Usage Parameter Control (UPC) in ATM networks. This paper is a contribution to this discussion.

The purpose of our work is to investigate four different UPC algorithms: Triggered Jumping Window, Exponentially Weighted Moving Average, Sliding Window and Leaky Bucket [4, 9]. Also we want to compare our results with the results which others have stated, for instance in [9].

We used a high level Petri Net language called Coloured Petri Nets (CPN) [6] to model the UPC algorithms and the CPN tool Design/CPN [7] to implement and simulate the models. Our approach is to simulate the UPC algorithms and expose them to realistic data traffic. A description of the models is found in [3].

The paper starts with an introduction to the UPC algorithms and the measures that we use to rank the UPC algorithms. After that we turn to an examination of our simulation results. Finally, we sum up our results in the conclusion. A description of the traffic data that is used as input to the model can be found in Appendix A.

2 Usage Parameter Control

The section contains a short description of the algorithms, the time consumption of each algorithm and the measures we use to rank the algorithms.

The UPC monitors traffic from a source on its way into the ATM network. The most important task of the UPC is to monitor if the source complies with the quality of the services that was agreed on when the connection was established.

When we evaluate an UPC algorithm several measures have to be taken into account. A UPC algorithm is appropriate (i.e. well suited for its purpose) if the following demands are met:

- The UPC algorithm detects and tags excess traffic.
- The UPC algorithm does not interfere with non-excess traffic.
- The UPC algorithm reacts quickly when an excess situation begins/ends.
- The UPC algorithm does not introduce a significant delay of the traffic.

The fact that a UPC algorithm must react quickly when an excess situation begins is sometimes relaxed a bit. A UPC algorithm is said to be flexible if it allows small excess peaks to pass without being tagged. The opposite reaction is called strict and it is an ongoing discusssion whether the UPC algorithm should be flexible or strict.

In the investigations we have used the mean bandwidth as a measure for the UPC algorithms. If the sources have large fluctuations this leads to a lot of tagged cells even though the sources do not exceed the mean bandwidth over a longer period of time. A choice of a higher bandwidth would result in fewer tagged cells but also poorer bandwidth utilisation. It is a complex question which traffic characteristics are necessary to balance optimal bandwidth utilisation and correct monitoring of the connections and this still has to be solved and will not be addressed further here.

2.1 The algorithms

A window is a time period during which we count the number of cell arrivals. The mechanism then compares this counter to a given limit. If the counter exceeds the limit an excess situation has arised. This limit is calculated on the basis of the bandwidth that is being monitored. Let D be the time period of the window in seconds and B the bandwidth in bits per second. The size of an ATM cell is 424 bits. The limit N is calculated by using equation 1.

$$N = \frac{D * B}{424} \tag{1}$$

Triggered Jumping Window (TJW) A TJW differs from a ordinary window mechanism only by one thing; when a window has expired a new one is not started until the moment when the next cell arrives.

The TJW algorithm is implemented by having two registers. One is used as the counter that counts the cells and one holds the fixed limit.

When a cell arrives the counter register is compared to the limit and if it exceeds the limit the CLP bit of the cell header is set to 1. Then it is checked whether the counter register is 0. If this is the case a timer is started at 0, which initiates a new window. In any case the last action on a cell arrival is to increment the counter register by 1.

When the timer reaches the time period of a window the counter register is reset to 0. This is the expiring of the window.

Exponentially Weighted Moving Average (EWMA) EWMA is a ordinary window mechanism, however, while the windows follow immediately after each other, the limit is no longer fixed. When a window expires a limit for the next window is computed as an exponentially weighted average of all previous arrivals. A limit N_i for the i'th window is computed by the equations in 2. N is the allowed mean number of cell arrivals while X_{i-1} is the actual number of arrived cells in the $(i-1)$'th window. S_{i-1} is the history of previous arrivals.

$$N_i = \frac{(N - \gamma S_{i-1})}{1-\gamma}\ , \quad S_{i-1} = (1-\gamma)X_{i-1} + \gamma S_{i-2}\ , \quad 0 \leq \gamma < 1 \tag{2}$$

The γ parameter is a value that determines the flexibility of the algorithm.

The algorithm is implemented by having three registers. The first register is used for the counter X_i. The second register contains the limit N_i which is no longer fixed and the third register contains the value of S_{i-1}.

When a cell arrives the counter X_i is compared to the limit N_i and if X_i exceeds N_i the CLP bit of the cell header is set to 1. Then X_i is incremented by 1.

When the timer reaches the time period of a window, the timer is started again from 0. The new limit N_{i+1} is computed and the counter X_{i+1} is set to 0. The new values of the limit N_{i+1} and history S_i are put in their respective registers.

Sliding Window (SW) The SW algorithm uses a window which slides continuously along the time axis. The cell arrival time is put into the window and it is removed again after the time period of the window. In this way a window always holds the number of arrived cells within the last time period and comparing this number to the limit gives a compliance measure.

The algorithm is implemented by using three registers containing the counter value, the counter limit and the buffer limit. A buffer is used to store the arrival time of each cell. Two timers are started simultaneously at the initialisation of the algorithm. The first timer is started from 0. The second timer is started from the time period of the window. Two timers are used to avoid unnecessary addition/subtraction.

When a cell arrives it has to be checked whether buffer capacity is available to store the arrival time. Then the counter and limit are compared and the counter is incremented like before. Finally, the value of the second timer at the cell arrival is stored in the buffer.

When the first timer reaches a value that matches a value from the buffer the counter is decremented by 1 and the buffer value is removed from the buffer. In this way the cell is in the window for exactly the duration of the window.

Leaky Bucket (LB) Imagine a bucket with holes. If you pour water in it the water will run out again through the holes in a steady stream. This is the idea of a LB algorithm. When the bucket is full new cell arrivals are tagged as overflow. Otherwise they just increase the water level. Cells are passed on and not actually put into the bucket since only a cell arrival is measured.

The algorithm is implemented using a register for the counter and a register for the limit. The limit is fixed.

When a cell arrives the counter is compared to the limit and if it exceeds the limit the CLP bit of the cell is set to 1. The counter is incremented by 1 only if the counter does not exceed the limit.

With fixed time intervals the counter is decremented by 1 given that the counter is not already 0. The time interval between decrements is determined by dividing the cell size by the allocated bandwidth.

2.2 Cell delay

We now discuss the time consumption of each algorithm, one at a time, but first we look at time consumption of some basic operations.

Operation	Time(ns)	Operation	Time(ns)
add	10	comparison	10
subtract	10	clock start	5
multiply	30	store	5
divide	60	zero check	0

Figure 1: Time consumption of basic operations.

In figure 1 the time consumption of eight basic operations are given. It is assumed that the mathematical operations and the comparison are performed on registers while the zero check is done by a flag check. The store operation is to move a value from a register into a storage. Using dedicated hardware implementations these time consumptions are set as ordinary values and not as values that would be possible to obtain with more sophisticated equipment. In the same spirit we define the transmission speed in a link to be 1.28 Gbps, which again is an ordinary value with fiber optic cables. In the following it is assumed that writing a 1 in the CLP field can be done on-line.

Turning to the algorithms, we can now define the time comsumption of each algorithm. This time consumption consists of three parts:

- **Time used before passing on the cell.** The CLP bit is located at the end of the third byte in the header of the ATM cell. This gives 23 bits that are allowed to pass through the UPC algorithm before it is necessary to delay the cell. At a speed of 1.28 Gbps this gives approximately 17.5 ns.
- **Time used to update the registers at a cell arrival.** To ensure that the registers are not incremented and decremented at the same time it is necessary to reserve the registers before using them.
- **Time used to update the registers according to the particular algorithm.** This value also describes a reservation time. This time is used, e.g. in EWMA to compute a new limit and reset the counter register.

From the suggested implementations of the algorithms each of the three values can be calculated for each algorithm. The results of the calculations are shown in figure 2. The number in parenthesis of LB line 2 is used at an excess situation (i.e. when the counter is not incremented) while the number in parenthesis of LB line 3 is used when the counter already has value 0.

Time used	TJW	EWMA	SW	LB
- before passing on the cell	0	0	5	0
- to update at cell arrival	20	20	35	20 (10)
- to update according to alg.	5	175	15	10 (0)
Total	25	195	55	30 (10)

Figure 2: Time consumption of each algorithm (Numbers in ns).

One might believe that our cell delay experiments depend strongly on the calculation of time consumption, while the rest of our work can not be affected by this. It could be argued that the operations could be done faster and that the network transmission speed could be higher. The question is how our model and our results would be influenced by this.

Comparison of the counter and limit registers are used by all four algorithms before the cell is passed on. A change in the time consumption for comparison could have the effect that SW would be able to perform the two comparisons within the 17.5 ns and thereby none of the algorithms would introduce a delay by performing the UPC check.

A change in the reservation time at a cell arrival would not affect anything. A cell is 53 bytes long and at a speed of 1,28 Gbps this gives a cell duration of approximately 323 ns. Therefore, all the operations can easily be made while the cell passes through the UPC.

A change in the reservation time at the register updates would have some effect. SW and LB use a subtraction while TJW uses a reset of a register. If the speed of subtraction could be increased the TJW, SW and LB algorithms would come very close to eachother in time consumption. The calculation time of EWMA is much larger than the other three algorithms so the calculation time of EWMA can not be made comparable to the calculation time of the other algorithms by simply increasing the calculation speed.

Finally, we have the assumed speed of 1.28 Gbps. Increasing this value could imply that all algorithms would delay every cell. However, the ranking of the algorithms is not influenced and the ranking is our immediate concern.

3 Validation

The first validation step is to investigate whether the UPC algorithms behave appropriately on an excess situation, i.e. if they tag the excess traffic and do not interfere with non-excess traffic.

Then we investigate how long time it takes before a UPC algorithm detects the start/stop of an excess situation. These times are denoted speed-of-reaction.

3.1 Reaction on excess

The traffic source Videophony is used as input to each of the algorithms (figure 3). The use of the Videophony source only serves as an example as the obtained results hold for the other sources too. Each experiment simulates three seconds of real time. For the Video source three seconds of real time is probably too little. However, three seconds are sufficient for the other sources and due to the

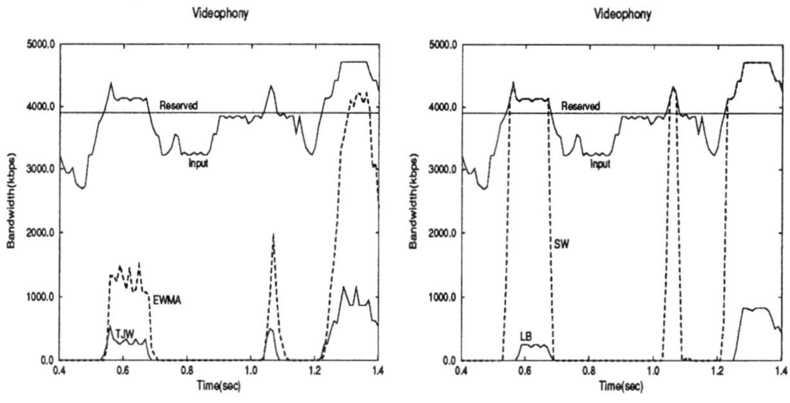

Figure 3: The algorithms used on a videophony source.

time required to simulate the Video source we choose as small a time periode as possible. Due to clarity only a part of the simulation time is shown in figure 3. The horizontal line in the figures indicates the reserved bandwidth. The input is shown by the solid line marked Input. The amount of tagged traffic is shown by the line marked with the name of the algorithm. The ideal algorithm only tags the traffic which is above the horizontal line.

In figure 3 the result of using the TJW and EWMA algorithms on the Videophony source is shown. The TJW algorithm tags the excess traffic. It does not seem to be sensitive neither to the size of the source nor to the length of the excess situation.

The result of using the EWMA algorithm shows that when the excess lasts for a longer period of time the EWMA starts tagging more and more. Also the size of the excess is of importance. The greater the traffic is the more non-excess traffic gets tagged. This is due to the fact that the EWMA algorithm recalculates its limit when a time period has expired. If the bit rate is high the new limit becomes smaller than the previous.

Also shown in figure 3 is the result of using the LB and SW algorithms. The SW algorithm tags far too much. In an excess situation the algorithm tags all cells that arrive at the UPC. This is because in an excess situation new cell arrivals are still saved in the buffer and the counter is incremented by one.

The LB algorithm tags most of the traffic that exceeds the reserved bandwidth. Small peaks are allowed as the second last in figure 3 (between 1.0 and 1.1 seconds). It might be argued that the algorithm ought to tag all of the excess traffic and therefore also small excess peaks. However, this is a discussion of how strictly/flexibilly the UPC algorithm has to monitor the input traffic.

The investigations above show that the SW algorithm tags too much when excess situations occur. The EWMA algorithm also tags too much, but not as much as the SW algorithm. Compared with these two algorithms the TJW algorithm is excellent. No more than the excess is tagged, but the algorithm on the other hand is very strict. All excess cells, even small peaks, are tagged. The LB algorithm also tags the excess traffic but the algorithm is more flexible than the TJW algorithm, since it allows small peaks to exceed the reserved bandwidth. A summary of the results is found in the first row of figure 4.

-	TJW	EWMA	SW	LB
Excess	**** Tags all excess cells, strict	** Tags too much	* Tags far too much	**** Tags all excess cells, flexible
Speed-of-reaction	*** Stops too late	* Starts too late Last to stop	*** Stops too late	** Starts and stops too late
Cell-delay	**** No delay	* Large delay, bandwidth dependent	** Small delays for all bandwidths	*** Small delays,bandwidth dependent
Summary	***	*	*	**

Excellent=* * **, Good=* * *, Average=**, Bad=*.

Figure 4: A comparison of the different algorithms.

3.2 Speed-of-reaction

Speed-of-reaction is measured by comparing the time when the excess situation occurs with the time when the UPC algorithm starts tagging excess cells. Similarly when the excess stops the speed-of-reaction is investigated. This is done in figure 5 where the first excess peak in figure 3 is shown. The numbers in the columns are in kbps. Zero indicates that the algorithm tags what it is supposed to. A positive number indicates that the algorithm tags too much while a negative number indicates that it tags too little.

Time (sec)	Excess	TJW	EWMA	SW	LB
0.53	0	0	0	0	0
0.54	33	8	-33	1167	-33
0.55	240	-116	-75	3900	-240
0.56	489	-75	836	3900	-489
0.57	240	-75	1085	3900	-240
0.58	199	-34	1043	3900	-117
0.59	240	-116	1250	3900	8
0.60	240	-116	877	3900	-33
⋮	⋮	⋮	⋮	⋮	⋮
0.67	240	-75	836	3900	-33
0.68	0	41	35	1705	82
0.69	0	0	165	0	0
0.70	0	0	0	0	0

Figure 5: Speed-of-reaction for the four UPC algorithms.

The figure shows that both the EWMA and the LB algorithms start tagging too late. This is not the case with SW and TJW which both start tagging when the traffic exceeds the reserved bandwidth. When the excess situation stops all the algorithms stop tagging too late. EWMA stops tagging later than the other algorithms.

On basis of this we conclude that SW and TJW have the best speed-of-reaction characteristics, followed by LB and EWMA. Similar results are found if other sources with other burstiness/excess qualities are considered. A summary of the investigations is found in the second row of figure 4. The investigation does not take into account the length of the time period and the bucket. Due to lack of space we do not report these investigations but interested readers are refered to [2].

4 Performance analysis

The delay introduced by the UPC is calculated by determining the difference in time for each cell as the cell enters and leaves the UPC. Below we use the Video source as an example. The results of the investigations are shown in figure 6. In the figure the cell delay is divided in time intervals. For each interval the number of cells in the interval is shown.

The TJW algorithm does not delay the cells. This is a major advantage of the TJW algorithm compared to the other UPC algorithms. However, as explained in section 2.2 the algorithm potentially delays the cells if the bandwidth is high enough. We do not find any delays because we do not use a bandwidth which is high enough in our test examples.

The EWMA algorithm does not introduce any notable delays for the large majority of the cells. However, a number of cells are delayed more than 12 ns.

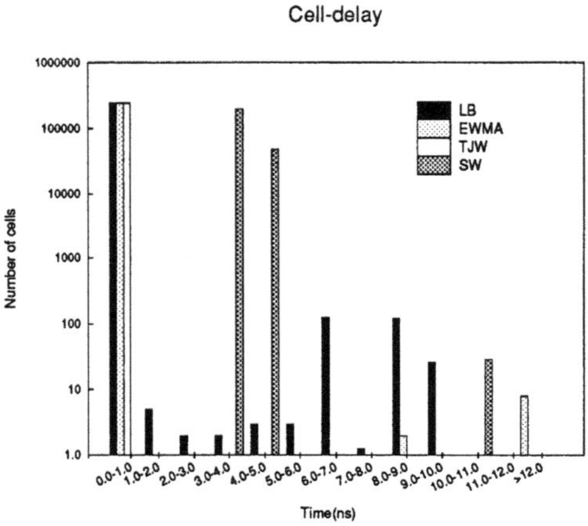

Figure 6: Cell delay for the four algorithms.

Actually one cell is delayed as much as 160 ns. The delay is caused by the rather complex functions needed in the calculations of the new limit of the window.

The SW algorithm delays the majority of the cells between 5.0 ns to 6.0 ns. The delay is introduced by the operations done each time a cells enters the UPC as explained in section 2.2. A few cells (29) are delayed even more (between 11 ns and 12 ns) due to updating of the window at the moment when the cell arrives at the UPC.

The LB algorithm delays the cells up to 10 ns. However, no cell is delayed more than 10 ns. Compared to EWMA and SW the LB algorithm is better as a UPC algorithm since most of the cells are not delayed and the maximum possible delay is rather small.

Our investigations have shown that most of the UPC algorithms delay the cells. This delay has to be added to other delays introduced by the ATM network, e.g. transmission- and switching time. Imagine that a 512 $\times$ 512 video frame is sent to update the picture. If a 1.28 Gbps fiber optical link is used the frame is transmitted in 1.6 ms. The delay introduced by the UPC is thus 6 orders of magnitude smaller than the transmission time of the video frame and as such irrelevant.

We have investigated the cell delay anyway because it shows the different behaviour of the UPC algorithms. Cell delay is then useful to determine which UPC algorithm is the most appropriate. A summary of the investigations is found in the third row of figure 4.

The number of delayed cells also has to be taken into consideration. Apart from the SW algorithm only a very small fraction of the total number of cells sent are delayed. For instance the EWMA algorithm where 242.176 cells are sent but only 8 cells are delayed. Hence a delayed cell is a rather rare event as we have already mentioned in section 2.2.

4.1 Discussion

The Triggered Jumping Window algorithm can detect excess traffic precisely and it does not interfere with non-excess traffic. It is a fast algorithm both with respect to speed-of-reaction and cell delay. We have not found any delays introduced by the TJW algorithm in our experiments and if a reasonable time period is chosen for the window the algorithm tags the excess traffic.

The Exponentially Weighted Moving Average algorithm can also detect excess traffic precisely. However, it interferes with non-excess traffic but this problem can be solved by a modification of the algorithm. When the excess situation has lasted for some time the limit gets so small that cells are tagged before the reserved bandwidth is exceeded. By changing the way the new limit is calculated (cf. formula 3) this problem is avoided.

$$N_i = Max(N, \frac{N - \gamma S_{i-1}}{1-\gamma}) \ , \ 0 \leq \gamma < 1 \tag{3}$$

The limit never gets below what is needed to allow the reserved bandwidth to pass without being tagged. The change does not affect the flexibility of the algorithm since the limit is allowed to be larger than N. By changing the algorithm it might be possible to overcome the tagging problem, but this does not change the delays introduced by the algorithm. We did not do any investigations of the modified algorithm.

The Sliding Window algorithm does not perform well as it tags too much. Again the problem can be solved by a modification in such a way that only compliant cell arrivals are stored. When the window is full, arriving cells are just tagged and passed on without storing their arrival time. However, this approach is a modification of the SW algorithm and again we did not do any investigations of this modified algorithm.

Delays introduced by the SW algorithm are very small even though every cell is delayed and not only some of them. The real problem with the algorithm is the buffer question. It is necessary to store cell arrivals in the time period of the window. If the bit rate is high the size of the buffer has to be correspondingly high. When the hardware implementation of the UPC algorithm is made the traffic characteristics of the sources using the network are still unknown. The buffer then has to be as large as the highest possible bit rate leading to a large buffer. A solution to this problem is not obvious if any exists at all.

The Leaky Bucket algorithm does not tag too much and has some good speed-of-reaction characteristics which are easily controlled by the size of the bucket. However, the algorithm has the drawback of introducing varying delays. By having knowledge of the implementation of the algorithm a fixed upper bound for the delay can be calculated and notified to the sources at connection setup as a part of the QOS parameters.

In the investigations we have used the mean bandwidth as a measure for the UPC algorithms. It is a complex question which traffic characteristics are necessary to balance optimal bandwidth utilisation and correct monitoring of the connections. Sources such as video sources with very large, long lasting peaks will require some kind of Fast Reservation Management (FRM) [1] if bandwidth is not to be wasted for long periods of time. By using a FRM more bandwidth is allocated or deallocated as needed and therefore the amount of bandwidth necessary to reserve at connection setup is reduced. The exact behaviour of and need for FRM is still a subject for further studies.

5 Conclusion

To use high level Petri Nets as a modelling and simulation tool is practical for the examination of UPC algorithms. The modelling and simulation on a machine by Coloured Petri Nets gives many possibilities of different analysis. We have seen that we could validate the algorithms and do performance analysis with respect to cell delay and speed-of-reaction. In [2] it is also shown that parameter variation can be investigated. Due to lack of space these results are not reported in this paper.

The TJW and LB algorithms are close to each other in performance, and it is hard to say which algorithm one should prefer. We prefer TJW because most of its calculations are done while a cell passes through the UPC and this prevents delaying cells. This is in contrast to the LB algorithm which has updates with fixed time intervals. On the other hand one might argue that the bucket size of the LB is easier to adjust to get a more flexible algorithm. We cannot support the statement in [9] that EWMA is one of the best algorithms. Only if the suggested modification is applied (and this is not done in [9]) one might tend to accept the delay problems with EWMA to get a nice flexible algorithm. However, the performance of the algorithm is still worse than the performance of the TJW and LB.

6 Acknowledgements

This paper has been written as a part of our master thesis at the Computer Science Department of Aarhus University, Denmark.

The modelling and simulation described has been made in cooperation with Søren Christensen, while the paper has been written together with Kurt Jensen and Jens Kristian Kjærgaard.

A Source traffic

Test data input to the model is generated according to a Markov Chain Process (MCP) [2, 5, 8].

Type	Application	Mean BW (kbps)	Burstiness
Continuous Bit Rate	Uncompressed telephone calls, Large file transfer	10000	Cont.
Dataflow	Resulting traffic from many different traffic types	2000	Low
Databurst	Reading/writing to a database, LAN-LAN interconnection	1000	High
Videophony	Compressed video conference	3900	Medium
Voice	Compressed telephone calls	1600	High
Video	HDTV, Multimedia applications	39457	Medium

Figure 7: The used source traffic.

By changing the parameters of the MCP we are able to generate different kinds of traffic types. The parameters are chosen such that: Firstly, there is a large variation in the values of the parameters and secondly, the resulting traffic resembles real traffic. Hereby, we have defined six traffic types shown in figure 7. Also shown in the figure are some basic characteristics of the traffic types such as the mean bandwidth requirement in kbps and the burstiness.

References

[1] CCITT SG XVIII: *Draft recommendations on Broadband ISDN.* Geneva, June 1992.

[2] H. Clausen, P.R. Jensen: *Usage Parameter Control Algorithms in High Speed Networks.* Master Thesis, Comp. Science Dep., Aarhus University, Denmark, May 1993.

[3] H. Clausen, P.R. Jensen: *Validation and Performance Analysis of Network Algorithms by Coloured Petri Nets.* Fifth International Workshop on Petri Nets and Performance Models, Toulouse, France, 1993.

[4] L. Dittman, S.B. Jacobsen, K. Moth: *Flow enforcement algorithms for ATM networks.* IEEE JSAC, vol. 9, no. 3, 1991.

[5] I.W. Habib, T.N. Saadawi: *Multimedia Traffic Characteristics in Broadband Networks.* IEEE Comm. Mag., vol. 30, no. 7, 1992.

[6] K. Jensen: *Coloured Petri Nets. Basic Concepts, Analysis Methods and Practical Use.* EATCS Monographs on Theoretical Computer Science, Springer Verlag, 1992.

[7] K. Jensen, S. Christensen, P. Huber, M. Holla: *Design/CPN. A reference Manual.* Meta Software Corporation, USA, 1991.

[8] B. Maglaris et al.: *Performance models of statistical multiplexing in packet video communications.* IEEE Trans. on Comm., vol. 36, no. 7, 1988, pp. 834-845.

[9] E. Rathgeb: *Modelling and performance comparisons of policing mechanisms for ATM-networks.* IEEE JSAC, vol. 9, no. 3, 1991.

Broadband Communications, II (C-24)
S. Tohmé and A. Casaca (Editors)
Elsevier Science B.V. (North-Holland)

Dimensioning a Traffic Control Device in an ATM network

Catherine Rosenberg[a] and Gérard Hébuterne[b]

[a]Ecole Polytechnique de Montréal, P.O. Box 6079, Station "A", Montréal, P.Q., Canada, H3C 3A7

[b]Centre National d'Etudes des Télécommunications, Route de Trégastel, BP 40 22301, Lannion Cedex, France

It is well known that a single Leaky Bucket cannot control all characteristics of a source. Especially, controlling burst length and mean rate may be incompatible, while a poor control of any of these characteristics impacts directly networks performances.

Given an ATM source, this paper addresses this issue and gives a general approach to choose the "best" Leaky Bucket parameters; the optimality criterion is the number of worst-case sources which the network can accomodate while ensuring the required Quality of Service.

Keyword Codes: C.2.1
Keywords: ATM (Asynchronous Transfer Mode), Network Control

1. INTRODUCTION

In Asynchronous Transfer Mode (ATM) based networks, resource allocation is based on a Traffic Contract negotiated between the user and the network during the connection establishment phase. The Traffic Contract, as specified in ITU-T Recommendation I.371 [1], is composed of the requested Quality of Service (QoS), the Cell Delay Variation clumping tolerance, and the Source Traffic Descriptor (a set of intrinsic traffic parameters). Up until now, only the Peak Rate of the source has been defined as a source traffic parameter and can be used to perform peak rate allocation in the network.

In the case of continuous bitrate (CBR) sources, Peak Rate Allocation allows optimal network operation: a simple dimensioning ensures that the network rejection probability is less than the Quality of Service figure (typical values being around $10^{-7} - 10^{-9}$).

The case of variable bitrate sources (VBR) has to be taken into account. For such sources – eg. video coders, database queries, etc – peak rate allocation leads to inadmissible waste in bandwidth capacity. Clearly, a major challenge in the near future for ATM networks will consist in performing some kind of *statistical multiplexing* while meeting QoS objectives (e.g., Cell Loss Ratios, burst blocking probability). For this purpose, information beyond the peak cell rate will be necessary to characterize traffic issued by an ATM source, especially a characterization of its stochastic fluctuations. In short, the source should characterize itself more precisely than with only its peak rate, eg. with more traffic parameters which should be enforceable by the network.

An example of a new traffic parameter proposal is to be found in the *Sustainable Ce* *Rate* (SCR) and the *Burst Tolerance* parameters [2]. The former is an upper bound on th mean cell rate which can be offered by the source, while the latter limits the size of a burs at the peak cell rate. The algorithmic definition of the SCR allows direct enforcement o these parameters by using algorithms identical to the Generic Cell Rate Algorithms (GCRA defined by ITU-T ([1], [2]) in order to monitor the source peak rate.

However, the choice remains among the possible values for the SCR parameters, or equiv alently the choice (among the many possible ones) of the GCRA the network has to use fo conformance testing. This is usually referred to as the dimensioning problem of the traffi control mechanism.

Defining efficient rules for dimensioning traffic control mechanisms happens to be an im portant issue, since they directly affect the network efficiency by allowing a certain level o statistical multiplexing. From the standpoint of this paper, we assume that the networ intends to control the traffic beyond the peak rate by using a second traffic control devic To simplify the mathematical analysis, we use the discrete state Leaky Bucket version [3] o the GCRA algorithm in the following for this second device. However, all the argumentatio could be rewritten easily in terms of the GCRA, due to the similarity between these tw algorithms.

In this paper, we survey the constraints on the parameters of this second device impose by the user's QOS objectives and by the network efficiency objectives (section 2). Then, w propose an approach for designing dimensioning rules for this device (section 3).

2. CONSTRAINTS ON THE PARAMETERS OF THE CONTROL DEVICE

Let us assume that the sources comply with their peak-rate requirements – for instance a Leaky Bucket or Virtual Scheduling Algorithm is in use, and a spacing function is in servic inside the User-Network Interface [4].

The "Discrete-State Leaky Bucket" used in this study conforms to the device initiall introduced in [3]. A copy of arriving cells is put into a queue of finite capacity N. Th queue is emptied at a rate R (the "leak rate"). If no space is available for the copy when th cell arrives, it is declared as non-conforming (and will be tagged or discarded, depending on network options). The Leaky Bucket only deals with copies of actual cells: a conforming cel incurs no additional delay in it.

It is well known that there exists no "universal Leaky Bucket", that is a Leaky Bucke able to control at the same time all statistical aspects of the traffic issued by a source. Fo instance, a Leaky Bucket efficiently controlling the mean rate of the connection is likely to provide a poor control on short term phenomena. On the other hand, an efficient control o burst length generally implies no tight control on the mean bitrate.

In other words, one can give various characterizations of the source, *depending on the tim scale observed*, while the Leaky Bucket can be used to properly control only a single tim scale at a time. This is to be accepted, as a basic limitation of such mechanisms [5]. Given this preliminary warning, one has nevertheless to choose the "optimal" Leaky Bucket in the set of all Leaky Buckets suitable to control the source beyond its peak rate. This optimalit can be viewed in many ways and we present one of them in this paper.

.1. The Source point of view

From the source point of view, once the traffic contract has been negotiated and the call ccepted, the end-to-end connection (control included) must offer a transparent interface, hich means loss probability and cell delay variation lower than negotiated. Especially, the ontrol device (eg. the Leaky bucket with parameters (R, N)) must not unduly reject too nany conforming cells.

The discrete-parameter Leaky Bucket can be modelled as a finite-capacity single-server ueue with deterministic service time equal to $1/R$ [5]. From the user's standpoint, a Leaky Bucket with parameters R and N would belong to the set of all Leaky Buckets suitable for ontrolling this source beyond its peak rate if and only if the actual rejection probability ncountered by the source in this Leaky Bucket is less than or equal to some prescribed value Pr (say around $10^{-7} - 10^{-9}$). Letting X denote the source process, this can be restated as ollows: *the rejection probability of the $X/D/1/N$ queue must be lower than Pr.*

In the following, we assume that the rejection probability can be computed using some ueueing model or simulation results, for a given source X and a given Leaky Bucket with arameters (R, N). In particular, we assume that the function giving the minimum value f N, for a given R such that the rejection probability is less than or equal to Pr has been omputed. Let this function be denoted as $N(R)$.

.2. The Network Management point of view

The Network Manager has to protect the network, that is to say all established calls, gainst non conforming connections. For that, we have assumed that it uses two Traffic Control devices, one devoted to Peak Rate enforcement and the second devoted to a control beyond the peak rate. This second device is a Leaky Bucket with parameters (R, N). Many ources would be considered as conforming with respect to these two parameters. Thus, the afer solution for the network is to rely on the *worst-case source*, that is the "worst" output rocess generated by the Leaky Bucket (worst from a network standpoint).

Let $C = 1$ be the network capacity, let $D \leq 1$ be the source peak rate and p its mean rate normalized with respect to C. The sources are assumed to conform with their negotiated peak rate. The network capacity is taken as unity, which means that time is measured in cell ransmission time units.

For a Leaky Bucket with parameters (R, N), the worst-case output source for the network behind the policing devices (eg. the worst-case source of peak rate equal to D assuming that he peak rate is enforced using a spacing device [4]) is the deterministic ON/OFF process, with period of activity at the peak rate D of length A (in cell transmission time units) during which B cells are sent: $(B-1)/D + 1 = A$ – see Figure 1. Since all cells are accepted, $N + R \times A - B \geq 0$. The silent period S is such that all credits are regenerated: $N = S \times R$. See Figure 1:

$$A = \frac{N + D - 1}{D - R} \qquad S = \frac{N}{R} \tag{1}$$

$$B = \frac{N.D - R + R.D}{D - R} \tag{2}$$

Remark: strictly speaking, since A, B and S are integer numbers, the worst-case process may consist in a series of 2 periods alternating between consecutive integers. For simplicity, we neglect these round-off problems.

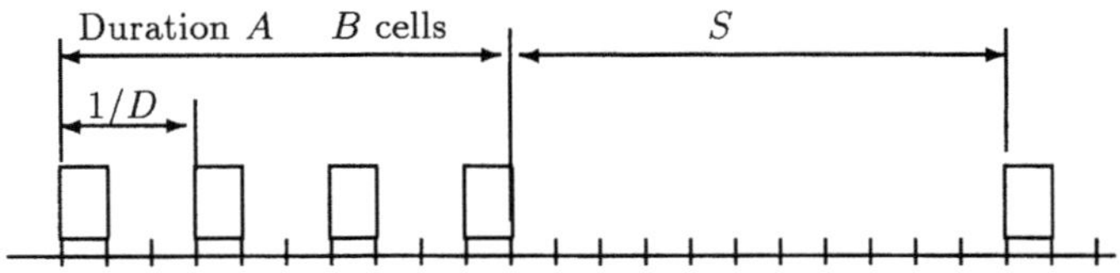

Figure 1. The Worst-Case Deterministic ON/OFF Process

The network point of view is to maximize its efficiency by maximizing the number of sources it can multiplex for its given capacity and a given QoS and thus to choose the parameters (R, N) of the above mentioned Leaky Bucket accordingly.

3. THE SELECTION RULES

3.1. Assumptions

We assume that the network supports n simultaneous identical connections (we return to this point in the Conclusion). The Network Manager point of view can thus be summarized as follows: *given the individual sources, among all the admissible Leaky buckets, the Leaky Bucket chosen to control any of them should have the parameter R (and $N = N(R)$) such that the number n_R of sources that it can multiplex with the same level of QoS is maximal.*

These n connections correspond to identical Leaky Buckets and to identical worst-case sources. The worst configuration for the network, with these n identical On/OFF sources, is to be observed if the activity periods of all sources are synchronized. This pattern produces a batch On/Off arrival process: during the ON period, a batch of n cells arrives every $1/D$ slots, an ON period being of length A and an OFF period being of length S.

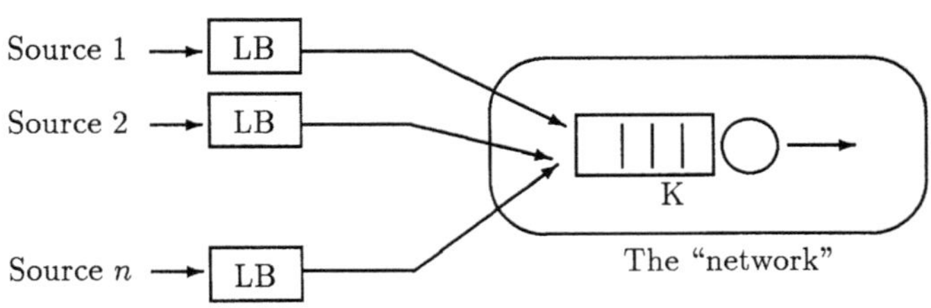

Figure 2. Statistical multiplexing of controlled sources in the ATM network

In a first approximation, one limits network occupancy ρ to figures around $0.7 - 0.8$: this ıitation accounts for end-to-end cell delay variation. Thus, one can allow a maximum $\supset = \rho \times C/D$ connections of peak rate D on a given multiplex with capacity C. For nplicity, we assume the network can be modelled as a finite-buffer queue with deterministic rver. Let K be the size of the buffer of this equivalent system.

2. Estimating the degree of multiplexing

It could be possible to develop elaborate queueing models dealing with the superposition ON/OFF processes and their interaction with the network-equivalent queue. We prefer ing a "fluid flow approximation" leading to simple results assuming all the sources are nchronized.

In the case where traffic control is performed with two devices, the so-called worst-case nfiguration corresponding to n identical sources, each one declaring itself according to a aky Bucket with parameters (R, N), yields an aggregated process corresponding to the tch arrival process described above.

Let us compute n_R for a given Leaky Bucket with parameters (R, N). n_R is the maximum mber of identical sources the network can multiplex with a rejection probability less than equal to Pr and without exceeding the load ρ.

During the ON period of the n individual sources, cells arrive at the rate nD, so that nB lls arrive on the whole. If $nD > 1$, they accumulate in the queue until the silence period. the end of the ON period, $nB - A$ cells are in the queue. They are eliminated during the ence period if $nB - A < S$, that is $n < 1/R$, and no rejection occurs iff:

$$B - A \leq K$$
$$n \leq \frac{K + A}{B}$$

The constraint on the network load can be written as $n \leq \rho/R$. Finally, n_R is the maximum satisfying the two above constraints, i.e.:

$$_R = \min\left\{\frac{\rho}{R}, \frac{K(D-R) + N - 1 + D}{N.D - R.(1-D)})\right\} \tag{3}$$

Recall that N (one should preferably write $N(R)$) is the function giving N (the "queue ze" of the Leaky Bucket) for a given R, for a prescribed rejection probability Pr. Under is form, the value of n_R of the above equation fulfills both requirements of the source uaranteed Quality of Service) and of the Network Manager (Protected network).

.3. The dimensioning rule

Using equation 3, obtaining R_{opt} is straightforward by maximizing n_R for R such that $< R < 1$:

$$\begin{aligned} n_{\text{opt}} &= \max_{p<R<1} \{n_R \mid n_R \text{ verifying } (3)\} \\ R_{\text{opt}} &= \min\{R \mid n_R = n_{\text{opt}}\} \end{aligned} \tag{4}$$

'his procedure calls for a few comments. It is quite general, in that it does not depend on any tatistical assumption about the sources. All the information about sources is summarized in the function $N(R)$. The form of the equation giving n_R relies only on the worst-c characterization of the Leaky Bucket algorithm, and does not require any assumption ab the source.

Obviously, the optimum value of R and n will depend on the network parameters ρ and and on the QOS constraint Pr. This point is illustrated on the curves shown below.

The inequalities used to derive n_R are obtained via a worst-case argument, assuming sources are synchronized. This means underestimating the load the network could actua handle with an acceptable QoS. For moderate burstinesses (the ratio D/p) and moderate assuming (more realistically) asynchronous sources would allow higher values for n_R. N however that the underestimation is the same for all possible (R, N) solutions. In ot words, the optimal solution given by (3) ought to remain the best, were sources assum asynchronous.

3.4. An example of dimensioning

As a simple example, let us assume Bernoulli sources with $p = 0.02$, and $D = 1$. T function $N(R)$ is obtained through classical queueing methods (see e.g. [6]). Figure 3 be shows the sample of possible controls. That is, each of the points on the curve represent Leaky Bucket through which the $p = 0.02$ source passes with a loss less than 10^{-7}.

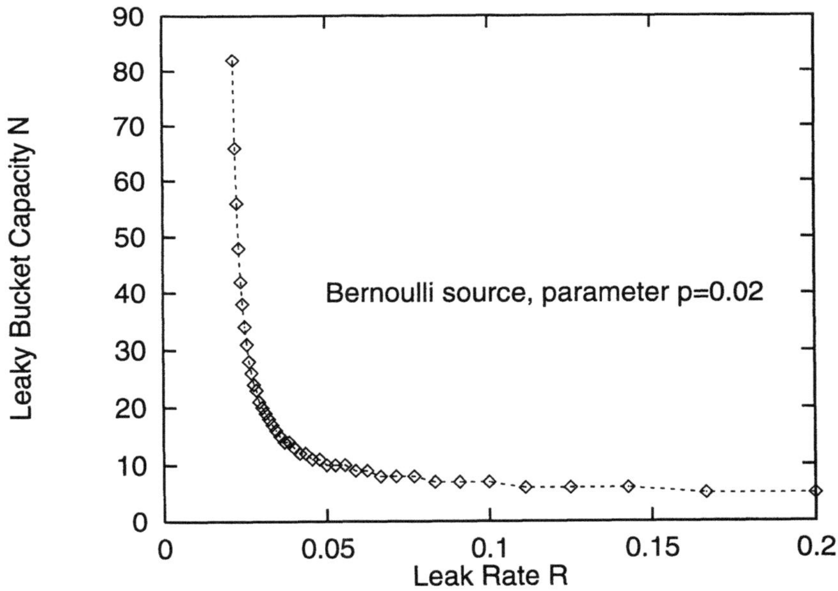

Figure 3. Possible (N, R) Leaky Buckets admissible for the Bernoulli source with $p = 0.0$

As above, the network is represented by a single-server queue with a limited capacity K, d the load is limited to $\rho = 0.7$.

The following table shows for various such admissible Leaky Buckets the resulting n_R and effective link occupancy (the examples assume $K = 60$).

R, N	Max burst In Nb of cells	n_R (Link load)	Comments
0.5, 3	6	1 (0.02)	max. burst length under control
0.1, 7	7	7 (0.14)	Optimal choice
0.0357, 15	16	3 (0.06)	
0.0222, 66	67	1 (0.02)	Mean rate under control

The curves on Figure 4 show the variation of the maximum number of sources the network n accept as a function of the leak rate R, for various values of K.

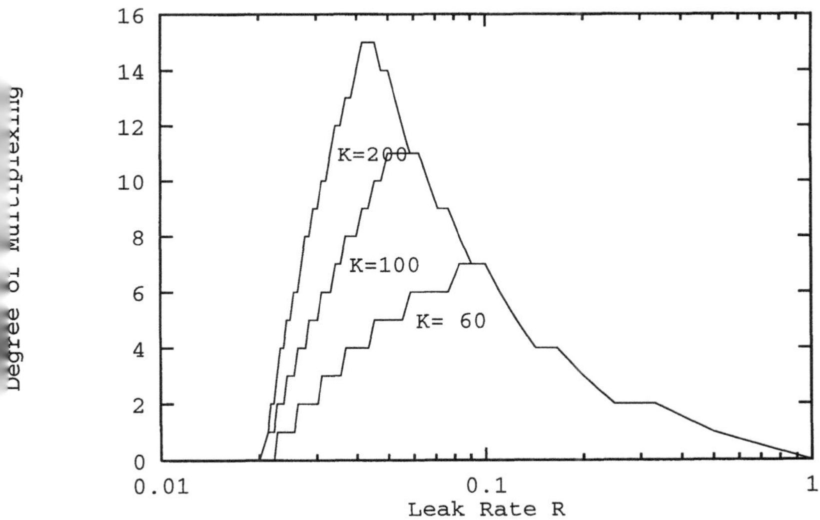

igure 4. Number of acceptable sources, as a function of R (the leak rate)

The curves illustrate known tendencies of the Leaky Bucket algorithm. As K increases, network is able to carry more bursty sources. It is then more advantageous to have R cl to the source mean bitrate, even if it implies tolerating more bursts (the worst-case is m bursty). At the same time, since the mean bitrate is better controlled, the utilisation rat the network increases.

The rather low value for n_{opt} is due to the strong constraint which is imposed on n (wo case with synchronous sources). More interesting than the brute result is the ratio betw all possible controls, allowing n_R to vary between 1 and 7.

4. CONCLUDING REMARKS

4.1. The negotiation

Two dual approaches can be taken to view the negotiation phase. Either the source decla itself completely and the network dimensions the device by computing the $N(R)$ curve a choosing R_{opt}, or the source declares itself in terms of the second traffic control devi parameters and thus performs the dimensioning by itself, sending to the network the tra parameters (R_{opt}, N_{opt}).

A major constraint for traffic characterization is that information exchanged between user and the network is conveyed by means of signalling. Thus the list of traffic paramet declared by the user at the connection set up should be minimal with respect to a cert level of traffic control complexity.

Dimensioning by the source seems a better choice as far as signalling has to be kept m imum, since it avoids passing the full characterization of the source. To push the source choose the optimal Leaky Bucket computed as described above, the network could use a ta based incentive. In that case, using identical sources to perform the computation of n_{op} the only judicious possibility since in any case a source has no way of knowing the traffic n already in the network.

Having the source characterizing itself in terms of the traffic control parameters is equival to the approach taken by the ATM-Forum with the Sustainable Cell Rate.

4.2. Further studies

In this paper, we have focused on a criterion for choosing actual parameters of a possi second source parameter control mechanism, in addition to the Peak Rate enforcement. T method proposed does not allow by itself any multiplexing gain: it only gives a way to sel the "best" control, that is the one which optimizes the admissible load by allowing the high multiplexing gain compatible with source requirements, network characteristics and with t complexity level the control can afford.

Note that, instead of selecting the best possible Leaky Bucket, the same approach may restricted to the case where the choice is to be made between, say, two possible devices.

For a given network dimensioning (i.e. a given K), this approach can be generalized for ca where the network is willing to take i traffic control devices for policing a source ($i = 1, ..., \infty$ To each "control level" i corresponds a "multiplexing level" L_i. $i = 1$ corresponds to a pe rate allocation policy for which the "multiplexing level" is $L_1 = \rho \times C/D$, $i = 2$ correspon to the configuration studied here (a kind of Sustainable Cell Rate), $i = \infty$ corresponds the case where the source gives a complete characterization of itself to the network. We ha $L_1 \leq L_2 \leq ... \leq L_\infty$. The case $i = 3$ is under study.

For example, in the case of Bernoulli sources of parameters $p = 0.02$ and $D = 1$ with a QoS responding to a rejection probability $Pr < 10^{-7}$ and $K = 60$ (note that $i = \infty$ corresponds dimension according to the Geo/D/1/K queue):

Number of Controls	Degree of Multiplexing
1	$L_1 = 1$
2	$L_2 = 7$
∞	$L_\infty = 35$

EFERENCES

CCITT Recommendation I.371 : *Traffic control and congestion control in B-ISDN.* Geneva, June 1992.

The ATM Forum : *UNI Specification. Version 2.2.* Boston, June 1993.

J. Turner : *New directions in communications (or which way in the information age ?).* Proceedings of the Zurich Seminar on Digital Communications, Zurich, March 1986, pp. 25-32.

P. Boyer, F. Guillemin, M. Servel, and J.P. Coudreuse : *Spacing cells protects and enhances utilization of ATM network links.* IEEE Network Magazine, Special Issue on "Switching and Congestion Control in ATM Networks", pp. 38-49, September 1992.

E. Rathgeb : *Modeling and performance analysis of policing mechanisms for ATM networks.* IEEE Journal on Selected Areas in Communications, 9(3), April 1991.

A. Gravey, J.R. Louvion, and P. Boyer : *On the Geo/D/1 and Geo/D/1/n queues.* Performance Evaluation 11, pp. 117-125, 1990.

RESOURCE MANAGEMENT AND NETWORK CONTROL

Broadband Communications, II (C-24)
S. Tohmé and A. Casaca (Editors)
Elsevier Science B.V. (North-Holland)

An Evaluation of Burst-level Bandwidth Reservation Methods in WAN Environments

Makiko Yoshida, Chinatsu Ikeda and Hiroshi Suzuki [a]

[a] C&C Research Labs, NEC Corp., 4-1-1, Miyazaki, Miyamae-ku, Kawasaki, 216 Japan
Fax:+81-44-856-2229 Tel:+81-44-856-2121
Email: myoshida@nwk.CL.nec.co.jp

This paper shows the evaluation results of fast bandwidth reservation (FRP) methods applied to bursty traffics, e.g. a large file transfer, in ATM networks with long propagation delay. These traffics require a large bandwidth for a short time, i.e., bursty characteristics and a strict cell loss quality. For these traffics, an FRP instead of a call level bandwidth reservation is effective to utilize network resources under guaranteed QoS. Comparison among FRP methods with peak rate controls in terms of a transmission completion time is carried out under a short and long propagation delay. We compare two types of FRP and FRPs with three adaptive peak rate controls.

Evaluation results show that confirmed type FRP is preferable to unconfirmed type FRP. In addition, we see from the results that FRP with peak rate control using network availability information provides the shortest transmission completion time under light load conditions. However, the results also show that FRP with simple peak rate control using ACK/NACK provides fair transmission completion time under heavy load and long propagation delay conditions.

1. Introduction

In ATM, to handle file transfer traffic is one of the important problems to be solved. The traffic should be dealt with efficiently over ATM networks, as demand for the interconnection among LANs increases.

File transfer traffic has the following properties: One is the bursty characteristics with a large bandwidth requirement for a short holding time. Second is an error free requirement which can be achieved either by retransmission or by no-cell-loss transmission. To achieve no-cell-loss transmission for the bursty traffic, fast reservation protocol (FRP) is proposed [2]. FRP tries to reserve a bandwidth for each burst transmission to guarantee cell loss quality. Since no-cell-loss is required, a peak rate bandwidth needs to be allocated.

One of the problems in FRP is a round trip delay for a request/confirmation of a bandwidth reservation. This could be a substantial problem, especially in wide area network (WAN) environments with a long propagation delay. If a bandwidth is reserved for each link before data transmission, the reserved bandwidth is wasted for one round trip time which is needed to confirm the availability on the route. Thus we need to study the performance, i.e., transmission completion time, of FRP methods under WAN environments.

In order to improve the performance, the combination of the adaptive peak rate control with the FRP is proposed [1]. The evaluation of an adaptive peak rate control applying on FRP under LAN environments is shown in [1]. In this paper, we focus on the evaluation which show the effect of the propagation delay on the reservation methods.

This paper is organized as follows: section 2 describes the evaluated bandwidth reservation method and section 3 shows performance comparison between several FRP methods. Finally, a summary of our conclusions is given in section 4.

2. The Evaluated Burst-level Bandwidth Reservation Methods

2.1. Confirmed / unconfirmed type FRPs

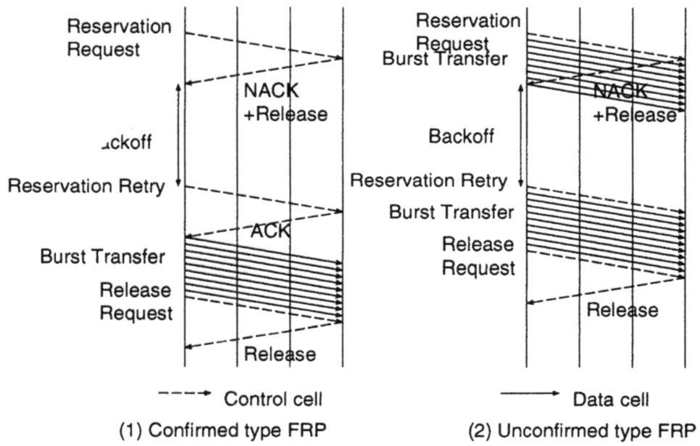

Figure 1. Confirmed type and unconfirmed type FRP

There are two FRPs with/without a round trip confirmation before sending data as shown in Fig. 1 [3][4][5]. Using FRP, a VC is established at call set up time, though a bandwidth for the VC is not provided at that time. The bandwidth is allocated for peak rate request of each burst.

When a source requests bandwidth reservation, the source indicates a desirable peak rate and a tolerable peak rate in the request. The source requires the desirable peak rate as the maximum bandwidth reserved and the tolerable peak rate as the minimum.

In confirmed type FRP, at first, a source sends a bandwidth reservation request cell to a destination. If less than request value of capacity is available, the destination sends a NACK cell back to the source. Then the source will try again after a random backoff time. If enough capacity is available, the destination sends an ACK cell back and the source starts to send the data. Using this, extra latency is introduced due to the confirmation. Though, it is easy to implement the confirmed type FRP. In addition, network information can be added on the route by using an ACK/NACK cell and the information can be used for peak rate control.

In unconfirmed type FRP, a source doesn't wait to an ACK/NACK cell and sends the data right after sending a bandwidth reservation request cell. Using this, no extra latency due to confirmation is introduced. However, additional mechanisms are necessary to selectively drop ineffective cells which comes into network during confirmation so as to protect effective cell transfer. The mechanisms are hard to be implemented. In addition, the network information which cannot be seen from the source is not able to be used for bandwidth reservation before sending the data.

2.2. Adaptive peak rate controls

We should evaluate the transmission completion time as the objective function as users demand a short transmission completion time for file transfer. To reduce the transmission completion time, [1] proposed the combination of the adaptive peak rate control with FRP. Reduction of peak rate is effective to reduce the blocking probability of bandwidth reservation for bursts. Nevertheless, transmission with a smaller bandwidth leads, of course, to longer transmission time. Thus we should consider trade-off between blocking probability, which leads to backoff time, and transmission time in peak rate control. In [1], a simple peak rate control method using ACK/NACKs is proposed for LAN environments. Peak rate controls with additional mechanisms could be more valid to make good use of network resources in WAN environments because a VC tend to use larger amount of resources than in LAN environments. Therefore, we propose two peak rate control methods using network information for WANs.

In this paper, we evaluate three methods, a method using ACK/NACK, a method using source information and a method using end-to-end information, to control peak rates corresponding to network situations, i.e., to decrease peak rates in heavy-load conditions and to increase peak rates in light-load conditions.

The method using ACK/NACK controls temporary peak rate to be increased /decreased corresponding to the number of consecutive ACK/NACKs[1]. The information of the number of consecutive ACK/NACKs shows the past global network conditions to some extent, but it does not show the current condition. As no additional information transfer in signalling is needed, it is simple to implement this control.

In the method using source information, a temporary peak rate request will be set to the value of available capacity of a link next to the source. If there is less available capacity than the temporary peak rate, the reservation will fail and the request will be tried again after a certain backoff time with the desirable peak rate request. Otherwise, the bandwidth is reserved and the source starts to send the data. To achieve this method, the source needs to see the available capacity which is next to each node. The information shows the current condition, but it could not show the global network conditions.

The method using end-to-end information is control using end-to-end information on the route. In this method, a temporary peak rate will be modified to the value of available capacity through the route. An ACK cell needs to bring available capacity information through the route in this method. If the available capacity through the route is less than tolerable bandwidth request, the request will fail and will be tried again after a backoff time with the desirable peak rate request. Otherwise the transmission will start. The source and intermediate nodes need to see the available capacity of a link which is next to each node. Though the method needs complex information transfer, the information shows the current global network conditions.

The peak rate requested, i.e. temporary peak rate, is limited in the range between desirable and tolerable peak rate, in any of these peak rate control methods.

3. Evaluation Results

3.1. Model

3.1.1. Network Model

To evaluate FRP methods from a performance viewpoint, i.e. transmission completion time, we studied models of hypothetical networks consisting of 4 switches, shown in Fig. 2.

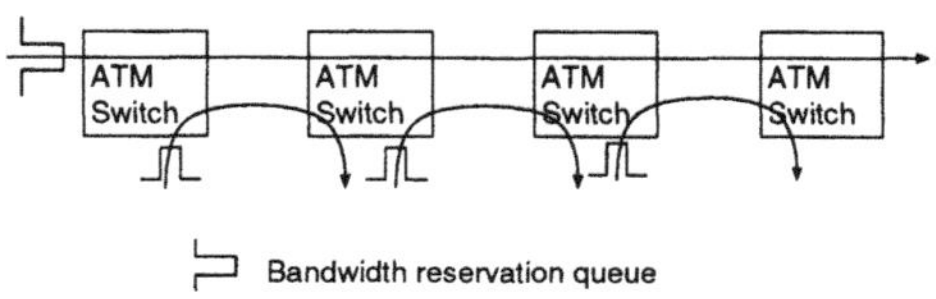

Figure 2. Network model

The lengths and propagation delays of all physical links are assumed to be the same. A propagation delay and transmission completion time are normalized by average service time. As FRP is suitable for transferring large size files, three kind of propagation delay parameter values, i.e., 0.001, 0.01 and 0.05, are used in the evaluations. These values correspond to 10ns, 100ns and 400ns for a 1M bit file transfer in 100Mbps of peak rate, and to 10ms, 100ms and 500ms for a 1G bit file transfer in 100Mbps of peak rate. Small propagation delay means either small networks, i.e., LANs, in geographical sense or a large file transfer which has relatively small overhead. On the other hand, long propagation delay means either large networks, i.e., WANs, or a small file transfer.

The backoff time parameter, which is a coefficient of the service time, is 3.0[6].

3.1.2. Traffic Model

In this evaluation, four kinds of traffic flows are used as shown in Fig. 2. Three of them are one-hop traffics and one is a three-hop traffics. FRP control is applied for any traffic. Each of the traffic has 50 VCs and all traffics of the VCs have unique characteristics. The inter burst arrival time and burst duration distributions are assumed to be geometrically distributed. The desirable and the tolerable peak rate are 0.8 and 0.1, which are values relative to the link capacity.

3.2. Comparison between confirmed and unconfirmed type FRP

Figure 3 shows the transmission completion time for confirmed / unconfirmed type FRPs as a function of throughput of total traffic, without any adaptive peak rate control.

Here, the throughput is defined as carried traffic normalized by the total network capacity. In the figure, p shows propagation delay parameter.

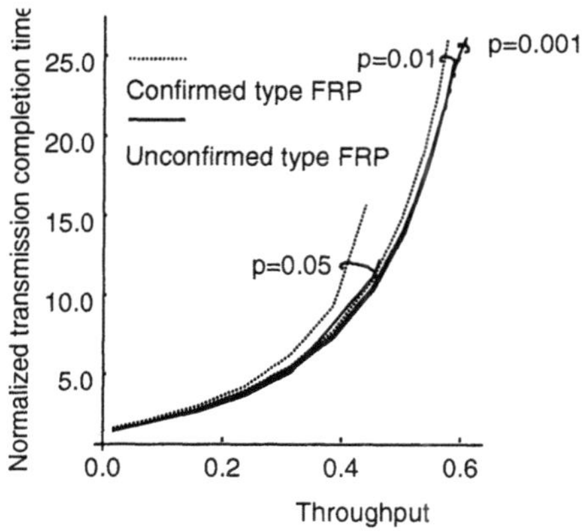

Figure 3. Comparison between confirmed / unconfirmed type FRP

As Fig. 3 shows, the difference in transmission completion time between two types of FRPs are less than 1 % and about 5 % for propagation delay 0.001 and 0.01, respectively. For propagation delay 0.05, while the difference is about 10 % at throughput 0.1, it increases to 30 % at throughput 0.4. Though, the transmission completion time rapidly increase at throughput 0.43 and 0.46 for each type FRP, so the range in which we see the advantage of unconfirmed type FRP is small.

The difference in performance between two types of FRP is in transmission overhead in successful transmission. In blocked transmission, the bandwidth is occupied for bandwidth reservation during a round trip delay in either type FRP. Therefore, the increase is a round trip delay itself in small throughput conditions.

When throughput increase, bandwidth occupation of overhead for the confirmation of successful reservation effect blocking probability. Thus, unconfirmed type FRP shows advantage over confirmed type. However, when blocking probability increase, either type FRP reach the maximum throughput. The same results are obtained using FRPs with adaptive peak rate controls.

If a round trip delay is longer or file size is smaller than that in this evaluation, the difference may not be negligible. To see this phenomena, we evaluated the performance with propagation delay parameter 0.1. The results show that the maximum throughput is about 0.15, so we see the range in which the advantage of unconfirmed type FRP is small even in large propagation delay case.

The results show that the difference in performance between the confirmed and the unconfirmed type FRPs are small. Since we need some mechanism in intermediate node to filter ineffective cells, the confirmed type FRP is preferable. Therefore we apply the confirmed type FRP in the following.

3.3. Comparison among adaptive peak rate controls

In this section, we show the comparison among three adaptive peak rate controls described in Section 2 and no peak rate control.

A. Primary FRP

This is an FRP without peak rate controls. Source nodes always request a fixed bandwidth, i.e., 0.8, in this method.

B. FRP with peak rate control using ACK/NACK

This is an FRP with the peak rate control using the number of consecutive ACK/NACKs. In the evaluation, a temporary peak rate to be requested b is given by the following exponential function. $b = b_0 \times 2^{\lfloor A/2 \rfloor} \times (\frac{1}{2})^{\lfloor N/2 \rfloor}$, where b_0 is the previous temporary peak rate, A is the number of consecutive ACKs and N is the number of consecutive received NACKs. b_0, A and N are reset when transition between receiving ACK/NACK to NACK/ACK occurs and when A or N crosses 2. $\lfloor x \rfloor$ means the largest integer which does not exceed the value of x. If b is more than desirable peak rate, b is set to be the desirable peak rate. If b is less than tolerable peak rate, b is set to be the tolerable peak rate. In [1], some other methods to change peak rate, such as transition between two values and step wise increase/decrease are examined. Since exponential function showed fairly good performance among these, we choose this function in the evaluation.

C. FRP with a peak rate control using source information

This is an FRP with the peak rate control which set temporary peak rate to be the available capacity seen from the source.

D. FRP with a peak rate control using end-to-end information

This is an FRP with the peak rate control which set temporary peak rate to be the available capacity through the route.

In each method, bandwidth is assumed to be reserved through the route during one round trip time for each reservation attempt, even if blocking is detected at a intermediate node. The reserved bandwidth which is blocked is assumed to be released after one round trip time.

3.3.1. The effect of propagation delay on each method

Figures 4 show the transmission completion time as a function of throughput of total traffic achieved, using the reservation method A (primary), method B (with ACK/NACK), method C (with src-info) and method D (with end-end info), with parameters of propagation delay.

In Fig. 4, we see that the transmission completion time increases according to the propagation delay increases in the results of methods C and D while there is small change in the results in methods A and B.

In order to examine the effect, we show Fig. 5 and 6. Figures 5 and 6 show the blocking probability and the peak rate as a function of throughput for method B and D, which show representative characteristics. The peak rate indicates the average of allocated rate in successful transmission.

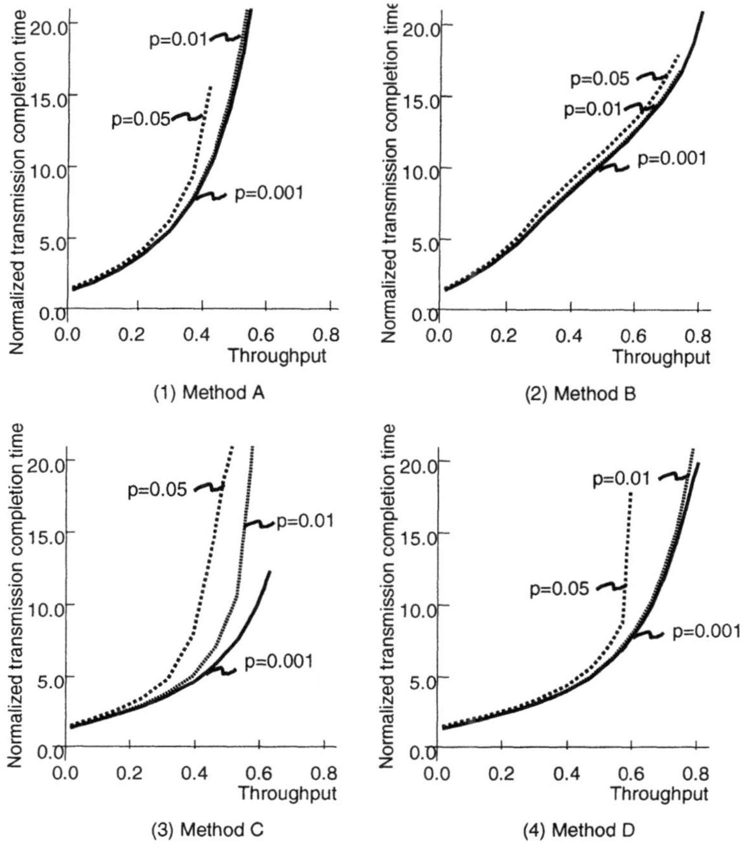

Figure 4. Transmission completion time vs. Throughput

Methods C and D tries to make full use of available capacity. Thus the blocking increases rapidly when the longer propagation delay causes the increase in offered load due to reserving the network resources for a round trip time. However, as using fixed peak rate, method A cannot get all the use out of available capacity. Since the method leaves some available capacity, blocking does not change much corresponding to increase in propagation delay.

Moreover, in method B, since the peak rate is controlled to be smaller than available

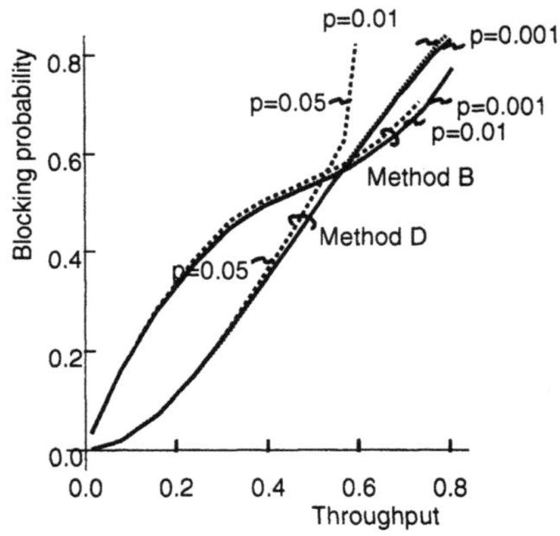

Figure 5. Blocking probability vs. Throughput

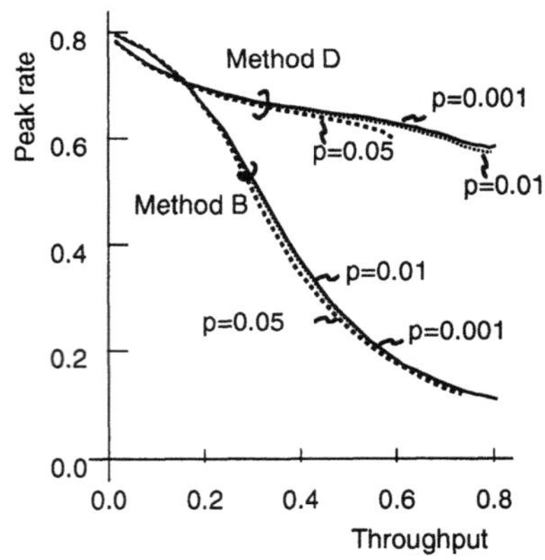

Figure 6. Peak rate vs. Throughput

capacity, which we can see from the comparison with method D, which sets peak rate to the available capacity on the route, in Fig 6. Thus the available capacity is left using method B, so the increase in blocking probability due to propagation delay is small, too.

As the performance degradation due to increase in propagation delay on method C and D is significant, we should be careful to apply these methods in WAN environments.

3.3.2. Comparison among the peak rate control methods

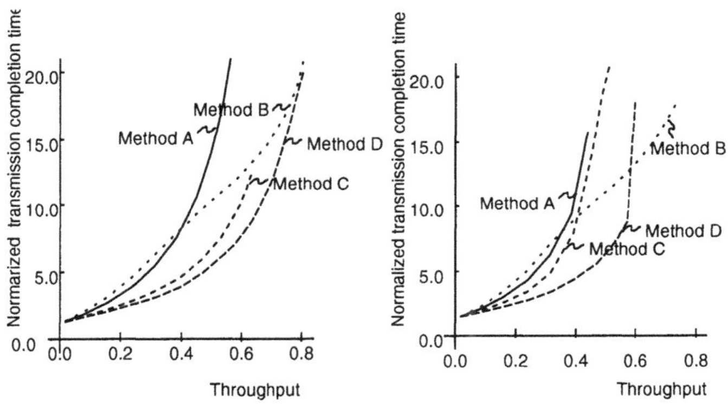

(1) Propagation delay=0.001 (2) Propagation delay=0.05

Figure 7. The comparison of transmission completion time among reservation methods

Figure 7 shows the comparison among peak rate control methods. As shown in Fig. 7, Method D shows the smallest transmission completion time in wide range of the throughput value. That is because of high utilization of network resources owing to end-to-end information.

On the other hand, for propagation delay 0.05, method B improves the maximum throughput. The reason is as follows. As shown in Fig. 6, method D can make good use of the capacity with a large peak rate allocated, while method B has a smaller peak rate allocated, compared with actual available capacity. However, as shown in Fig. 5, the blocking probability for method D rapidly increased at throughput of 0.6, for long propagation delay. Method D, which fully utilizes network resources, improves performance in light load conditions. On the contrary, in heavy load condition, if the network resource are fully utilized, the blocking probability will rapidly increase as will transmission completion time.

Therefore, we should apply method D if we require good transmission completion time performance, such as a few times multiplied by service time, particularly under relatively short propagation delay conditions. However, if we require wide range operation under long propagation delay conditions and/or heavy load conditions, we should apply method B to achieve good performance.

4. Conclusion

This paper has described evaluations on FRPs which provide an efficient transmission for bursty traffics such as file transfers in ATM networks. We evaluated two types of FRPs and three adaptive peak rate controls in short and long propagation delay environments. The two types of FRPs are confirmed and unconfirmed type primary FRP, without peak rate controls. In addition, we compared the primary FRP with three adaptive peak rate controls, i.e., controls using ACK/NACK, using source information and using end-to-end information.

The evaluation results showed that the difference in transmission completion time between confirmed and unconfirmed type FRP is small even in long propagation delay conditions. Thus the confirmed type FRP, which is easy to be implemented, is favorable to be used.

We see that the transmission completion times for FRP methods using source or end-to-end information deteriorate significantly for long propagation delay while these for a primary method or a method using ACK/NACK show small change according to increase in the propagation delay.

The results also showed that FRP with peak rate control using end-to-end information improves performance for relatively small networks or large-size bursts transmission under light load conditions, because of its full utilization of network resources using network information. However, FRP with the simple peak rate control using ACK/NACK shows larger maximum throughput than the method using end-to-end information, specifically under longer propagation (WANs) and/or heavy load conditions. It indicates that we should apply an appropriate FRP method with a peak rate control according to the condition of offered traffic, network and file size and performance required.

ACKNOWLEDGMENT

The authors would like to thank Professor Domenico Ferrari, Dr. Masahiro Yamamoto, Mr. Takao Takeuchi and Mr. Hiroyuki Okazaki for giving the authors the chance to work on this research. Thanks are also to Tenet members in U.C. Berkeley and ICSI for the discussion.

REFERENCES

1. C. Ikeda and H. Suzuki, "Adaptive Congestion Control Schemes for ATMLANs," submitted to INFOCOM'94.
2. P. Boyer,"A congestion control for ATM," 7th ITC seminar Oct. 1990.
3. C. Ikeda and H. Suzuki,"Comparison of Fast Bandwidth Reservation Schemes in ATM Networks," 1992 IEICE Fall Conference, Sep. 1992 in Japanese.
4. H. Suzuki and C. Ikeda, "Performance Comparison of Fast Bandwidth Reservation Schemes in ATM Networks," 1992 IEICE Fall Conference, Sep. 1992 in Japanese.
5. P. Boyer and D.. Tranchier,"A reservation principle with applications to the ATM traffic control," Computer Network and ISDN systems 24, pp. 321-334, 1992.
6. N.F. Maxemchuk, "Dispersity routing in high-speed networks," The 2nd annual workshop on very high speed networks, Mar. 1991.

Broadband Communications, II (C-24)
S. Tohmé and A. Casaca (Editors)
Elsevier Science B.V. (North-Holland)

Bandwidth Allocation By Pricing In ATM Networks

John Murphy[a] and Liam Murphy[b]

[a]Electronics Engineering Department
Dublin City University
Ireland

[b]EECS Department
University of California, Berkeley
USA

Admission control and bandwidth allocation are important issues in telecommunications networks, especially when there are random fluctuating demands for service and variations in the service rates. In the emerging broadband communications environment these services are likely to be offered via an ATM network. In order to make ATM future safe, methods for controlling the network should not be based on the properties of present services. We propose one bandwidth allocation method which has this property. Our proposed approach is based on pricing bandwidth to reflect network utilization, with users competing for resources according to their individual bandwidth valuations. The prices may be components of an actual tariff or they may be used as control signals, as in a private network. Simulation results show the improvement possible with our scheme versus a leaky bucket method in terms of cell loss probability, and confirm that a small queue with pricing can be efficient to multiplex heterogeneous sources.
Keyword Codes: C.2.1; C.2.4
Keywords:Network Architecture and Design; Distributed Systems

1. INTRODUCTION

Asynchronous Transfer Mode (ATM) is a form of fast packet switching which uses short fixed–length packets [1]. ATM has been adopted by CCITT as the transfer mode for the Broadband Integrated Services Digital Network (BISDN), a service–independent network capable of supporting all the communication services that users now require or may require in the future. ATM is also emerging as a local area networking technology, as it provides flexible bandwidth–on–demand and internetworking capabilities for conventional data communications. Although ATM was originally developed with high bandwidth optical fiber as the intended transmission medium, there is increasing interest in using ATM for lower bandwidth non–fiber environments such as satellite or wireless networks [2], [3]. If ATM can in fact transport any service – irrespective of its bit rate, bursty nature or quality requirements – it offers the possibility of a unifying communication paradigm, allowing not simply integrated applications but also integrated operating environments and administrative domains [4].

ATM is connection–oriented and provides no link–by–link error protection or flow control within a connection, called a virtual circuit. The fixed size and reduced header functionality of ATM cells is intended to facilitate the very high–speed switching necessary in networks with link speeds of 150–600 Mbps. The short cell length (48 byte information field + 5 byte header) is intended to keep the delays for real–time services such as voice within acceptable limits. However ATM is a relatively new concept and its implementation is still a very dynamic area, with a wide range of proposed solutions and little practical experience with actual users. Whether ATM can live up to its promise as a universal transfer mode is unknown and many issues have yet to be decided [5].

CALL ADMISSION CONTROL

One of the fundamental issues which has not yet been satisfactorily resolved is the mechanism by which user traffic will be accepted into the network, usually referred to as Call Admission Control (CAC). When a new connection is requested by a user, the network must decide whether or not to accept the call; and if so, how to route it through the network and what resources to reserve for its virtual circuit. Packet–switched networks use higher–layer protocols to guarantee acceptable packet delivery but these are not expected to scale well to broadband speeds. In circuit–switched networks (such as most telephone networks) the CAC mechanism results in call blocking when the bandwidth of a requested connection exceeds the available bandwidth [6]. But in an integrated–services network the traffic source may be *bursty*, so the required bandwidth of its virtual circuit varies with time during the call. The nature of this time–varying behavior and the mean bandwidth requirement vary widely among different sources. Therefore it is difficult to characterize the bandwidth of a requested connection. This difficulty has led to proposals to reserve the peak bandwidth of the call (deterministic multiplexing), as required for constant bit rate (CBR) sources. However the gain in efficiency possible by taking advantage of the statistical nature of variable bit rate (VBR) sources has led to many schemes for statistical multiplexing. Such schemes assign less than the peak bandwidth required, and therefore may introduce cell loss and/or delay. The extent to which these service degradations occur is measured by the quality of service (QOS) offered to the call.

The aim of a preventive CAC scheme is to balance the QOS offered to admitted calls against network utilization by limiting the number of calls using the network. Many schemes described in the literature decide whether or not to accept a call based on knowledge of the *call behavior*, the user's *quality requirements*, and the *current state* of the network. Ideally a user requesting a call would give a complete statistical description of the call, but in practice only a limited indication of expected call behavior is feasible [7]. Call behavior is described by a set of parameters called traffic descriptors, such as mean bit rate, peak bit rate, maximum burst length, probability of cell arrival in a fixed interval, and so on. User quality requirements are usually expressed in terms of acceptable cell loss, delay and jitter. Based on these requirements, traffic sources are divided into classes and each class is provided with a different QOS guarantee, eg. [8]. The current state of the network can be determined by monitoring the utilization of network resources and/or by characterizing the behavior of calls already admitted. For example, traffic models based on fluid flow approximations have been used in analyzing network bandwidth and buffer utilization [9]. Based on the above knowledge, CAC schemes have been developed in which each source is assigned an effective bandwidth [10] in order to meet its QOS while still permitting a statistical multiplexing gain.

There are many problems with these approaches. Users may not know enough about the statistics of their calls to provide accurate traffic descriptors to the network. This is particularly true for certain data transfers with long holding times. For example, in browsing a remote image database, it may not be possible to give a reasonable estimate of the mean bit rate except over a long period of time. Even if users know their traffic descriptors they may be unwilling to reveal them, or may try to misrepresent them. This implies the need for some form of traffic policing and enforcement at the network access points, in order to ensure that each source is complying with what it declared at call setup. Assuming accurate traffic descriptors, a complicated analysis is required to determine whether sufficient resources are available to support the requested call's QOS while not degrading the QOS of admitted calls. Current computational methods cannot do this analysis on the short timescale in which CAC decisions must be made [7]. Consequently either simplifying assumptions about call and resource interactions are introduced, or calls are grouped into classes whose members are assumed to have similar characteristics. Making simplifying assumptions can lead to an over–controlled network, or produce unreliable results. For example, recent studies have shown that LAN and VBR video traffic is self–similar [11], which is very different to the behavior predicted by Poisson–based or fluid flow models. There is a tradeoff between simplicity, which implies a small number of traffic classes, and the large number of classes needed to cover a wide range of different traffics and services. There is likely to be a wide variation in user requirements even within the same traffic class, due to individual preferences and internetworking considerations. Even if accurate traffic and network models were available, technological advances may quickly render their assumptions invalid. For example, the bandwidth required for video or image transfer applications may continue to decrease due to improvements in compression / decompression algorithms and technology [12]. In addition, it is impossible to predict the characteristics of all future applications, or the application mix when users become familiar with new network capabilities. Hence **it may be unwise to base CAC and bandwidth allocation schemes on the requirements and properties of current services.**

The area of CAC and bandwidth allocation is the subject of intensive study and work is in progress addressing the above problems. Reactive control and feedback schemes are also under investigation (eg. [13], [14]) and may be useful in certain network environments or on longer timescales, eg. call level rather than burst or cell level. Experience with real users will show which schemes are feasible in an actual network as opposed to a research or test environment. The problem is complex and is likely to require a multilevel solution approach [15]. It is fair to say that how bandwidth will be allocated in ATM networks is still an open question.

INTEGRATED–SERVICES NETWORKS

The exact form of integrated–services networks of the future is also unclear, but based on current trends in communications and computing it is possible to predict some of its features with reasonable confidence. Multimedia services offering storage and transfer of voice, video, images and data will be supported, allowing remote conferencing and collaboration and replacing text processing with multimedia 'document' processing. Networks will be interconnected and will allow users and organizations to set up virtual private networks embedded in the physical network [16], [17]. Intelligent network interfaces will be needed at internetwork boundaries and at user access points to hide the interface details and present the image of a single network to the users. Users will demand a wide spectrum of application requirements as they become familiar with the network. Some may continue to opt for network–defined services or choose from a library of predetermined services, but others may want to customize their calls and vary call parameters dynamically. Computational power of various types may be offered for rent by the network or third–party providers, enabling users from different fields to access the type of processing appropriate to their needs.

We take the view that current trends and future considerations in the communications and computing areas point to the need for a change in the traditional user–network relationship. The traditional model is centralized with passive users : the network provides well–defined services and users choose from this limited range. User feedback is long–term and inferred from the aggregate demands for services. This is appropriate for large–scale provision of a single service, as in the phone network, but may be unsuitable for the kind of integrated multi–service network discussed above. The relationship implied by many of the proposed CAC schemes is more of a 'contract' : users describe their traffic and make quality demands, and the network provides a stated level of service while enforcing the user commitments. Some problems with this model were outlined above, although it is a step in the right direction. Taking this process further, our view is that

- the network should be a provider of **resources** rather than services;
- the responsibility for packaging these resources into services should lie with the **users** (or in real–time their interface equipment);
- the network's primary function should be to **coordinate requests** for its resources. The goal of this coordination could be to optimize some measure of network performance, or to maximize a suitably–defined global user satisfaction, or to ensure some degree of fairness, or some other objective.

Thus we view the network simply as a high–speed cell relay network in which the central issue is transporting cells rather than implementing services defined at higher levels [4]. Our view is consistent with that expressed in [18] that an ATM virtual circuit is similar to a virtual wire, since the service delivered by the virtual circuit is similar to that of the physical layer in the OSI reference model. Users may actually be working with higher layer protocols such as TCP/IP running on top of an ATM network, which also argues for keeping the ATM network as simple as possible. One proposal along these lines is to move call admission and bandwidth allocation decisions to the terminal equipment at the network access points, and somehow ensure that the combined user rates do not exceed the network capacities [19]. In our view, users take responsibility for requesting sufficient network resources to meet their QOS requirements. At call setup this involves specifying some parameters of the requested connection, either manually or in a menu-driven environment. If users want to adjust connection parameters during a call then these parameter specifications may be automated. This kind of in-call negotiation is desirable for bursty calls where the actual resource usage is of interest [20]. One traffic management scheme for burst-level resource reservation is described in [7], and a form of in-call negotiation is now commercially available in some network interface equipment. Even for static connection requests, the call setup process may involve a negotiation between the user and the network in which the user modifies their request to conform to the current level of resource utilization. For example [21] outlines a real-time channel setup procedure in which detailed information is sent to the user if a connection request is refused, allowing the user to take this feedback into account in a revised request. Some work is underway on modifying user traffic inputs via pricing in the Internet [22] and in integrated-services networks [23], [24]. A contract-based CAC scheme is proposed in [25] in which the charge to users is related to how accurately they declared their traffic, rewarding users who provide better information on their call characteristics.

This view of the network may require a new approach to the CAC and bandwidth allocation problems. We have developed one solution approach which is compatible with this new user-network relationship. Our approach is based on users regulating their traffic inputs in response to bandwidth prices calculated by the network. The network and users are considered to form a system and we apply economic principles of pricing for resource allocation to this system. Another traffic management scheme similar in spirit to our approach is discussed in [26].

PRICING BANDWIDTH

The system made up of the users together with the network has various resources that can be used to meet service demands. However in all realistic systems these resources are limited and some method of allocating them is needed when total demand is greater than the resource limit. In this paper the resources are the capacities of the ATM connections. The bandwidth allocated to a user is considered to be a commodity which is sold by the network to the user [27], [28]. We view the users as placing a benefit, or willingness-to-pay, on the bandwidth they are allocated. Given a price per unit of bandwidth, a user's benefit function completely determines that user's traffic input. Users are assumed to act in their own best interests and to be capable of responding to changes in the price for bandwidth.

In our formulation, network constraints such as virtual path capacities [29] are translated into cost functions on the bandwidth demands. This is because one user's consumption of bandwidth gives rise to a social 'cost' (in terms of longer delays, less available bandwidth, etc) which is borne by all users. The network operator sets the prices so that the marginal benefit the users place on their bandwidths is equal to the marginal cost of handling that traffic in the network[1]. The network operator dynamically adjusts the prices based on monitored network conditions. We assume that when the price of bandwidth increases, users reduce their demands. There is no need for enforcement of call parameters in our scheme because users do not declare anything about their traffic. The mathematical formulation of the scheme is in an Appendix.

A typical configuration that we envisage is shown in Figure 1. The user programs the terminal equipment with their benefit function. The bandwidth in cells per second allocated to the user by the network depends on this benefit function and the price that is sent from the network to the terminal equipment.

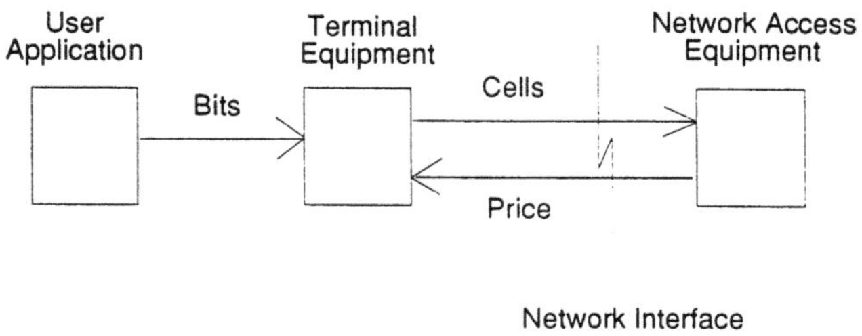

Figure 1. Proposed configuration for access

Our approach is intended to be applicable to general ATM networks and to the full range of user and service types, such as voice, video, real–time data and non–time–sensitive data. However a special case that is particularly suitable is that of a Virtual Private Network (VPN) in which one operator controls the network and all applications running on it. The relative importance of various applications could then be varied over time and from node to node. For example, the operator may place a high priority on operational data transfer and low priority on other data transfers; or interactive video and voice might get higher priority than other traffic types.

However, assigning dynamic priorities is difficult. If the real–time applications such as voice and video are given priority to ensure timely delivery, then data traffic may suffer higher loss though it may not be able to tolerate cell loss as well as voice. On the other hand, if priority is given to data and large buffering is employed, then real–time applications may suffer large variable delays.

What is needed is a dynamic adaptive intertemporal priority scheme. The priorities should change to track changes in the network state or in the application requirements over multiple time periods. Rather than have a complicated priority scheme, a pricing scheme like the one proposed in this paper could be used. The operator would set the benefit functions for the different applications, and could also set different benefit functions for

[1]We address only variable costs corresponding to network constraints. The recovery of fixed network costs could be done by an access charge or some other method.

applications of the same type. Each application would then input traffic according to its assigned benefit function and the current state of the network, as reflected in the prices. We have simulated a VPN and implemented a simplified pricing scheme to investigate this possibility. In the rest of the paper we describe our simulation model and the results we have obtained with it so far.

2. SIMULATION MODEL

To investigate our proposed algorithm and its efficiency, we built a model of a single link which represents a single virtual path. Typically there could be diverse services using this path and so a number of models of sources were needed. We implemented three basic source models, representing some of the services that will be demanded in an ATM network : voice, video and data. For simulation purposes we investigated a link with a bit rate of 2.048 Mbps, although our results are expected to scale to higher link speeds.

The voice model that we use is a standard model with two states, speaking and silence [30], [31]. During speaking periods, cells are produced by the source at the bit rate of the voice. During silent periods no cells are produced. The mean duration of the speaking (or ON state) is 0.352 seconds. If we count the number of cells produced in this period it is found that this number is geometrically distributed. In our case this geometric distribution will have a mean of 59 cells. The silence is made up of two parts, one from the silences when listening and the other from the silences between words or sentences. Therefore to properly represent the silence distribution there should be two distributions. However it has been found that a hyper–exponential distribution will accurately model the silences, with the lower exponential for the inter–word silences between and the upper exponential for the listening silences. The average silence is 0.65 seconds long. The probability of being in the lower exponential (rather than the upper exponential) is 0.8746. From these parameters it is possible to fully characterize the hyper–exponential silence distribution. The two means of the silence distribution are 0.372 seconds and 2.592 seconds. The standard deviation of the distribution is then 1.22525. The resulting model assumes an input of 64 kbps voice which if placed directly in the ATM layer would require a bit rate of 71 kbps due to the 5 byte overhead for every 48 bytes of data. However as only the speaking portion is sent the required bit rate is 24.8 kbps. This gives a peak–to–mean ratio of 2.8. The model outlined here assumes 64 kbps voice, but this can be dynamically changed during the simulation if desired.

One of the major intended uses of ATM networks is for video applications. The video source model that we use here is a standard one for video conferencing [32]. The codec is a one layer model with a compressed bit rate of 222 kbps, which adheres to the H.261 standard for video telephony. The model used is a discrete markov model with nine states. These states capture the transitions from one average bit rate to another. The maximum bit rate delivered by the codec is 750 kbps. Within a state the bit rate is assumed to be uniformly distributed. The transition probability was engineered so that the matrix only had diagonal and one off diagonal entries. This allows a birth–death queueing structure to be formed. Another possible model is based on two layer codecs. In this model the base layer is approximately CBR and if this bit rate can be reserved the base layer delivers a minimum level of quality. The enhancement layer is VBR with a peak–to–mean ratio of maybe 40, and is used to enhance the picture. The advantage of using a two layer codec is that the enhancement layer data can be treated as best effort traffic. However it is unlikely that the same compression can be achieved with two layer codecs as with one layer codecs.

A data source is one of the most difficult sources to model as the source type depends on what applications are being run and on what systems. In this paper we model a file transfer application. This captures the bursty nature of data communications as well as its looser delay requirements relative to voice and video. A model was built based on transferring files from one computer to another. An empirical distribution for file size ranges was obtained from actual files stored on one of our computers. In the simulations a range was chosen according to this empirical distribution. and then a file size was chosen from a uniform distribution within this range. The amount of data transfer can be varied depending on how many files are to be transferred. For one file per second the average bit rate is 197 kbps. The peak–to–mean ratio of this source can be high with values up around 1000. The user's quality requirement is that the file will be transfered at a rate determined by the network. A rate is given to the data source for a time interval and the rest of the file is stored awaiting transfer.

Simulation Environment

The network and source models were simulated using SES/workbench [33], a discrete-event simulator that allows hardware and software simulation. The models in this paper were mainly created by use of its graphical user interface. SES/workbench compiles the graphical code to C and creates an executable. The simulation execution platform was a cluster of Sparc-10 workstations. The run time for the simulations was 1000 seconds, or 5 million cells. Sharing the 2.048 Mbps link were 20 voice sources, 2 video sources and a data source. A schematic of the simulation setup is shown in figure 2 below, where the data and video sources are shown with leaky bucket implementations.

The simulation model is made up of submodules. each of which performs a well defined function. The sources generate cells which are input to a network interface submodule. The network interface takes the source bit stream and forms ATM cells. For voice cells 5 bytes of header are added to the 48 byte information load. For data and video cells we implemented AAL 3/4 which uses 44 bytes of information with a 4 byte AAL header and trailer in addition to the 5 byte ATM header. The cell stream from an interface is then input to the ATM switch buffer submodule. This submodule smooths the arrival of cells to the ATM network and so takes care of cell scale congestion. This buffer is the limited resource critical to the operation of the model. In a typical implementation the buffer is managed by an input control mechanism, such as a leaky bucket scheme.

The leaky bucket is a method for policing the mean rate of a source and still allowing bursts to occur. The mean rate is the rate at which tokens are given to allow cells past the network interface and into the network. If there were no storage of tokens then this would be a pure mean rate policing function. If the source can store tokens, it could burst all those tokens in one go. If a source has no remaining token allowance then rather than dropping that cell usually it is merely marked. If there is congestion in the network and cells have to be discarded the marked ones are disacarded first. The leaky bucket does

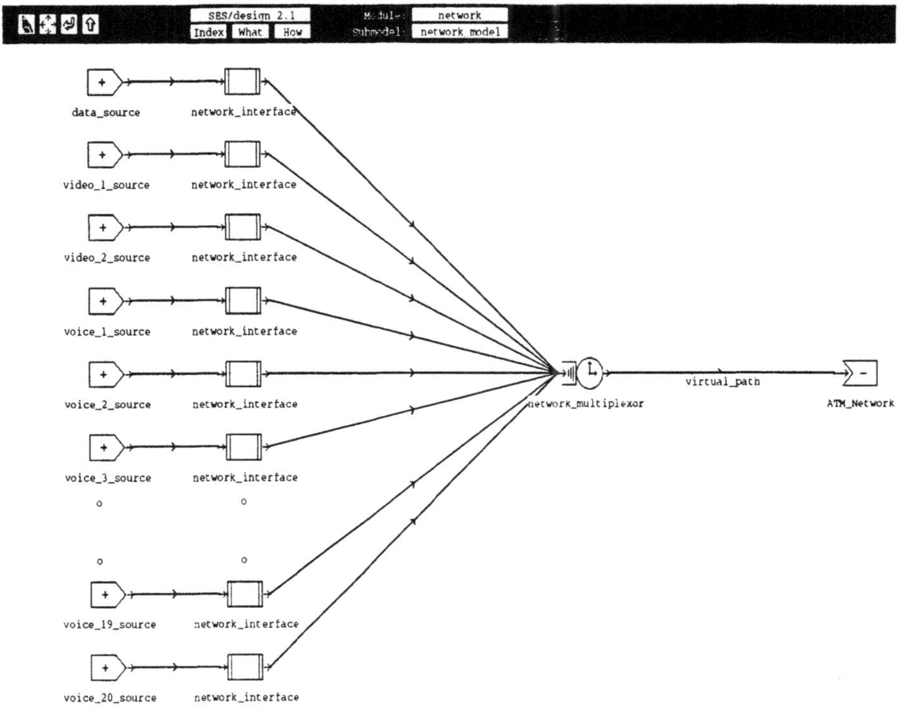

Figure 2. Simulation Schematic

not lower the cell loss in a buffer; it merely selects which ones to lose. Our pricing scheme actually lowers the probability of cell loss by smoothing the traffic adaptively based on the buffer occupancy to avoid congestion.

In our proposed scheme a price is generated by the network based on the present state of the network buffer, and the sources adapt their demands based on this. The leaky bucket can also be implemented on top of our scheme so that if there is cell loss we can discard the marked ones first. The model takes in cells over an interval of about 50 msec and gives a price to all the sources sharing the link. The price reflects the congestion in the buffer (if any) and hence on the virtual path. The higher the buffer occupancy, the higher the price sent back to the sources. This price is received by the network interfaces which use it to calculate how many cells they *should* input in the next interval. For example, a data source may calculate that at the given price it should input 100 cells to the network, but if no cells are actually generated by the data source in that time interval then obviously there is no input. The outline of the distributed pricing algorithm we used is contained in the Appendix.

In an interval of about 50 msec the link can serve about 250 cells and the buffer size we choose is 250 cells. This means that when the buffer is full there should be no more than 250 input cells over the next interval. The input sources are shaped as much as possible by the time the cells are formed at the network interfaces, so the cells arrive approximately equally spaced in time. It is standard to guarantee real–time services that if their cells get through then the maximum access delay they experience is the buffer queueing delay which in this case is small (50 msec).

3. RESULTS

We investigated two cases, first where the leaky bucket approach has cell loss but our pricing scheme does not, and secondly with a link utilization close to 100 percent to see how both schemes perform in heavy traffic.

In practice voice and video calls may be inflexible with respect to bandwidth, and even if they are flexible they do not permit a huge statistical multiplexing gain (except for the enhancement layer of a two layer codec) compared to data. In our simulations we assumed that the voice and video have inflexible bandwidth demands. We found the buffer never gets close to capacity for the first case, so the prices are low and even if voice and video were price–sensitive it would have a negligble effect on the performance of the system. For this case we used 20 voice sources which have a combined mean of 492 kbps and a potential maximum of 1413 kbps. The video has two channels, both one layer codecs, with a combined mean of 507 kbps and a potential maximum of 1773 kbps. Even without a data source there is clearly a nonzero probability of cell loss. The data source has a mean of 237 kbps and a potential maximum equal to bit rate of the link, 2048 kbps.

When a leaky bucket scheme alone was implemented the results in Figure 3 was obtained for the sampled buffer occupancy.

From an analysis of the results, there were 32.268 marked cells and 191,044 unmarked cells lost out of a total of 2,914,191 generated over the 1000 seconds. This give a loss probability for the marked cells of 1.1 percent and for the unmarked cells of 6.6 percent. In Figure 4 we show the plot of buffer occupancy over the interval when our pricing scheme was used.

The buffer is not as occupied as in Figure 3. This is apparent when we compare the loss probabilities. The pricing scheme produced no cell loss, either of marked or unmarked cells[2]. There is a small decrease in the quality of service provided to the data source to gain this improvement. Figure 5 shows the sampled number of waiting data cells of data over the interval when the leaky bucket is used.

By comparing Figure 5 to Figure 6 below for the pricing scheme, we see that the mean waiting time for the data source has increased.

Next we investigated increasing the data source bandwidth demand to see the effect that would have on the loss and delays. In the first instance the data source input was increased by a factor of 3 and then applied to the leaky bucket. The resulting graph of buffer occupancy is shown in Figure 7.

The loss probabilities for this case were calculated to be 1.7 percent for the marked cells and 16.1 percent for the unmarked cells. The results for our pricing scheme with a

[2]It should be noted that there could be cell loss if a longer interval was chosen.

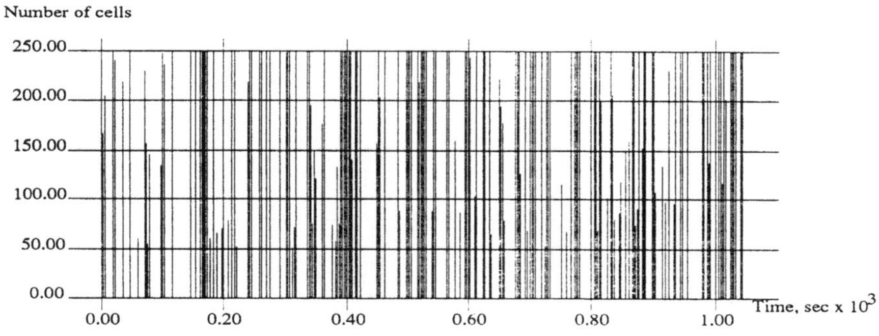

Figure 3. Buffer Occupancy for First Case with Leaky Bucket.

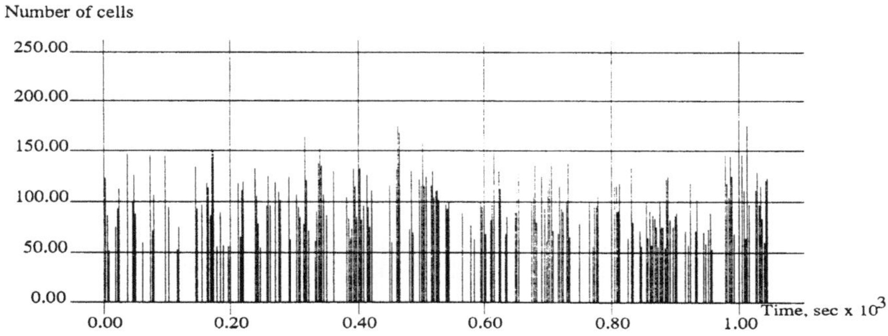

Figure 4. Buffer Occupancy for First Case with Pricing.

loaded link are shown in Figure 8.

Again a much lower buffer occupancy is observed, and the loss probabilities are 0.29 percent for the marked cells and 1.5 percent for the unmarked cells. These figures are almost an order of magnitude lower than for the leaky bucket case.

4. DISCUSSION AND CONCLUSIONS

This paper suggests pricing and user self-regulation as a means of allocating bandwidth in ATM networks. The pricing scheme is independent of the traffic arrival model and does not require the network to police and enforce user commitments in real-time as cells arrive from the sources. The pricing algorithm is distributed and local to the ATM access nodes, suggesting that it may be able to meet the tight time requirements of an admission control

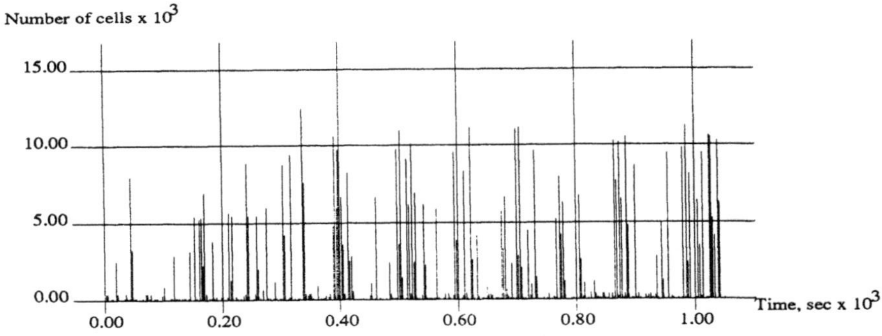

Figure 5. Waiting Times for Data Cells with Leaky Bucket for First Case.

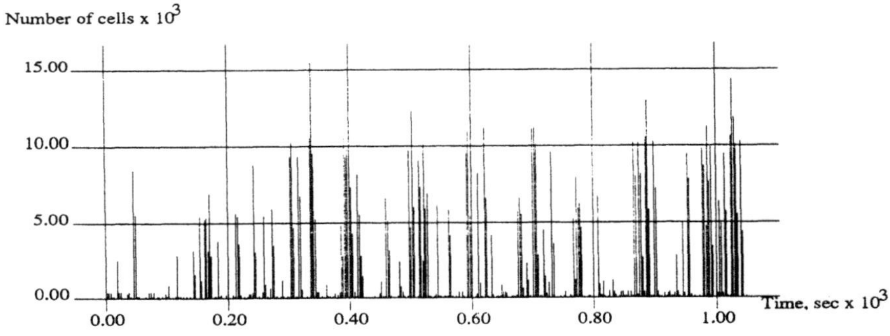

Figure 6. Waiting Times for Data Cells with Pricing for First Case.

scheme. There is no need for the network to define and support traffic classes. The users have more flexible access to network resources than in other CAC schemes, but they take the responsibility of determining how much network resources to acquire in order to meet their particular needs. It should be noted that this view of the user–network relationship should not be judged based on the feasibility or otherwise of our pricing scheme. Other CAC schemes may also be suitable and it remains to be seen whether our scheme is compatible with other approaches, perhaps in a multi–level solution approach. Similarly our pricing scheme offers users the possibility of taking more control over their service than they have under traditional system models, but it does not *require* them to. Users whose applications have inflexible bandwidth demands, or who set a threshold price above which they demand zero bandwidth, can be accomodated in our formulation. In order to pursue an inflexible strategy, users may have to pay premium prices at times of heavy

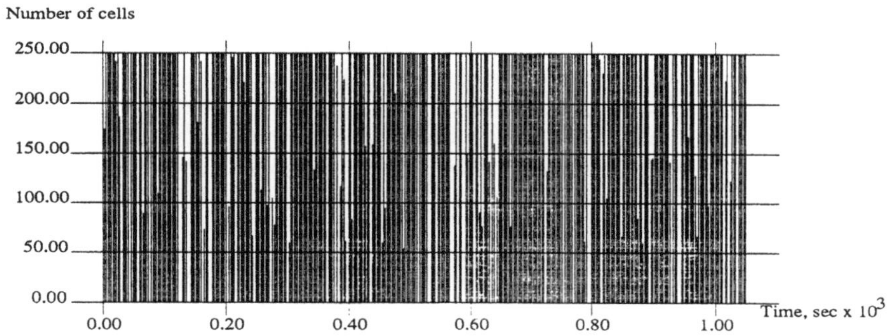

Figure 7. Buffer Occupancy for Loaded Link with Leaky Bucket.

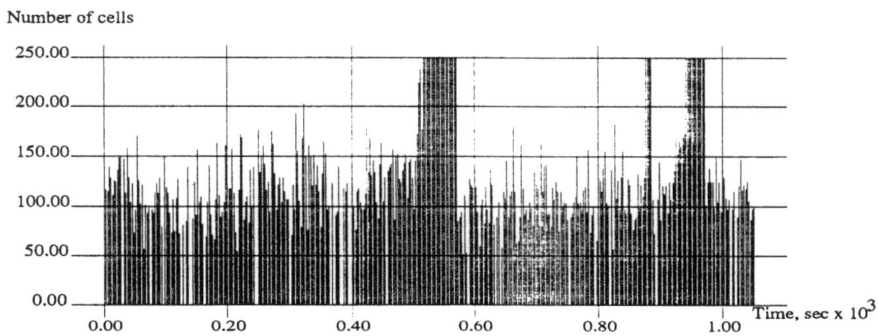

Figure 8. Buffer Occupancy for Loaded Link with Pricing.

demand; but it is our belief that simply being forced to examine the relative importance of their usage will give rise to a variation in user valuations that a scheme like ours can exploit. The success of our approach depends on having a sufficient number and range of flexible applications, although how much is enough is as yet unknown.

In VPNs pricing can be used to choose the desired application mix and vary this mix dynamically, as demonstrated in a simple case by our results. Another use of pricing is in selectively dropping cells during congestion. Suppose that an application uses Ethernet frames and one cell is dropped from a frame. Then the benefit of the rest of the frame should be set to zero if the whole frame has to be retransmitted. Similarly, if there are a number of video calls in progress, it may be better to drop all the cells from one call during congestion rather than to spread out the loss. The feasibility of using our pricing approach as a dynamic adaptive priority scheme in this way is one avenue for further research.

Our pricing scheme also applies to general ATM networks in which the user benefit functions are unknown to the network operator. There are many issues which would have to be resolved in a practical implementation. For example, our assumption here is that the benefit of making a call is associated entirely with the sender. But in practice the benefit is shared, and some pre-call negotiation may be necessary to agree on call parameters acceptable to all parties. If intermediate ATM nodes switch on the VCI as well as on the VPI [1], users at different ATM nodes have to compete for shared resources, which complicates the algorithm and may increase the minimum length of a pricing interval. Whether the prices used in our scheme could be used as part of an ATM network tariff is not clear. However we believe that the paradigm suggested in this paper is an interesting one and may be helpful in solving these difficult network management problems.

Acknowledgements

We are grateful for the help and advice given by the late Ed Posner of Caltech and JPL in the initial stages of this work. We thank Pravin Varaiya, Gustavo de Veciana, Mike Wong, Padmanabhan Srinagesh, Mike Honig and Jerry Teahan for helpful discussions about this work. The first author acknowledges Charles McCorkell of Dublin City University for his continuing support.

APPENDIX

We briefly describe the mathematical formulation of our bandwidth allocation algorithm here. A fuller treatment can be found in [34].

We assume that a virtual path is set up for every source–destination pair, and all virtual circuits using a virtual path are for the same source–destination pair. Thus when a call arrives at the access to the network it is assigned a virtual circuit, which is amalgamated on one virtual path with the virtual circuits for other calls from that source to destination, as shown in Figure 9.

The virtual paths are carried by the physical network trunks, whose capacities are assumed to be fixed. We assume that statistical multiplexing of the sources is done in the sense of allocating a capacity to a virtual path which is less than the sum of the virtual circuit peak bandwidths using this virtual path. The network decides the capacities of the virtual paths and these satisfy the physical trunk capacity constraints. For example, the network could simply do deterministic multiplexing at the virtual path level, as in Figure 9. In this paper the virtual path capacities are assumed to be fixed, although this assumption can be relaxed [34].

The users connected to an ATM switch wish to send traffic to various destinations. Their benefit versus bandwidth curve for a particular call is assumed to be concave increasing in general, as shown in Figure 10. This follows the usual economic assumption that users value the first part of a commodity the most, with diminishing incremental valuation as they get more of the commodity [27],[35].

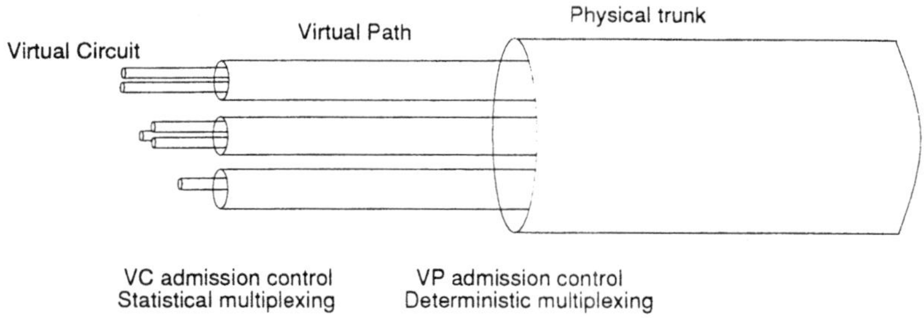

Figure 9. Virtual Path/Virtual Circuit connections

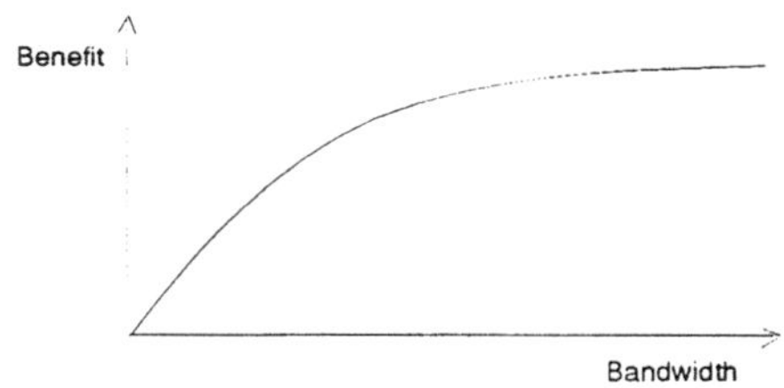

Figure 10. User Benefit as a function of Traffic Rate

Consider the problem faced by a user attached to ATM node i in deciding their bandwidth demand for a particular call to destination ATM node j. Let the virtual path corresponding to this source–destination pair be denoted V_q. The network quotes all users a price per unit of bandwidth on V_q, π_q. Each user then uses their private benefit function ben_{rq} for call r to decide their bandwidth at that price, b_{rq}. This decision process can be formulated mathematically as follows :

$$maximize \quad ben_{rq}(b_{rq}) \; - \; \pi_q \cdot b_{rq}$$

The user solves a maximization problem of this form for each call to destination ATM node j, for each j. The optimality condition for this maximization is [36]

$$\frac{\partial ben_{rq}}{\partial b_{rq}} - \pi_q \; = \; 0 \, , \quad \forall \; V_q \tag{1}$$

Consider now the system made up of all the users and the ATM network, and suppose for the moment that there is a single 'system manager' who knows all the problem data. We will see later on that no such 'system manager' is needed.

The problem facing this (hypothetical) system manager is to choose the bandwidth demands to maximize aggregate benefit, subject to the virtual path capacity constraints.

Here we replace the virtual path capacity constraint by a corresponding buffer capacity constraint, which ensures that the ATM switch buffer occupancy b_q is less than the buffer size B_q but allows the input rates to temporarily exceed the output rate of the virtual path :

$$b_q \leq B_q , \quad \forall\ V_q$$

We drop this constraint but add a barrier function term to the system objective function corresponding to it, $Buffcost_q$. For example.

$$Buffcost_q := \frac{1}{B_q - b_q} - \frac{1}{B_q}$$

The shape of this cost term is shown in Figure 11, and ensures that as the buffer capacity is approached, the 'cost' of assigning bandwidths increases to infinity.

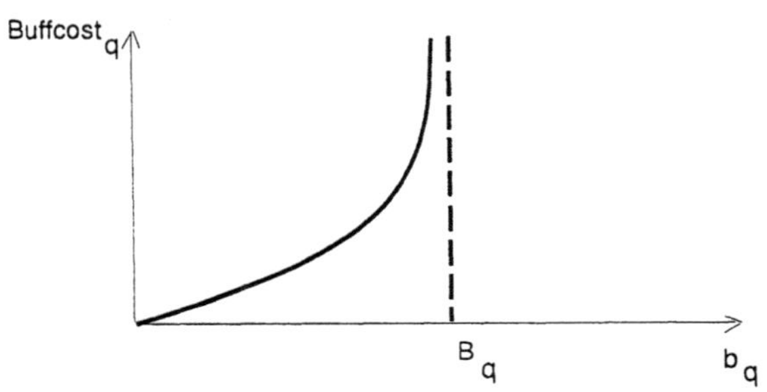

Figure 11. Barrier Function For q^{th} Buffer Capacity Constraint

The system problem can be formulated as follows :

$$maximize \quad \{\sum_{all\ V_q} [\sum_{r} ben_{rq}(b_{rq})] - Buffcost_q\}$$

The optimality conditions for the system problem are [36]

$$\frac{\partial ben_{rq}}{\partial b_{rq}} - \frac{\partial Buffcost_q}{\partial b_q} \cdot \frac{\partial b_q}{\partial b_{rq}} = 0 , \quad \forall\ r, V_q \tag{2}$$

Comparison of equations 1 and 2 shows that by setting

$$\pi_q = \frac{1}{(B_q - b_q)^2} \cdot \frac{\partial b_q}{\partial b_{rq}} \tag{3}$$

individual user benefit maximization (equation 1) coincides with system optimality (equation 2). Therefore if the network operator sets the prices according to equation 3, the users will self-select their optimal bandwidth demands while competing for the virtual path capacity. There is no need to have a 'system manager' who knows all the user benefit functions as well as the network data. The network operator can induce system-optimal user behavior by setting the prices as in equation 3 without any knowledge of the user benefit functions.

Distributed Pricing Algorithm

Let time be divided into successive pricing intervals of length T. Within each interval the benefit functions of all the users are assumed to be fixed; however, these benefit functions may change from one interval to the next. The price per unit of bandwidth on virtual path V_q, π_q, is announced by the network at the start of each pricing interval, and remains fixed for the duration of the interval. Hence each user solves a maximization problem to decide their optimal average bandwidth b_{rq}, and transmits exactly $b_{rq} \cdot T$ cells during the interval. At each ATM node, the resulting buffer occupancies are measured, enabling the network operator to calculate the marginal cost of each buffer with respect to its total traffic input rate, as given by the right-hand side of equation 3.

If these marginal costs are equal to the prices that were announced, then the system optimality conditions in equation 2 are satisfied. If not, then the prices must be updated to correct for the difference. In this simulation study we used an update rule based on the method of False Position.

We summarize this negotiation between the network and the users by the following Bandwidth Allocation Algorithm :

Step 0. network operator chooses initial values for the prices $\{\pi_q\}$

Step 1. network operator announces these prices to the users at the start of the current pricing interval

Step 2. users respond by choosing their $\{b_{rq}\}$ according to equation 1

Step 3. network operator calculates the marginal costs of the buffers with respect to the bandwidths, as on the right-hand side of equation 3

Step 4. network operator adjusts the prices $\{\pi_q\}$ to reduce the difference between the terms in equation 3$\Rightarrow$go to Step 1

Note that the Bandwidth Allocation Algorithm is purely local to an ATM node and so can be done at the ATM node to which the users are attached. This is because we have assumed above that the virtual path capacities are fixed.

REFERENCES

1. M. de Prycker, *Asynchronous Transfer Mode : Solution for Broadband ISDN*, Ellis Horwood, 1991.
2. P. Komisarczuk, M. H. Hadjitheodosiou, F. Coakley and C. Smythe, 'B-ISDN Implementation Via Satellite : The Catalyst Project', *10th UK Teletraffic Symp.*, BT Laboratories, Martlesham Heath, April 1993.
3. J. Murphy, E. Chow and R. Markley, 'ATM Service-Based Selective Retransmission Over DSN Satellite Links', to be presented at the 15th AIAA International Communication Satellite Systems Conference, San Diego, USA, March 1994.

4. I. M. Leslie, D. R. McAuley and D. L. Tennenhouse. 'ATM Everywhere ?', *IEEE Network*, Vol. 7, No. 2, pp 40-46, March 1993.
5. I. Nikolaidis and R. O. Onvural, 'A Bibliography on Performance Issues in ATM Networks' *Computer Communication Review*, Vol. 22, No. 5, pp 8-23, Oct. 1992.
6. V. E. Benes, 'Programming and control problems arising from optimal routing in telephone networks'. *Bell Syst. Tech. J.*, Vol. 45, pp 1373-1438. 1966.
7. J. S. Turner, 'Managing Bandwidth in ATM Networks with Bursty Traffic', *IEEE Network*, Vol. 6, No. 5, pp 50-58, Sept. 1992.
8. L. Trajkovic and S. J. Golestani, 'Congestion Control for Multimedia Services'. *IEEE Network*, Vol. 6, No. 5, pp 20-26, Sept. 1992.
9. S. Low and P. Varaiya, 'A Simple Theory of Traffic and Resource Allocation in ATM', *Proc. of Globecom '91*, pp 1633-1637, Dec. 1991.
10. F. P. Kelly, 'Effective Bandwidths at Multi-Class Queues', *Queueing Systems Theory and Applications*, Vol. 9, No. 1-2, pp 5-15, Oct. 1991.
11. W. E. Leland, M. S. Taqqu, W. Willinger and D. V. Wilson, 'On the Self-Similar Nature of Ethernet Traffic', *Computer Communication Review*, Vol. 23, No. 4, pp 183-193, Oct. 1993.
12. G. J. Armitage and K. M. Adams, 'Using the Common LAN to Introduce ATM Connectivity', *IEEE 18th Conf. on Local Computer Networks*, Sept. 1993.
13. H. Kanakia, P. P. Mishra and A. Reibman, 'An Adaptive Congestion Control Scheme for Real-Time Packet Video Transport', *Computer Communication Review*, Vol. 23, No. 4, pp 20-31, Oct. 1993.
14. H. Gilbert, O. Aboul-Magd and V. Phung, 'Developing a Cohesive Traffic Management Strategy for ATM Networks', *IEEE Commun. Mag.*, Vol. 29, No. 10, pp 36-45, Oct. 1991.
15. S. Yazid and H. T. Mouftah, 'Congestion Control Methods for BISDN', *IEEE Commun. Mag.*, Vol. 30, No. 7, pp 42-47, July 1992.
16. R. Ahmad, F. Halsall, K. Lam and J. Zhang, 'Performance Evaluation of Private ATM Networks', *10th UK Teletraffic Symp.*, BT Laboratories, Martlesham Heath, April 1993.
17. J. M. Schneider, T. Preuss and P. S. Nielsen, 'Management of Virtual Private Networks for Integrated Broadband Communication', *Computer Communication Review*, Vol. 23, No. 4, pp 224-237, Oct. 1993.
18. M. de Prycker, R. Peschi and T. van Landegem, 'B-ISDN and the OSI Protocol Reference Model', *IEEE Network*, Vol. 7, No. 2, pp 10-18, March 1993.
19. S. M. Walters, 'A New Direction for Broadband ISDN', *IEEE Commun. Mag.*, Vol. 29, No. 9, pp 39-42, Sept. 1991.
20. J. W. Roberts, 'Variable-Bit-Rate Traffic Control in B-ISDN', *IEEE Commun. Mag.*, Vol. 29, No. 9, pp 50-56, Sept. 1991.
21. D. Ferrari, J. Ramaekers and G. Ventre, 'Client-Network Interactions in Quality of Service Communication Environments', *IFIP Transactions C (Communication Systems)*, Vol. C-14, pp 221-234, 1993.
22. J. K. MacKie-Mason and H. R. Varian, 'Pricing the Internet', presented at *Public Access to the Internet*, JFK School of Government, Harvard University, May 26-27, 1993.
23. S. Shenker, 'Service Models and Pricing Policies for an Integrated Services Internet', presented at *Public Access to the Internet*, JFK School of Government, Harvard University, May 26-27, 1993.
24. C. Parris, S. Keshav and D. Ferrari, 'A Framework for the Study of Pricing in Integrated Networks', Technical Report TR-92-016, International Computer Science Institute, Berkeley, CA, March 1992.
25. F. P. Kelly, 'Tariffs, Policing and Admission Control for Multiservice Networks'. *10th UK Teletraffic Symp.*, BT Laboratories, Martlesham Heath, April 1993.
26. S. H. Low and P. P. Varaiya, 'A New Approach to Service Provisioning in ATM Networks', to appear in *IEEE Trans. on Networking*.
27. A. de Fontenay, M.H. Shugard and D. S. Sibley, *Telecommunications Demand Modeling : An Integrated View*, North-Holland, 1990.
28. J. R. Meyer, R. W. Wilson, M. A. Baughcum, E. Burton and L. Caouette, *The Economics of Competition in the Telecommunications Industry*, Oelgeschlager, Gunn and Hain, 1980.
29. J. Burgin and D. Dorman, 'Broadband ISDN Resource Management : The Role Of Virtual Paths', *IEEE Commun. Mag.*, Vol. 29, No. 9, pp 44-48, Sept. 1991.
30. J. G. Gruber, 'A Comparison of Measured and Calculated Speech Temporal Parameters Relevant to

Speech Activity Detection', *IEEE Trans. Communs.*, Vol. COM-30, No. 4, pp 728-738, 1982.
31. P. T. Brady, 'A Statistical Analysis of On-Off Patterns in 16 Conversations', *Bell Syst. Tech. J.*, pp 73-91, 1968.
32. 'Redundancy Reduction Techniques for Coding of Video Signals in Multi-Media Services', COST 211ter, Compendium ATM, Coding Procedures, Simulation Subgroup, pp 1-83, May 1993.
33. SES, *SES/workbench Reference Manual*, Release 2.1, Feb. 1992.
34. J. Murphy, L. Murphy and E. C. Posner, 'Distributed Pricing For Embedded ATM Networks', to be presented at *International Teletraffic Congress ITC-14*, Antibes, France, June 1994.
35. US Department of Commerce, *Telecommunications in the Age of Information*, Oct. 1991.
36. D. G. Luenberger, *Linear and Nonlinear Programming*, 2nd Ed., Addison-Wesley, 1984.

Broadband Communications, II (C-24)
S. Tohmé and A. Casaca (Editors)
Elsevier Science B.V. (North-Holland)

A congestion control framework for broadband ISDN using selective window control

Ke-Qiang Liao[a] and Lorne Mason[b]

[a]Département d'informatique, Université du Québec à Hull
P.O. Box 1250, Station B, Hull, Québec, Canada J8X 3X7

[b]INRS-Télécommunications, 16, Place du Commerce,
Ile-des-Soeurs, Verdun, Québec, Canada H3E 1H6

Keywords: Network architecture and design, ATM (Asynchronous Transfer Mode) networks

Abstract - In this paper, we consider a congestion control framework which allows the support of different services while providing high utilization rate and performance guarantees to all types. We divide the services into two performance classes: isochronous (real-time) services and anisochronous (data) services and deal with the congestion control of delay based traffic. An important problem is how to prevent the congestion for data traffic from impacting on the performance of isochronous traffic to achieve control separation and effectively control the data flow in the network as a function of the traffic level of isochronous services. In this paper we propose to use a selective windowing control scheme together with a priority mechanism. Theoretical considerations and simulation results are presented to show that the optimal selective window size under heavy load condition is independent of the isochronous traffic level and a priority scheme prevents this control from having any effect on the performance of isochronous services. Numerical results for a network case show that the selective windowing scheme performs better than a random discard scheme since the former regulates the data traffic as a function of the instantaneous traffic level of isochronous services.

1. INTRODUCTION

Fast packet switching, called Asynchronous Transfer Mode (ATM) in the broadband ISDN area, is one of the most promising means to integrate the transport of services with a wide spectrum of traffic statistics such as average bit rate and peak bit rate. This transport flexibility can be achieved with management and control schemes which provide different grades of service to different services or with general congestion controls, which must provide adequate uniform transport performance for all services carried in the network. In the latter case, the simplicity in control implies a decrease in the efficiency since the performance must be tightly controlled in order to satisfy the most stringent requirements [23]. Consequently if we want to maintain a certain efficiency in the heterogeneous environment, we require fairly complicated algorithms to resolve the contention of different traffic streams for common network resources.

Let us consider first the traffic types that integrated services broadband networks have to support. In this paper, we define two performance classes with associated cell delay and loss objectives, into which any given broadband service can be mapped. The first class, the isochronous traffic (Classes A and B in CCITT standards), includes traffic generated by voice and video sources and traffic from existing circuit switched systems which have stringent performance requirements in terms of delay and delay jitter. The second traffic class, the data traffic (Classes C and D in CCITT standards), has less stringent performance requirements in terms of delay compared to isochronous traffic. In other words, the first traffic class is delay sensitive and may be loss sensitive while the second one is not delay sensitive.

The congestion control strategies can be divided into two categories: reactive control and preventive control [9]. Reactive control allows the network to congest and then takes corrections to relieve congestion while preventive control does not wait until congestion occurs, but rather tries to prevent the network from reaching an unacceptable congestion level. It is frequently argued that because of the domination of propagation delay over queueing delay in high speed networks, the reactive control may not be suitable. However Mitra demonstrated in [17] that in a high speed data network with window flow control, most packets are in the process of being propagated and only a small proportion are in the node buffers. It follows that a large window size does not imply large node buffers and we could consider using window flow control in ATM networks. Other researchers [1, 4, 14] have also considered the use of windowing flow control as a congestion control mechanism and error recovery scheme in high speed networks.

The two main components of preventive control in an ATM network are admission control at the call level and bandwidth enforcement (source policing) at the cell level. In admission control, the decision to accept a new call is based on the predicted network performance taking into account the declared traffic characteristics. Since users may deliberately or inadvertently exceed the traffic volume declared at the call set-up, after a connection is established, one must monitor the traffic flow to ensure that the actual traffic flow conforms with that specified at call set-up. The essential point of preventive control is that the performance requirements of an admitted call which conforms with the traffic characteristics declared at call set-up must be satisfied. It is the viewpoint of the authors of this paper (see also [22]) that a preventive control mechanism should be implemented for isochronous traffic since: i) it is difficult for an isochronous application to accept a service degradation once accepted; ii) the isochronous cells can not be delayed beyond a certain limit.

On the other hand, the data traffic is not delay sensitive and can accept a relatively longer delay than isochronous traffic. When the network is congested, the data traffic cells can be delayed or discarded and then retransmitted. This is the reason why reactive control can be applied to the data traffic. In [3], Boyer proposed a control scheme, Fast Reservation Protocol, which performs the admission control not only at the call level but also at the burst level. The fast Reservation Protocol was proposed to control the traffic with less severe delay constraints. It will be interesting to compare the performance of the scheme we propose in this paper with that of the Fast Reservation Protocol.

In this paper we concentrate on the congestion control of data traffic and we assume the existence of an appropriate congestion control mechanism for isochronous traffic [9, 12]. An important problem is how to prevent the congestion for data traffic from having a significant effect on the performance of isochronous traffic to achieve control separation and effectively control the data flow in the network as a function of the traffic level of isochronous services. Control separation provides a simplification for global congestion control in ATM networks. The method we propose is called selective window control and it can be implemented for either virtual circuit control or network access control (isarithmic scheme [8]). The window is called selective because it controls only the data traffic (it will be detailed in Section 3). Since the isochronous traffic has a more stringent performance requirement in terms of delay, we propose a priority mechanism to give non-preemptive priority to isochronous traffic cells. For the time being, an explicit priority indication in the ATM cell header has been recently introduced by CCITT [5] for the discarding process at buffer overflow occurrences. This specific bit could be used to distinguish different performances classes (e.g. data services from isochronous services). Consequently, the effect of data traffic on the performance of isochronous traffic is negligible and the performance requirements of isochronous services could be satisfied by employing an appropriate preventive control mechanism without considering the presence of data traffic.

One might argue that as the transmission speed of high speed networks scales up, the window size becomes larger and larger and it would create an implementation problem since a large window size implies increased signalling and control traffic. However as discussed in [22], one could use a permit to control a group of cells instead of individual cell. This will reduce the signalling and control traffic.

As the proposed method controls only the data traffic and the effect of data traffic on the performance of isochronous traffic is negligible due to the priority mechanism, we consider in this paper only the performance of data traffic.

The paper is organized as follows. In Section 2 we briefly describe the performance metric used in the paper and a detailed description of the congestion control framework is given in Section 3. Theoretical considerations and simulations results are presented in Section 4 to show that the optimal selective window size in the heavy load case is independent of the isochronous traffic level. In Section 5, we present simulation results to show that the selective control scheme performs better than rate regulation based on random discard. Section 6 concludes.

2. PERFORMANCE METRIC

The power function, proposed by Kleinrock [10] and defined as the ratio of average throughput over average delay, has been used in the context of flow control as a means to unify the antithetic objectives of high throughput and low delay. If we consider only a virtual circuit, we use the power function as performance metric. On the other hand, if we consider the network on the whole, Bharath-Kumar and Jaffe pointed out in [2] that the product of powers is a more appropriate optimization criterion since the maximum power operating point can be unfair in that certain users have zero throughput. By using a game theoretic framework, Mazumdar et al. [16] provided a proof of why the product of powers is indeed a reasonable design criterion. The product of powers is defined as

$$P = \prod_{i,j \mid \Lambda_{i,j} > 0} \frac{\lambda_{i,j}}{T_{i,j}} \tag{1}$$

where

$\Lambda_{i,j}$: offered traffic from node i to node j,

$\lambda_{i,j}$: throughput of traffic from node i to node j,

$T_{i,j}$: average end-to-end delay for carried i-j traffic.

Note: In our model we use a single permit queue per source node. This implies that the loss probabilities for all traffic originating at the same node are equal. By removing this constraint, i.e., by allowing independent control for each origin-destination pair, the product of powers can be increased. However this implies N-1 (N: number of nodes) permit queues per source node, resulting in increased complexity.

3. METHOD AND DESCRIPTION

As one moves toward high speed networks, the effect of propagation delays on congestion control become increasingly consequential. Several researchers have studied the effect of propagation delays on congestion control in high speed data networks [7,17-19]. Mitra [17] formulated and solved the problem of optimal design of sliding windows for the congestion control of virtual circuits. Cotton and Mason [7] dealt with the network access level control by using the isarithmic scheme, first proposed by Davies [8], to control the total number of packets in transit within the network.

Since no window size is optimal for all loads, several researchers (see [21] and the references therein) have proposed methods for adaptive window size adjustment. In the context of high speed packet switching networks, Mitra and Seery [19-20] proposed a method to optimize the window size for a virtual circuit based on an explicit expression relating the optimal window size to the mean round trip response time; Liao and Mason [13] studied a self-optimizing method to optimize the window size.

The window size optimization in ATM networks is much more complicated because of the presence of isochronous traffic. In fact, the window size which maximizes the performance metric (power in the virtual circuit case or product of powers in the network access control case) is a function of the isochronous traffic level. In a changing environment in an ATM network, we need a fast algorithm to adapt the window size to the changes in the isochronous traffic level which could be difficult to realize considering the possible variations in a future ATM network.

In this paper, we propose an alternative to avoid the aforementioned adaptation problem during periods of congestion (the network is heavily or over loaded) where we need congestion control. The method we propose uses a counter with a certain initial value, called window size or permit population, to control the number of cells of data traffic in transit as a function of the state of isochronous traffic. In real implementation we should have a permit queue (or a counter) at each node to control the traffic streams originating at that node. The selective window controls only the cells of data traffic and has no effect on cells of isochronous traffic although an admitted cell of both traffic classes decreases the value of the counter by 1 and on the other hand, a cell departure increases the value by 1. In other words, a cell of isochronous traffic is admitted irrespective of the value of the counter while a data traffic cell gets access only when the value is positive. Evidently, the value of the counter can be negative (a negative credit). This is the major difference between the selective window and the traditional windowing schemes.

The selective window control scheme can be implemented at various levels: virtual circuit control or network access control (isarithmic scheme). In the latter case, we can employ a centralized controller to allocate the released permits. Another approach is decentralized where we have a controller associated with each destination [11]. For both centralized and decentralized cases, Mason and Gu [15] proposed several learning automata control schemes to make controllers adaptive to network conditions. Cotton and Mason [7] studied the centralized scheme in the context of high speed data networks.

On each link we can have a shared large buffer with a head-of-line priority given to isochronous services or two separate buffers: one for each service class and isochronous services have non-preemptive priority over data services. In the latter case, the buffer for isochronous traffic can be finite (in a similar way to the former case we can have a threshold to limit the number cells of isochronous services in the buffer. One of the advantages of using finite buffers in the interior of the network is that we can avoid the phenomenon that some isochronous cells arrive at the destination but are of no use because of their excessive delay. This can improve the network efficiency since we do not waste network resources carrying useless cells. However we have isochronous cell loss because of buffer overflow. Throughout this paper we assume that a multiplexor is able to detect a cell loss and consequently the cell loss is taken into account by the window counter.

It is worth noting that in a slotted ATM system, where cells can arrive only at slot marks, the data traffic will have no impact on isochronous traffic. We will show that in the continuous time case, the influence of data traffic on isochronous traffic is negligible.

4. SELECTIVE WINDOWING SCHEME

We might ask the following question: Does the optimal window size remain constant for different isochronous traffic levels in the heavy load case? In this section, we present some simple justifications for the proposed selective window scheme. Because of the complexity of the network access control case, we consider only the virtual circuit case. However we believe that the conclusions hold in the more complicated network case where the isarithmic control is employed. We will first present a simple theoretical consideration where cell length is assumed to be exponentially distributed and in the second subsection, some simulation results for constant cell length to show that the optimal window size (value of the counter) is independent of the isochronous traffic level.

4.1 Theoretical Consideration

Mitra [17] presented an elegant analysis of sliding windows for congestion control of virtual circuits in high speed data networks and solved the problem of designing the optimal window size. In order to apply the results obtained by Mitra, we assume that cell length is exponentially distributed rather than of constant size in ATM networks. We will use the same notations as in [17] and in order to simplify the presentation we consider only the case with one node. (The same conclusion can be obtained in the same way for cases with many nodes.)

Let μ denote the service rate of the node; it has dimensions of cells/time. The time unit is chosen to be the mean round trip propagation delay. The node service rate μ gives the average number of cells which are in the process of being propagated in the heavy load condition. Evidently, the higher the transmission rate, the larger μ. Let us now define the reduced nodal processing rate

$$\lambda = \mu - \nu \tag{2}$$

where ν is the rate of exogenous, cross traffic in [17] and the isochronous traffic rate in our considered case.

In case of large λ (high speed), Mitra gave in [17] the following asymptotic solution for the optimal window size of the considered virtual circuit:

$$\mathbf{K}^* \sim \lambda + \mathbf{0{,}55}\sqrt{\lambda} \tag{3}$$

For high speed networks, since $\sqrt{\lambda}$ is much smaller than λ, to leading order, the optimal window size K* is approximately λ. In our considered case, although the counter (selective window) does not control the isochronous traffic, it takes into account the total number of cells in transit (being propagated or served at the node). Hence the optimal window size W* should be the sum of the optimal window size for data traffic K* and the number of isochronous cells in transit. Since the number of isochronous cells in transit is approximately ν, we can see that the optimal window size W* is approximately μ and is essentially independent of the isochronous traffic level.

4.2 Constant Cell Length Case

In the previous analysis of virtual circuit windows in a network, the statistics of a virtual circuit are directly obtained from the analysis of a network which is identical to the original network except that (i) all cross traffic classes (isochronous traffic in our considered case) are deleted, and (ii) the nodal service rate is redefined to be the reduced service rate. This is evidently an approximation and one may very well argue that because of the presence of variations in isochronous traffic, the optimal window size W* could depends on the isochronous traffic level. On the other hand, in order to apply Mitra's results, we assumed in the previous subsection that cell length was exponentially distributed rather than constant as in ATM networks. However simulation results of realistic cases, which are to be presented in this subsection, show that the conclusion that the optimal window size W* is independent of the isochronous traffic level still holds although the conclusion was obtained in the precedent subsection with the assumption of exponentially distributed cell length and the approximation of reduced nodal processing rate.

In the following, the time unit is ms and the throughput is in bits/ms. It follows that the power has dimensions of bits/(ms)^2.

We first consider the well-known packetized voice system where we have a multiplexor of capacity 1.536 Mbits/s. The voice packet length is 64 bytes which results in a fixed voice packet transmission time of 1/3 ms. The round trip propagation delay is assumed to be 47 ms corresponding roughly the distance across the continental USA divided by the speed of light. This gives $\mu = 141$. The packet stream from a single voice source is modelled by arrivals at fixed intervals during talk spurts and no arrivals during silences. The talk spurt and silent period are both exponentially distributed with means 352 ms and 650 ms, respectively. The data packets are assumed to arrive according to Poisson process and the packet length distribution is the same as voice traffic.

Fig. 1 shows the power as a function of the window size for the two cases with 60 and 90 voice sources respectively. We can see that the optimal window size (143) remains unchanged for both cases although the traffic level of isochronous services is changed (the utilization rates of isochronous traffic are respectively 0.439 and 0.659 for 60 and 90 sources).

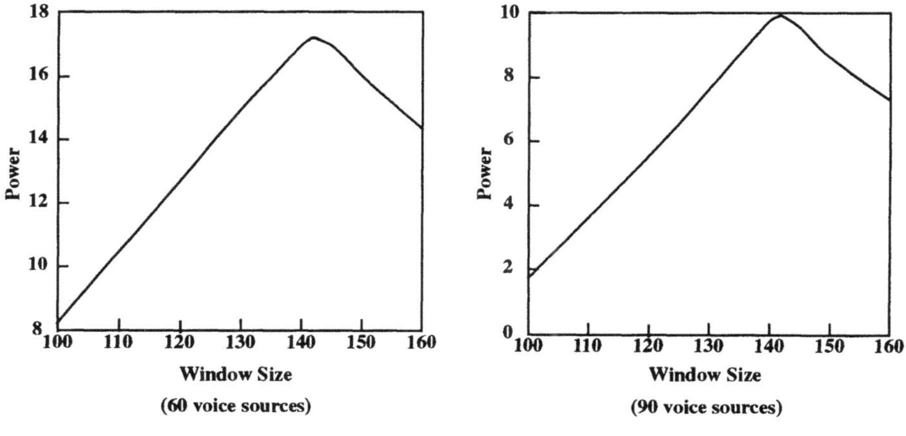

Fig. 1 Power vs. Window Size (Packetized Voice)

In the following, we deal with the case where the cell arrival process of isochronous services is modelled by two-state Markov Modulated Poisson Process (MMPP) which is commonly used to approximate the superposition process of bursty sources. The transmission rate is 1.7 Gbits/s which is a standard for optical transmission. The round trip delay is assumed to be 0.117 ms which corresponds approximately to a metropolitan network. The cell length is chosen to be 53 bytes (ATM standard) which gives $\mu = 470$. We consider two cases where we have different parameters for the two-state MMPP:

	λ_1	λ_2	r_1	r_2
Case 1	$4.8\text{x}10^6$	$2\text{x}10^6$	1/150	1/150
Case 2	$4.4\text{x}10^6$	$1\text{x}10^6$	0.001	0.001

Table 1 Parameters of MMPP

λ_1 and λ_2 are respectively the arrival rates (cells per second) during states 1 and 2 and r_1^{-1} and r_2^{-1} are mean sojourn times (in cell transmission time) in states 1 and 2.

Fig.2 presents the power as a function of the window size for the two cases. It can be seen that the optimal window size (466) is indeed independent of the traffic level of the isochronous services. In the simulation, we have separate buffers for isochronous traffic and data traffic. The buffer for isochronous traffic is finite with 100 waiting places while that for data traffic is assumed to infinite for modelling purposes. Simulation results also show that the effect window size has on performance of isochronous traffic is negligible because of the priority scheme. In fact the loss probabilities for isochronous traffic are 2.135% and 3.5% for cases 1 and 2 respectively and they remain almost the same for all window sizes.

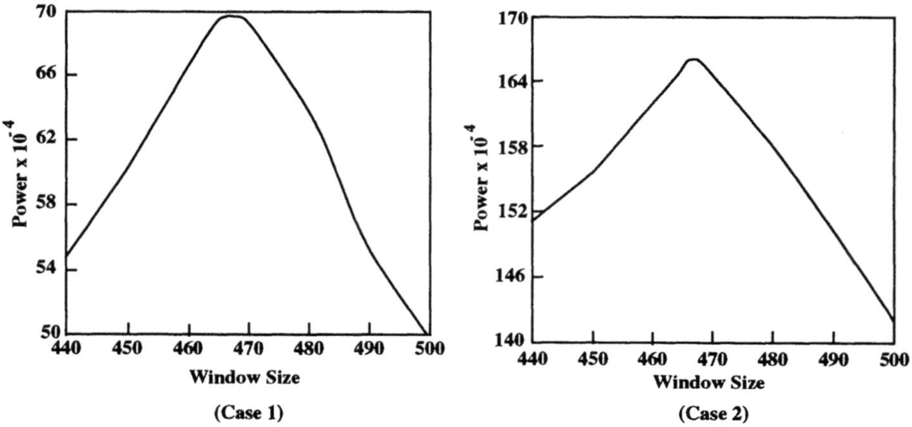

Fig. 2 Power vs. Window Size (MMPP)

It is worth noting that the optimal window sizes, for deterministic service time case, obtained by simulation are very close to the results of Mitra [17], obtained for the exponential service time case. This leads us to conjecture that in high speed networks, the optimal window size is a function of mean service time but not sensitive to service time distribution.

On the other hand, we can see that the power, as a function of window size, is reasonably well behaved (quasi-convex). This means that hill-climbing techniques can be employed to locate the global maximum [13].

5. NETWORK CASE

In this section, we consider the isarithmic flow control at the network access level. The flow control optimization involves two subproblems: the window size and the redistribution algorithm for released permits. In this paper, we concentrate on the problem of selecting window size for the sake of simplicity. An interested reader is referred to [20] where Pelletier considers the global optimization problem in a homogeneous data network. The centralized redistribution algorithm for released permits employed in this paper is not adaptive. Therefore the performance of the control scheme can be further improved by using an adaptive permit allocation algorithm.

5.1 Description of the simulated network

The particular network used in these studies had 4 nodes and its configuration is shown in Fig. 3.

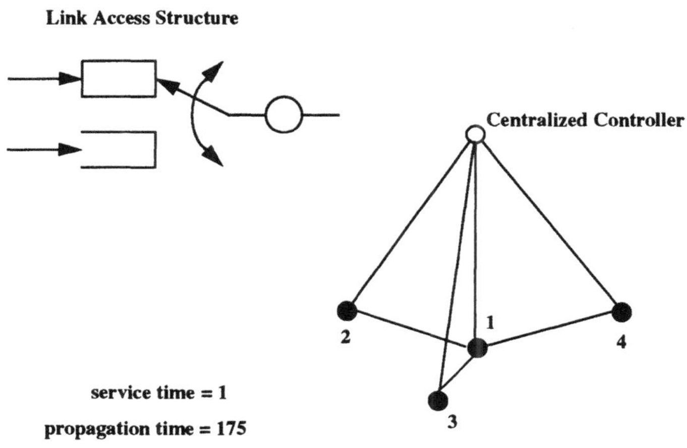

Fig. 3 Configuration of the simulated network

In this network we assumed the same transmission speed for each link and the time unit was the time required to transmit a cell. The value of the time unit is arbitrary but the time unit was chosen to be the time required to transmit a cell of 53 bytes on a link of bandwidth 1.7 Gbits/s which is a standard for optical transmission. The propagation delay was chosen to be 175 time units, corresponding roughly to a metropolitan-scale broadband network. (For comparison, the transcontinental end-to-end propagation delay for the current ARPANET, with larger packets and much narrower bandwidth, is only 4 to 5 time units.) All transmission lines were assumed to be error-free, so that the cell losses were from buffer overflow. We modelled the propagation delays from a node to the centralized controller and vice versa and all were assumed to be 175 time units.

For each Origin Destination pair, the cell arrival process of isochronous services was assumed to be a Markov Modulated Poisson Process with a certain burstiness while that of data services was modelled by a Poisson process. There was no traffic between the OD pairs with node 1 as destination. We had the same cell arrival pattern for all other OD pairs. It follows that the network was symmetric as far as the total traffic offered to a link is concerned. For the considered network, the stream with node 1 as origin had a better grade of service, since the cells of these streams needed to pass through only one link. In the simulation, we assumed that all links were bidirectional. The results obtained may be slightly optimistic but the conclusions hold in the case with unidirectional links.

5.2 Simulation Results

We have simulated the network described above, with total isochronous traffic λ_1= 1.1025 (each MMPP has the following parameters: λ_1=0.2, λ_2=0.045, r_1=r_2=0.005). For different window sizes, we present in Table 2 the throughput, the average sojourn time, and the product of powers for data services as a function of the offered data traffic λ_2. As expected, we can see that by using different window sizes we can increase (decrease) the throughput of data traffic but at the same time the average sojourn time is increased (decreased). We have observed that for a certain traffic level of isochronous services, the optimal window size is a function of the traffic level of data services. When the traffic level is low, it is better to do nothing (infinite window). Coderre [6] has observed the same phenomenon in a data network.

We have performed some more simulations to determine the optimal window size when the network is heavily loaded. The results are presented in Table 3. The optimal operating point is at 1100 and the network utilization rate is 0.924. From this table, it can be seen that the product of powers, as a function of window size, is reasonably well behaved. The uniqueness of the maximizing point and the good behavior (increasing function when the window size is smaller than the optimal one and decreasing function when it is larger than the optimal operating point) mean that hill-climbing methods can be used to find the optimal operating point.

In order to see the effectiveness of the selective window control scheme under heavy load condition, we have compared it with an open-loop control scheme: random discard. The traffic level of data services after random discarding was chosen to be the throughput in the case of optimal window size 1100. The average sojourn time and the product of powers are respectively 476.4 and $1.74x10^{-35}$ for the random discard scheme. We can see that the selective window control scheme gives a better performance. The reason why the selective window control scheme performs better is that the window control scheme regulates the amount of data services as a function of the instantaneous traffic of isochronous services while the random discard scheme does not. In the random discard case, if we model the arrival processes of isochronous traffic by Poisson processes rather than MMPPs but with the same arrival rates, the average sojourn time and the product of powers are respectively 329.9 and $4.73x10^{-34}$. The results are even better than those obtained with window control scheme but with MMPP arrival processes. However this comparison is unfair to the windowing scheme, since the pattern of arrival processes of isochronous traffic evidently has effect on the performance of data traffic. In fact, the windowing scheme regulates the data traffic as a function of the isochronous traffic level but for each level of isochronous traffic, we can not reach the optimal operating point immediately. As mentioned in the introduction, it will be interesting to compare this scheme with others like the rate based flow control or the Fast Reservation Protocol.

It is worth noting that the redistribution algorithm of released permits in the simulation is static and we can improve the performance of the windowing scheme by implementing an adaptive algorithm for the centralized controller.

λ_1= 1.1025		Window Size			
λ_2		750	1000	1500	1750
	Throughput	**9.94x10-02**	**1.59x10-01**	**2.00x10-01**	**2.05x10-01**
0.225	av. S.T.	**3.10x10+02**	**3.12x10+02**	**3.16x10+02**	**3.18x10+02**
	pr. of powers	**1.58x10-40**	**9.41x10-39**	**6.46x10-38**	**8.05x10-38**
	Throughput	**1.54x10-01**	**3.03x10-01**	**4.09x10-01**	**4.29x10-01**
0.450	av. S.T.	**3.07x10+02**	**3.22x10+02**	**3.50x10+02**	**3.60x10+02**
	pr. of powers	**7.74x10-39**	**2.34x10-36**	**1.64x10-35**	**1.97x10-35**
	Throughput	**1.81x10-01**	**4.24x10-01**	**6.49x10-01**	**6.66x10-01**
0.900	av. S.T.	**3.06x10+02**	**3.30x10+02**	**7.61x10+02**	**1.09x10+03**
	pr. of powers	**3.53x10-38**	**3.83x10-35**	**9.50x10-37**	**4.89x10-38**
	Throughput	**1.88x10-01**	**4.58x10-01**	**6.64x10-01**	**6.70x10-01**
4.500	av. S.T.	**3.07x10+02**	**3.28x10+02**	**8.53x10+02**	**1.21x10+03**
	pr. of powers	**5.13x10-38**	**8.11x10-35**	**4.31x10-37**	**2.31x10-38**

Table 2 Simulation Results

Window Size	Throughput	Av. S.T.	Pr. of Powers
750	0.188	307.0	5.13x10-38
900	0.339	315.7	8.38x10-36
1000	0.458	328.0	8.11x10-35
1050	0.514	340.2	1.67x10-34
1100	0.561	365.7	1.90x10-34
1150	0.597	402.7	1.41x10-34
1200	0.624	449.1	7.87x10-35
1250	0.640	505.8	3.35x10-35
1500	0.664	853.0	4.31x10-37
1750	0.670	1211.	2.31x10-38

Table 3 Optimal Window Size in Heavy Load Case ($\lambda_1 = 1.1025$, $\lambda_2 = 4.5$)

6. CONCLUSIONS

The main objective of this paper is to synthesize a congestion control scheme for data traffic in ATM networks where a preventive control is employed for isochronous traffic. An important problem is how to separate the two controls in particular, the congestion control for data traffic should not have a significant effect on the performance of isochronous services which are considered to have higher priority because of their stringent performance requirements in terms of delay and delay jitter. The importance of control separation is evident; with separation we can first provide performance guarantees to isochronous traffic and then consider the control problem for data traffic. This makes the global congestion control much simpler.

We have studied a congestion control framework using a selective window control for broadband ISDN. This control together with a priority scheme provides performance guarantees to both isochronous and data service classes while maintaining a high utilization. Performance guarantee is a very important requirement in ATM networks where we have many services with a wide spectrum of traffic characteristics and performance requirements. Theoretical considerations and simulations have been presented to show that the optimal selective window size in the heavy load case is independent of the isochronous traffic level and a priority scheme prevents this control from having any significant effect on the performance of isochronous services. Numerical results for a network case have shown that the selective windowing scheme performs better than a random discard scheme since the former regulates the data traffic as a function of the instantaneous traffic level of isochronous services.

Although we have considered in this paper only the case where isochronous traffic has access to all network resources, it is easy to extend the presented control framework to include the possibility of reserving a certain amount of capacity on each link for data traffic to avoid the case where data services have a very poor performance because of a high traffic level of isochronous services. In order that isochronous traffic carried on each link does not exceed the allocated link bandwidth, we could implement a spacer (peak rate enforcer) on each link. It is worth noting the similarities between this mechanism and the well known movable boundary scheme employed to integrate voice and data.

ACKNOWLEDGMENT

The authors would like to acknowledge Prof. Zbigniew Dziong for many helpful comments and suggestions and Anne Pelletier for the simulation work in Section 4. This work was mainly supported by the NSERC Strategic Grant No. STR 0045121 and partly by an internal grant from Université du Québec à Hull.

References

1 O. Aboul-Magd, H. Gilbert and M. Wernik, "Flow and congestion control in broadband packet networks", ITC-13, Copenhagen, Denmark, June 1991.

2 K. Bharath-Kumar and J.M. Jaffe, "A new approach to performance oriented flow control", *IEEE Trans. Communications*, vol. COM-29, pp. 427-435, April 1981.

3 P. Boyer, "A congestion control for the ATM", 7th ITC Specialists Seminar, paper 4.3, New Jersey, Oct. 1990.

4 T.T. Bradley and T. Suda, "Performance of error recovery schemes in a fast packet switching network", IEEE ICC'90, Altlanta, April 1990.

5 CCITT Draft Recommendation I.356: "ATM layer specification for B-ISDN", Study Group XVIII, Geneva, Jan. 1990.

6 J.R.J. Coderre, "Modèle optimal d'un contrôle global de flux appliqué aux réseaux à commutation par paquets", INRS-Télécommunications Technical Report, no. 89-15, 1989.

7 M. Cotton and L.G. Mason, "Adaptive isarithmic flow control in fast packet switching networks - Heavy traffic case", ITC Specialists Seminar, Cracow, Poland, April 1991.

8 D.M. Davies, "The control of congestion in packet switching networks", *IEEE Trans. Communications*, vol. COM-20, pp. 546-550, June 1972.

9 M. Decina, "Open issues regarding the universal application of ATM for multiplexing and switching in the B-ISDN", IEEE ICC'91, Denver, June 1991.

10 L. Kleinrock, "Power and deterministic rules of thumb for probabilistic problems in computer communications", IEEE ICC'79, June 1979.

11 F. Vasquez and L. Mason, "Decentralized isarithmic flow control in packet switched networks", Canadian Conference on Electrical and Computer Engineering, Toronto, Sept. 1992.

12 K.-Q. Liao, Z. Dziong, L.G. Mason and N. Tetreault, "Effectiveness of leaky bucket policing mechanism", IEEE SUPERCOMM/ICC'92, paper 341.4, Chicago, June 1992.

13 K.-Q. Liao and L. Mason, "Self-optimizing window flow control in high speed networks", 3th Int. Conf. on Advances in communication and control systems, COMCON'3, Victoria, British-Colombia, Canada, Oct. 1991. (also in *Computer Communications,* vol. 16, no. 11, pp. 706-716, Nov. 1993)

14 D.T. Luan and D.M. Lucantoni, "Throughput analysis of a window-based flow control scheme subject to bandwidth management", IEEE INFOCOM'88, New Orleans, March, 1988.

15 L.G. Mason and X.D. Gu, "Learning automata models for adaptive flow control in packet switching networks", in Adaptive and Learning Systems - Theory and Applications, K.S. Narrenda, Plenum Press, pp. 213-227, 1986.

16 R. Mazumdar, L.G. Mason and C. Douligeris, "Fairness in network optimal flow control: optimality of product forms", *IEEE Trans. Communications*, vol. COM-39, pp. 775-782, May 1991.

17 D. Mitra, "Asymptotically optimal design of congestion control for high speed data networks", *IEEE Trans. Communications*, vol. COM-40, pp. 301-311, Feb. 1992.

18 D. Mitra and J.B. Seery, "Dynamic adaptive windows for high speed data networks: Theory and simulation", *ACM Comp. Comm. Rev.*, vol. 20, no. 4, Oct. 1990.

19 D. Mitra and J.B. Seery, "Dynamic adaptive windows for high speed data networks with multiple paths and propagation delays", IEEE INFOCOM'91, Miami, April 1991.

20 A. Pelletier, "Contrôle de flux et acheminement adaptatif combinés dans les réseaux à commutation par paquets rapide", INRS-Telecommunications, Technical Report no. 92-11.

21 S. Pingali, D. Tipper and J. Hammond, "The performance of adaptive window flow controls in a dynamic environment", IEEE INFOCOM'90, pp. 55-62, San Francisco, June 1990.

22 T. Toniatti and F. Trombetta, "Performance simulation of end-to-end windowing in ATM networks", IEEE INFOCOM'92,

23 G.M. Woodruff and R. Kositpaiboon, "Multimedia traffic management principles for guaranteed ATM network performance", *IEEE J. Selected Areas in Communications*, Vol.8, No. 3, April 1990.

Broadband Communications, II (C-24)
S. Tohmé and A. Casaca (Editors)
Elsevier Science B.V. (North-Holland)

"The control of network resources for complex B-ISDN services"

A. Paglialunga (CSELT) - L. Ronchetti (ERICSSON-FATME),
L. Faglia (ITALTEL) - S. Minzer (BELLCORE)

ABSTRACT

Today we are witnessing the realisation of ATM Networks, initially in the form of national field trials consisting of ATM Cross-Connects which will eventually be connected to one another, with the scope to verify the feasibility of the ATM switching and begin to offer basic services to users. The next phase will be to introduce signalling to offer connections on demand, with the final objective in mind of offering advanced multimedia multiconnection services which will utilise to the full the bandwidth and speed offered by ATM.

The introduction of complex multimedia multiparty services introduces an added complexity in terms of signalling (i.e. for the resource control of the network). Resources such as bridges, combiners, distributers and converters will be especially precious in ATM networks. Already a step has been taken to best utilise these resources by separating the once monolithic signalling protocols to permit virtual ATM connections to be established independently of the call. In this paper we examine the necessity to optimise the use of special resources and identify the need for an additional level of separation between Call Control and Bearer Control, which results in the need for a separate protocol which we call Resource Control. A protocol architecture and stack is presented and it is proposed that this new protocol be based on the atomic action principles of the CCR (Commitment, Concurrency and Recovery) protocol.

Finally a short discussion is given on the impact of managing resources within an exchange proposing some techniques to be adopted to best use the limited resources.

Keywords Codes: C.2.2

Keywords: Network Protocols

1. INTRODUCTION

In the long term, the objective in many countries is an intelligent integrated broadband network for all services - the Broadband ISDN. It is based on the Asynchronous Transfer Mode (ATM) which offers the user the possibility of supporting services requiring a wide range of bit rates and several "virtual" connections via the network access, flexibly depending on instantaneous requirements. ATM, in addition to transporting continuous bit streams of any desired rate, complies with the requirements of burst-oriented data transfer, motion-video with variable bit rates, and versatile multimedia communication.

Also growing in importance is multimedia communication, which is combined use of data, text, graphics, image motion video and audio (voice and sound). Multimedia communication is very close to the human being's perception, communication and way of acting, but also makes new technical demands.

Video represents another category of potential broadband services. The ability to support interactive video services such as "video-on-demand", broadband videotext, and distribution services offers many opportunities. The ability to offer extended-quality TV and High-Definition TV (HDTV) services to the residential market will also be a key ingredient of broadband networks.

There is general agreement that end users will derive definite benefits from the Broadband ISDN. Desktop video teleconferencing appears to have growing acceptance among business users. So, as consumers begin to share this vision, computer vendors are working to make it possible.

2. OVERVIEW ON SERVICES

Requirements and demand in broadband communication differ on one hand in the location size and sector of businesses, or according to the income of households. On the other hand, they differ by region, such as in the inner cities and in the industrial, mixed-users and residential areas. The introduction and distribution of new technologies and applications will be segmented correspondingly. For reasons of utility and outlay, the demand for additional broadband communication will initially arise in large companies (e.g., publishing houses, banks, insurance companies, and industry), in research institutes, and public authorities. As networks spread their coverage, and as equipment costs and charges decline, medium-sized and smaller organizations and private households will also come to enjoy the extended range of facilities of broadband communication.

For example, today, many workstations are attached to LANs. However, in the future, high-speed workstations may be connected directly to a public broadband network in order to communicate with other directly attached workstations as well as with PCs, workstations, and servers attached to LANs served by the broadband network. Host computers could be connected directly together or to high-speed peripherals (such as front-end processors or disks) via a direct channel connection. High-speed distributed host computing can be supported by a broadband network in conjunction with appropriate channel extension equipment. Besides these traditional data applications, there is a strong emerging market for digital imaging services: production, storing, and transport of high-resolution medical images; electronic image capture and transfer of financial and insurance documents; production and sharing of graphic design files; and electronic library services. In addition, multimedia services appear at this moment attractive for the future of telecommunications. Because of the nature of the traffic (high-speed, large files, and burst transmission) as well as the potentially large community of interest, a shared public broadband data network offers an economically attractive solution.

2.1. Requirements for signalling

The above presented scenarios demand high-speed transmission and *switching* within the network, and many services require new *signalling* capabilities. For example, one can imagine asymmetric or unidirectional connections for the transport of some types of information, re-negotiation of the basic attributes of the various connections making up a call and the addition of new connections to an existing call when an additional medium is invoked and adding new legs to a call. This evolution has strong impact on the signalling systems, since high flexibility and efficiency are required to handle complex services in a broadband environment.

Demand will arise for Distribution Services (e.g. TV), which are the point-to-multipoint (in particular multicast or broadcast) transfer of information from a distribution center to distribution service users. The point-to-multipoint configuration is in this case a distribution tree. The distribution service users may or may not select the source by a selection/control path and even the order of the information of the source may or may not be user controlled.

Retrieval services (e.g. videolibrary) provide unidirectional transfer of information to a calling user from one or more other users under the control of the calling user via a bi-directional selection path offering selection of the actual retrieved information.

In addition, the conference services (e.g. videoconference) seem attractive, since they can provide interactive real-time multipoint-to-multipoint transfer of information between multiple users.

All the above services will require specialized resources (such as bridging/branching nodes, code/protocol converters), which could be located within the network and dynamically and efficiently allocated according to the instantaneous requests by means of an intelligent control system.

3. NETWORK ARCHITECTURE

Today there are some projects already implemented, or being placed in the field, to create a B-ISDN pilot network.

Broadband ISDN pilot networks are supported by ATM on a stage-by-stage basis, first locally, then nationally and finally on an international basis. This serves not only to clarify technical and operational questions, but also to stimulate incorporation of suitable applications and to speed-up development of the necessary low-cost terminal facilities. Because of the high development and investment costs involved, the Broadband ISDN will not be able to attain full area coverage on an ATM basis in the short or medium term. The aim must therefore be to extend the existing networks step-by-step with suitable technical concepts, which could mean expanding the telephone network via ISDN into a B-ISDN, and to design new networks in such a way that they can economically cope with existing demand from the outset and can later be incorporated in the B-ISDN.

To satisfy this step-by-step approach two large groups of applications have been identified that can be supported using either the Virtual Path (VP) or Virtual Connection (VC) concept. A first set of applications uses only semi-permanent VP connections. In a second phase, fully "on demand" VC connections are offered to the subscribers.

VP connections are semi-permanently allocated between the two endpoints of the virtual connection. The establishment and release of these VP connections is performed by means of management procedures. The systems used to offer this end-to-end Broadband ATM connectivity are often called Cross-Connect.

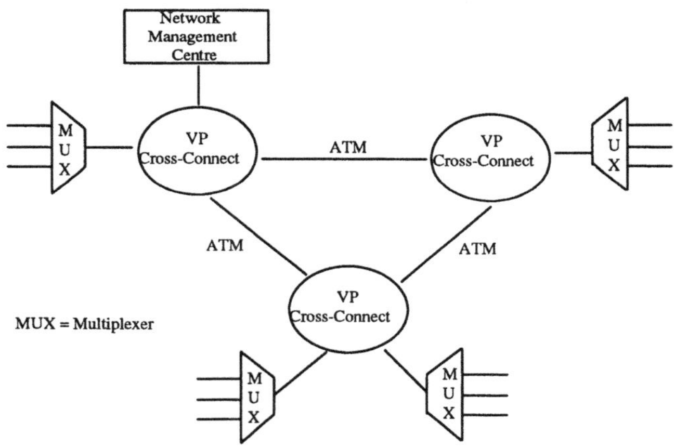

Figure 1. VP Cross-Connect network architecture

The next evolutionary step is the transition from Cross-Connect to full Exchange capability. This implies that to allocate VC connections demanded by the user, signalling procedures are needed. With respect to the procedure used for permanent and semi-permanent connections, the signalling procedures allow every terminal to request the network to establish and release a call and the related virtual ATM connection(s), therefore only when really needed by the service and with the parameters (e.g. bit rate, Quality of Service) as required by the application.

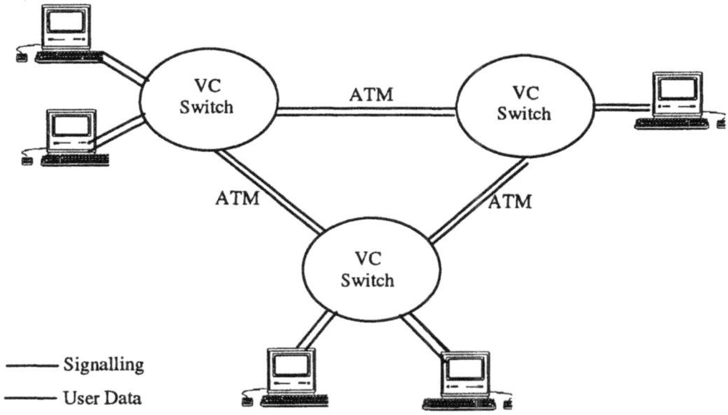

Figure 2. VC Switched network architecture

It has to be noted that the two network architectures are not mutually exclusive. In fact they can co-exist to form a unique B-ISDN network. To this purpose the Cross-Connects could be used either as transit between Switch nodes or as access to the Switch itself.

4. SWITCHING AND SIGNALLING

4.1 Evolutionary steps

In N-ISDN the User Plane and the Control Plane share only the Physical layer, in the sense that they make use of different protocol stacks from the Layer 2 to the Application/User Parts. In B-ISDN, also the ATM layer and the AAL layer will be able to satisfy the requirements of both planes. However, it must be said that the functionality and the services offered by the AAL will be slightly different, because the requirements are different. Anyway, in this paper the focus is on the User Parts of the Control Plane.

The signalling system of a modern telecommunication network can be seen as the infrastructure in charge of controlling and co-ordinating the functions of the network. For example, it can enable stored program control exchanges, network databases, and other "intelligent" nodes of the network to exchange messages related to call/connection control and supervision management.

Signalling must be considered as separated in two parts: access signalling, between the user and the network node, and network signalling, between network nodes. The current N-ISDN signalling systems are the DSS.1 (Digital Subscriber Signalling system No.1) for the access side, and the SS7 (Signalling System No.7) for the network side.

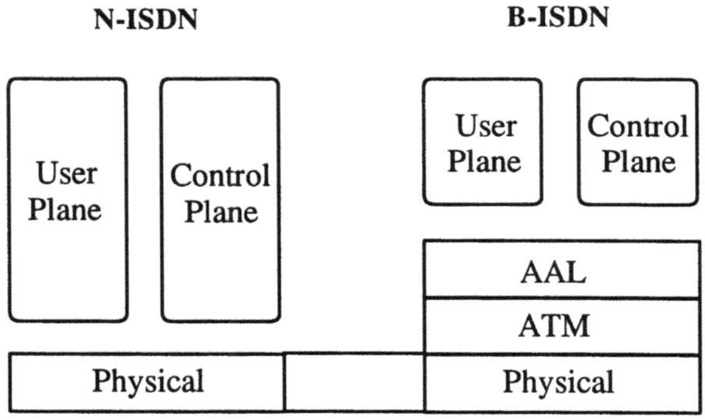

Figure 3. The N-ISDN and B-ISDN stacks

Because of the high development and investment costs involved, and of the high technological gap from the current N-ISDN, the Broadband ISDN will be developed through successive phases (Releases). The use of ATM will initially (Release 1) allow the support of the existing telecommunication services, such as voice telephony and constant bit rate video services. Then (Release 2 and beyond), the B-ISDN will cover a wide spectrum of new and enhanced services.

Two main bearer services will be supported by the Release 1 B-ISDN: circuit emulation service (connection-oriented with constant bit rate and end-to-end timing), and connectionless data service (connectionless with variable bit rate and no end-to-end timing). ATM connections bandwidth will be assigned according to peak requirements only. As regards connection configurations, only point to point connections will be allowed, with one connection for each call. To control this network and service features it has been decided to modify the present Narrowband ISDN signalling systems. To this aim the Q.93B (i.e. the User Part of DSS.1) and the B-ISUP (i.e. the User Part of SS7) are being specified in the standardization bodies respectively for the access and the network sides. The protocol architecture characteristics have been retained, but the parameters used to control switched circuits have been replaced by those for ATM connections control.

A big jump will be made in the network by the introduction of Release 2 and 3. In fact, new enhanced bearer services (e.g. to support multimedia services) and connection configurations (such as point-to-multipoint connections) will be supported by the network. For example, two connection-oriented bearer services will be introduced: variable rate services delivered in "real time" end-to-end (variable bit rate and end-to-end timing) and variable rate data services (variable bit rate and no end-to-end timing). ATM connections bandwidth will be specified according to the traffic characteristic parameters, such as average bandwidth, burstiness and peak duration. As regards the connection configurations, simple point-to-multipoint connections will be allowed, including the ability to add and drop connection branches. In addition more than one connection per call will be supported with the ability to add and drop connections during the lifetime of the call.

In B-ISDN Release 2 and 3, the main innovative concept will be the introduction of the separation of control functions in contrast with the current N-ISDN where call and bearer controls are handled in a "monolithic" way. With this paper we support the performing of control separation by means of three levels. A call is a logical relation between two end-users to perform Call Control (CC) functions and it has end-to-end significance. The Bearer Control (BC) is concerned with the functions needed to control a "physical" connection between two adjacent nodes and it has link-by-link significance. Aspects more related to the handling of special resources (e.g. bridging nodes, converters) are covered by Resource Control (RC).

The control functions separation leads to a great flexibility in the handling of network resources. For example, inside a Call, a flexible managing of the associated connections will be possible: each connection might have its own routing and connections might be added or deleted without the need for the Call to be released or repeated. In addition it offers the possibility of checking the state of the called terminal during the Call establishment phase without reserving bearer resources. In other words, the bearer connections are set-up only after the signalling system has verified that the call will be successful (called user not busy, terminal compatible, proposed service accepted, etc.), decreasing the waste of resources due to unsuccessful calls. In future, some services could require large amount of bandwidth to be supported (e.g. HDTV distribution); therefore, bandwidth might be a precious resource: to prevent waste is an important goal!

4.2. The Functional Model

In the previous paragraph the separation of Control functions has been introduced. Within a RACE (R&D Advanced Communication-Technologies in Europe) project called MAGIC (Multi-service Applications Governing Integrated Control) studies have been made to identify how to design the signalling protocol, based on the separation concepts, to provide advanced B-ISDN services as identified by the Standard Bodies for the Release 2 and 3.

A first result of the studies has been the specification of a functional model consisting of three levels: Call Control, Resource Control and Bearer Control. The essential function of CC is to co-ordinate and to handle the negotiation of telecommunication services among a set of users. RC is responsible for 'implementing' the service, using special resources and the service of BC which only handles simple point-to-point connections.

The introduction of RC allows separated control of special resources inside the network, such as multiparty multimedia conferencing units, audio bridges, and branching circuits for point-to-multipoint connections. Separation of RC from CC and BC allows the exchange of RC information separately from CC and BC information. CC, RC, and BC can each then have information flows suited to the requirements of their specific functions. This functional separation gives the network greater freedom in its distribution and use of special resources.

The reason that brought to the definition of a dedicated level for the special resources handling, came out from different considerations. In particular, the control over special resources can not be put in CC, since CC entities only exist at the edges of the network, and special resources could reside inside the network. The control over branching circuits could be put in BC. However, if this functionality is put in BC, the BC protocol should also be capable of handling point-to-multipoint connections (not only set-up, but also adding branches, failure of branches). For compatibility reasons, the decision was made to have the BC protocol without inclusion of the multicast capability.

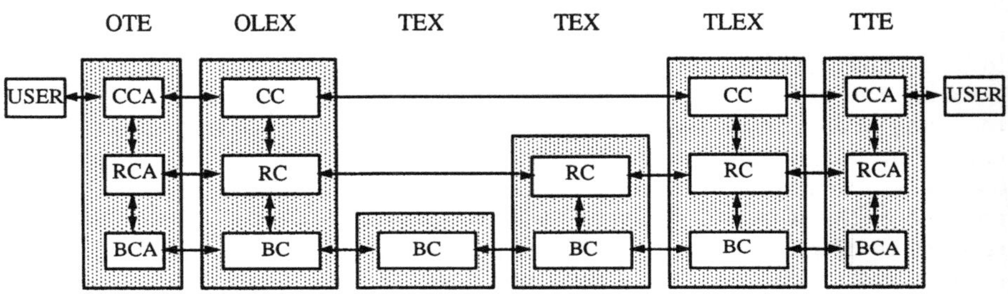

Figure 4. The three levels functional model

4.3. The protocol stack

As we have seen Release 1 protocols will be based on the existing Narrowband protocols with the appropriate ATM parameters. Now in order for the network to support more advanced multimedia, multiconnection services we require new functionalities in the protocols. In accordance with Recommendation CCITT Q.1400, at least two approaches are possible for introducing these new requirements into the signalling protocols:

- extend the Release 1 protocols taking into consideration backward compatibility;
- develop new OSI based protocols.

Studies within the RACE MAGIC project on the second approach show that basing the protocols on the OSI ALS, while still reusing much of the Release 1 protocols, seems to be the most promising way of developing the Release 2 and 3 protocols.

Very briefly the OSI ALS is the highest layer of the OSI protocol stack, layer 7, and it proposes that an Application Process (AP) communicates with another AP by means of an Application Entity Invocation (AEI). The AEI contains all the functionality to permit the two applications to communicate, the functionality of the AEI is contained in the Application Service Elements (ASEs) which compose the AEI. ASEs can be seen as blocks which perform a certain set of well defined functions, the ASEs are coordinated by a Control Function (CF). The set of ASEs and CF in a particular AEI is known as the Application Context (AC). In every AEI there is a special common ASE called Association Control Service Element (ACSE) which has the responsibility of negotiating the AC (i.e. the set of ASEs) required for the two processes to communicate. Figure 5 shows diagrammatically the structure of the ALS.

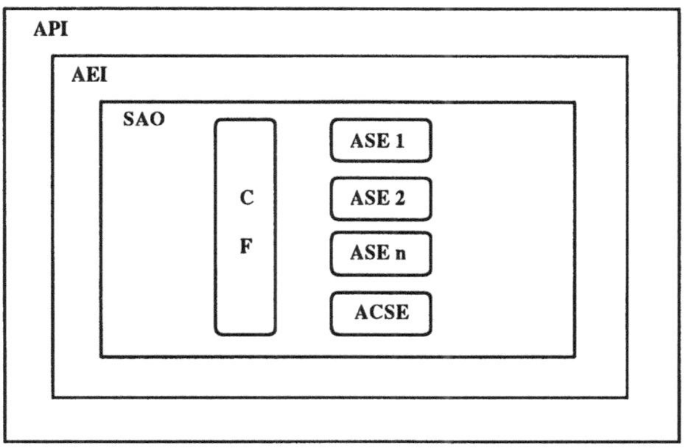

Figure 5. Architecture of the OSI ALS

The benefits of basing signalling protocols on the OSI ALS are:

i) *flexibility*, ASEs can be combined to suit the particular needs of the application;
ii) *modularity* permits ASEs to be altered without effecting other ASEs;
iii) *extendibility*, new functionalities can be easily supported by adding a new ASE to support the additional protocol requirements;
iv) *efficiency*, nodes need only to contain those part of the protocols for which functionality need to be provided e.g. a pure transit node may not need to support call control functionality.

The requirements that have been identified for separation between CC and BC can be easily supported using the OSI ALS by designing an ASE for CC and an ASE for BC, in a similar way the RC can be implemented as a separate ASE.

Given the particular nature of the application, some of the requirements of the ALS can be relaxed, i.e. at the NNI nodes are generally aware of the signalling capabilities of the nodes to which they are directly connected, adjacent nodes, thus it is not necessary to negotiate the BC ASE between nodes. Furthermore, a full implementation of an ACSE as specified in the recommendations would not be efficient for the real time requirements of telecommunications; a better solution is to reuse the capabilities of TCAP to carry the Application Protocol Data Units (APDUs) of other ASEs. By reusing the Release 1 protocols it is possible to base the BC ASE on the Release 1 B-ISUP.

At the UNI we can make a further simplification by noting that routing is not relevant, thus the APDUs of BC, RC, and CC are mapped directly onto the S-AAL. Therefore TCAP might not required in this case.

Thus possible protocol stacks at the UNI and NNI which satisfy this architecture is shown in the following figure.

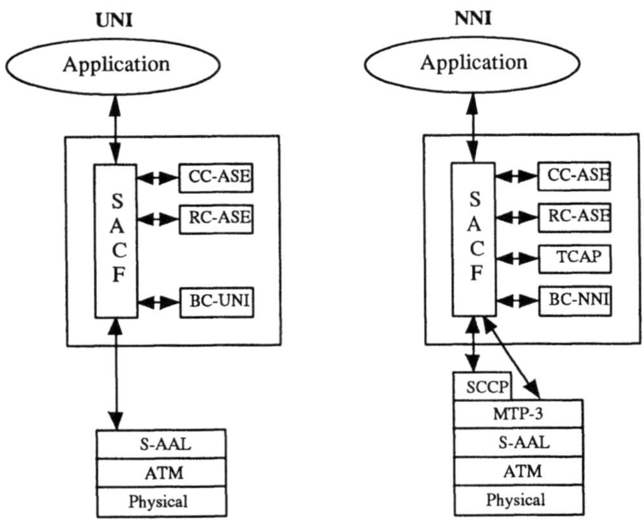

Figure 6. Protocol stack at the UNI and NNI

Note at the NNI that the BC-ASE APDUs map directly onto the MTP3, which provides routing capabilities (OSI layer 3) whereas the ASEs that require the support of the ACSE (TCAP in this case) map APDUs onto the SCCP for routing, the SACF (Single Association Control Function) provides the co-ordination functions between the various ASEs and decides whether to map the APDUs onto the SCCP or MTP3 layer. Notice also that at both UNI and NNI it is possible to use the same CC and RC ASEs - the advantage is that interworking between UNI and NNI protocols is greatly simplified compared to interworking for Release 1 protocols based i.e. Q.93B at the UNI and B-ISUP at the NNI.

5. THE HANDLING OF SPECIAL RESOURCES

A way to understand the reasons why a flexible handling of special resources (in terms of allocation and de-allocation) is needed is to consider an example concerning a multiparty service. For this type of service, to avoid the waste of resources (e.g. connections and bandwidth), it is very important to localize within the network the most appropriate Branching node taking into account the call scenario.

Consider a scenario in which a user living in Italy who wants to establish a call with two users who reside in the USA. It would be very inefficient to perform the Branching functions at the originating exchange (i.e. in Italy) because it would require two connection legs from Italy to the USA (see Figure 7).

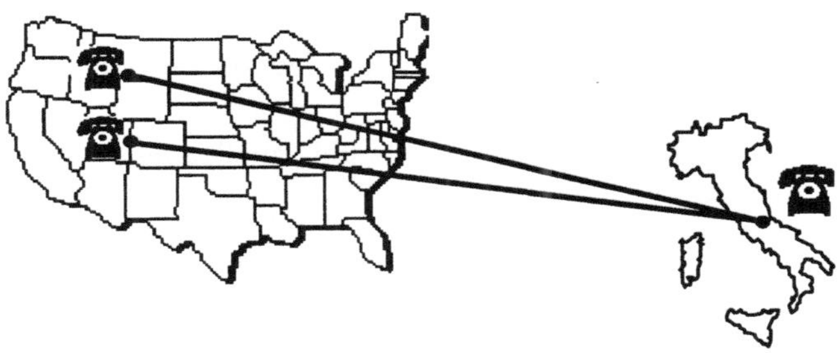

Figure 7. Multiparty call: an inefficient scenario

A more appropriate solution would be to perform the Branching functions in a node located in the USA closer to the two American users as depicted in Figure 8. In this example, the Resource Control functions are therefore used to find the most appropriate location of the Branching node; these functions must be able to adapt themselves dynamically for every configuration of the call taking also into account that not all the nodes could be able to perform "special resource" functions (e.g. branching , video/audio conversion, etc.).

The need for these Resource Control functions is more evident when we consider an already existing point-to-point call that has to be upgraded to a multiparty call. Again, 'how can the signalling system know where the most appropriate location of Branching node is?' The answer to this question is 'by means of the Resource Control'.

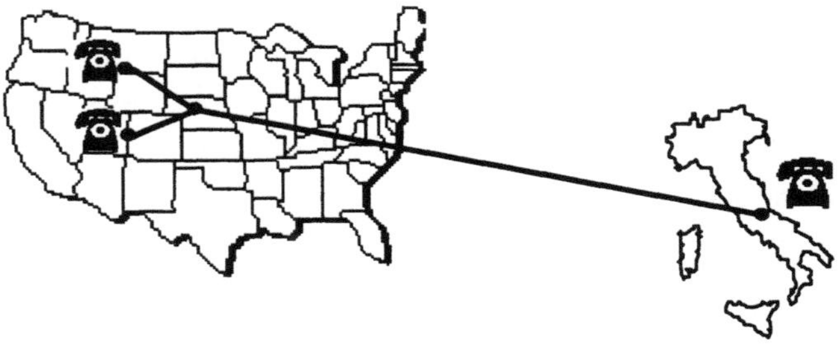

Figure 8. Multiparty call: an efficient scenario

6. ATOMIC ACTION

6.1. The basic principles

To facilitate understanding of the Call Control (CC) and Resource Control (RC) signalling application, this section first briefly introduces CCR (Commitment, Concurrency and Recovery) concepts. It then shows how these concepts can help to define a generic model for multiparty information flows applicable for Call Control.

CCR provides a mechanism for distributed applications to maintain the reliability of dynamically modified data. For CC, this data is the state and attributes of component objects of the call model. CCR calls this bound data. Multiple nodes can engage in an atomic action changing bound data that is distributed among these entities.

Usually, CCR is used for data base applications. But the atomic action mechanism can also be used to support multiparty call control negotiations. No changes have to be made to the CCR mechanism, though the relationship between the users of the mechanism is a little different. In CC applications entity relationships are best characterized as peer-to-peer, whereas data base entities normally function in a client-server relationships.

An atomic action provides a mechanism for a distributed application to develop a system with the following characteristics:

- *Atomicity* - a set of related operations on the bound data are either all performed or none are performed (commit or rollback)
- *Consistency* - the effect of related operations in each node is performed correctly with respect to application semantics
- *Isolation* - partial results are not accessible and do not affect other atomic actions
- *Durability* - effects of operations are not altered by node failure.

Not all of these characteristics are important for CC and RC entities. Atomicity has a slightly different meaning when some operations are mandatory and others are optional. Durability is not so significant if the nodes are protected against failure or if there is little purpose in recovering state after a node is restarted. But consistency and isolation are important for signalling. A protocol that is fundamentally based on a single phase two-party peer-to-peer exchange cannot be "extended" to guarantee isolation and consistency when negotiations involve more than two parties. Different procedures are required.

Figure 9 depicts the basic idea underlying CCR atomic actions. The direct interaction between a pair of function entities (CCs in our example) within a specific instance of an atomic action is defined in terms of a superior/subordinate relationship. This relationship defines the role and functions performed by each entity within an instance of an atomic action. A superior can have more than one subordinate, but a subordinate only interacts with one superior.

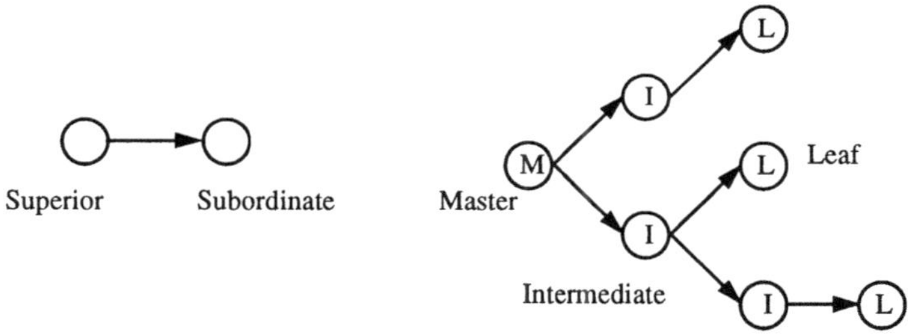

Figure 9. Atomic Action Structure

An atomic action involves two phases. In the first, a superior issues a BEGIN create a branch of an atomic action tree to a subordinate. In response, if it is wants to accept the atomic action, a subordinate replies to its superior with a READY. If not, its response is a REFUSE, rejecting the atomic action. A REFUSE causes the atomic action to fail, causing superiors to generate a ROLLBACK to all subordinates in the atomic action.

For the CC signalling application, a subordinate can REFUSE an atomic action only when it must confirm a mandatory operation contained in a BEGIN. Note that the type of operations within the BEGIN is distinct from the BEGIN function itself. A BEGIN can conceivably contain operations creating, deleting and/or modifying the bound data or "objects" used to define telecommunications services.

In the second phase, the superior instruct its subordinates to COMMIT or ROLLBACK, based on the responses from subordinates in phase one.

In a CC application, decisions to issue a READY or a COMMIT may take into account interactions with other functional entities besides CC. CC interaction with RC and indirectly BC entities can cause a atomic action that might otherwise have succeeded, based purely on negotiation among CCs, to fail. This happens when RC and/or BC cannot provide necessary User-Plane resources for the telecommunications service.

So far, we have focused mainly on the superior/subordinate relationship. These relationships can be extended to create an atomic action tree, as depicted in Figure 9, by allowing a node to have both superior and subordinate relationships.

The root of an atomic tree is called the Master. The Master is a node that is a superior with one or more subordinates, but no superior of its own. An Intermediate is the subordinate of another node, but is also the superior of one or more subordinates. A Leaf is the subordinate of a node and superior to none. Unlike Masters and Intermediates, a Leaf only has a single CC relationship, making their functionality easier to implement.

Figure 10 applies the action tree concept to the CC functional entities. Note that Call Control Agents (CCAs) are always Leaves. All of the other nodes are CCs in the network.

- A user node (square) is a leaf.
- A network node (circle) is a user node's superior.

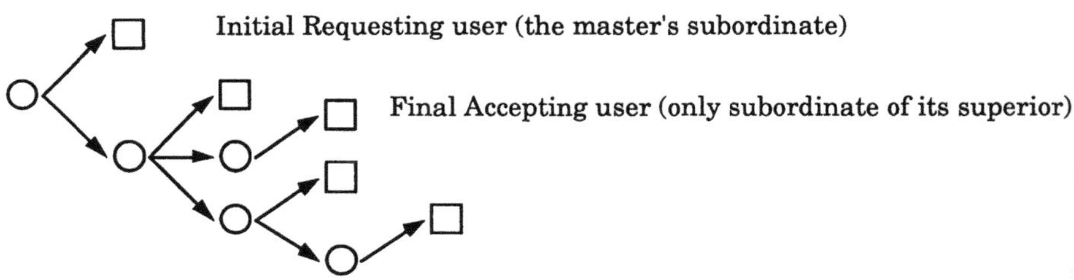

- Intermediate Accepting user (all other user nodes in tree).
- Affected users get information on success, but not in tree.

Figure 10. Atomic Action Structure for CC Entities.

6.2. The application for Special Resource handling

To illustrate better the concept of the Atomic Action for the special Resource handling purpose, it's worth considering some scenarios that could be realistic in a B-ISDN environment.

The following examples show a point-to-point scenario requiring a special resource, for example a converter for video signals. The same examples are applicable also to a point-to-multipoint scenario, since the mastership is owned by the local exchange at the originating side, which is able to manage a multiple parties call.

The first example shows a pre-allocation scenario; it means that the resources needed to support the requested service are allocated before the called terminal(s) is(are) alerted. This strategy could be used when a human user is involved (for instance, for speech services to avoid clipping). In fact, the user answering the call finds the connection already established.

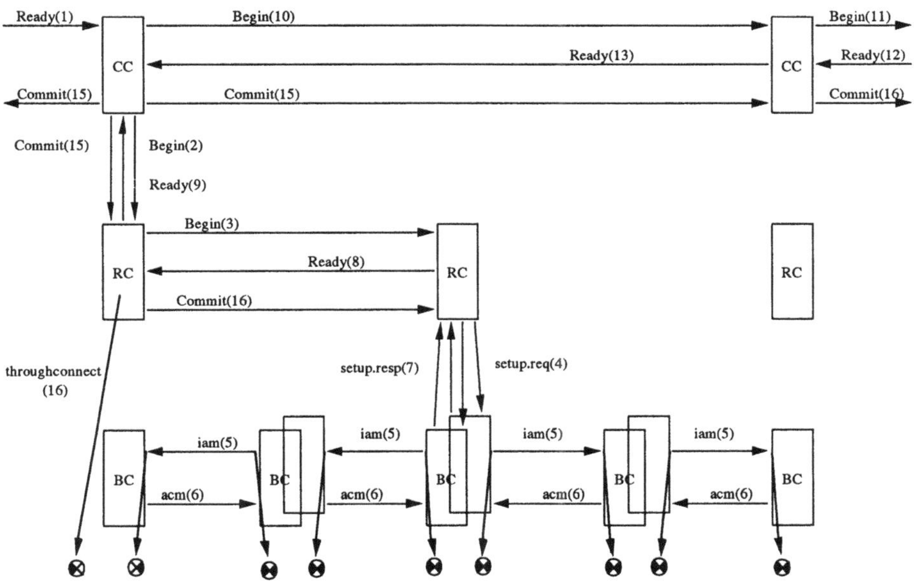

Figure 11. Pre-allocation of resources

The second example shows a post-allocation scenario; it means that the resources needed to support the requested service are allocated after the called terminal(s) is(are) alerted. This strategy could be used when saving resources might appear important, for instance when large bandwidth connections or very specialized resources are requested. In this case, it could be useful to verify the compatibility or the terminal equipments and the availability of the user to accept the proposed service.

The pictures show the message flows and some internal triggers for the two cases. The boxes are not to be intended as actual protocol entities but as logical entities in the sense that for simplifying the drawing, the primitives between the protocols and the Application Process are not explicitly shown. The BC modules that are duplicated in the transit nodes indicate the logical entities for the incoming and outgoing side.

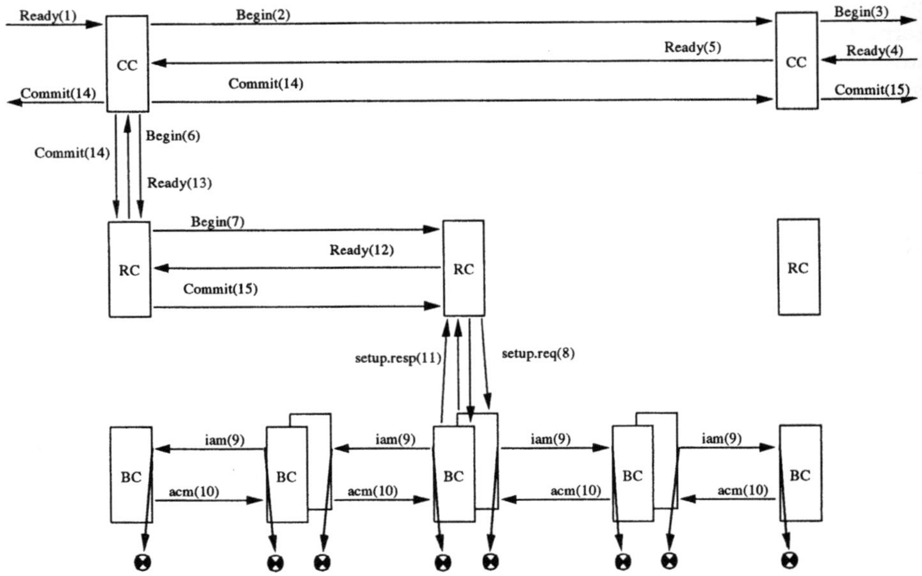

Figure 12. Post-allocation of resources

7. THE INTERFACE BETWEEN THE SIGNALLING PROCESSING AND THE SWITCH

In the evolutionary scenario of B-ISDN, there are a lot of possible configurations for an exchange. We can imagine a complex example in which a local exchange is provided with all the control capabilities, that is CC functionalities, RC governing special resources and BC for connection handling.

In the switch, there are both access signalling protocols and network signalling protocols; interworking can be required between peer level protocols if they are not compatible. Having similar and compatible solutions for UNI and NNI protocol stack avoids the need to provide interworking between the two interfaces. Moreover, there could be the need for interworking with signalling protocols of previous networks (i.e. N_ISDN, B-ISDN Release 1). Now we will give some examples of the impact of the separation of control functions in the switch.

7.1. Impact of RC actions on the special resources located in the Switch

As depicted in Figure 4, the Functional Model shows the Resource Control entity at both UNI and NNI. Therefore, if this solution is adopted, interworking between access side and network side is not required. We can distinguish the involvement of the switch in resource handling, according to the presence or not of the special resources requested by an incoming call. If these resources are managed by the switch, the switch itself is directly responsible of their allocation to a particular call. However, the exchange can request the coordination of other network nodes, that are devoted to RC, because the resources are not owned by a single node, but they have to be shared by the whole network.

This aspect highlights the importance for every node to know the topology of the special resources and their availability. In fact, if the switch can not satisfy the request for special objects, it must be able to re-route the request itself to an appropriate different node. The Resource Control Protocol is activated by the Call Control Protocol and it delegates the Bearer Control Protocol to set up the appropriate connections. The resource allocation phase and connection setting up has to be hidden to Call Control. Only in case of unavailability of the requested resources, the release of the call is notified to Call Control.

7.2. Impact of BC actions on the Switch

As we have proposed in this paper, the BC protocol is able to manage simple point-to-point connections. In this section, we underline three aspects that are of relevance to the switch:

- Connection establishment
- Asymmetrical handling of resources
- Control of the grade of service

With regards to connection establishment, every service can have a different scheme of allocation strategy (resource allocation and through connect). Once the configuration of the Call has been negotiated among the parties, the CC enables the BC via RC to set up the required connections immediately (pre-allocation), or after all the involved parties have been alerted (post-allocation). In both cases, BC must verify, for each connection, the availability of the requested bandwidth and the routing path between the incoming and the outgoing termination.

There can be the necessity to consider a group of connections belonging to a same Call that need a common routing. This function is identified as "Common Route Connection Group" (CRCG). In this case the switch must assure that the outlets of the connections follow the same path in the network; this means that BC must guarantee that all connections are routed to the same adjacent node.

Another function that must be foreseen in a switch, is the handling of unbalanced bandwidth assignment in the opposite transmission directions. For example, video-on-demand requires a small bandwidth in the user-network direction and a very large bandwidth in the opposite one. If this kind of traffic is significant in the switch, it may be necessary to adapt the switching fabric to assure a good Grade of Service (GoS), that is the target performance of the exchange, assuming it is possible to deduce it from the end-to-end parameters of Quality of Service.

The last aspect we consider, is the capability of the switch to guarantee the agreed quality of service and to check that any connection respects the declared traffic parameters after the establishment phase. To satisfy this requirement, the switch must be provided with special units devoted to the periodic detection of parameters of quality of service. If it is possible to control directly the information flows of the sources, the switch could be equipped with policing units. If any violation are detected, the upper layers could manage the situation.

8. CONCLUSION

The next evolutionary step of Broadband ISDN network is the transition from Cross-Connect to full Exchange capability. This implies that to allocate VC connections demanded by the user, signalling procedures are needed. With respect to the procedure used for permanent and semi-permanent connections, the signalling procedures allow every terminal to request the network to establish and release a call and the related virtual ATM connection, therefore only when really needed by the service and with the parameters (e.g. bit rate, Quality of Service) as required by the application.

With the introduction of new and complex on demand services, a signalling systems to allow high flexibility and efficiency on the service handling is required.

This article has focused on Signalling aspects with partcular attention for the handling of special resources since some of the envisaged services will require specialized resources (such as bridging/branching nodes, code/protocol converters), which could be located within the network and dynamically and efficiently allocated according to the instantaneous requests by means of an intelligent control system.

After an overview on signalling requirements and network architecture, a protocol architecture based on three levels has been proposed: Call Control, Resource Control and Bearer Control. The Resource Control and the Call Control protocols are based on Atomic Action principles for which a description have been given. The Bearer Control protocols developed for the UNI and the NNI re-use part of the Q.93B and B-ISUP protocols specified for the Release 1.

To illustrate better the concept of the Atomic Action for the special Resource handling purpose, two examples are proposed. The first shows a point-to-point scenario requiring a special resource (e.g. a converter for video signals) with pre-allocation of resources while the second example shows a case of post-allocation of resources.

Finally we briefly examine the impact of supporting new complex services on the resource management of a switch and propose ways of effeciently utilising the resources of the switch.

9. ACKNOWLEDGMENTS

The authors wish to thank the members of the RACE MAGIC project and in particular M. Siviero (CSELT), G. Marino (ITALTEL), H. Bussey (BELLCORE) and G. Reyniers (ALCATEL-BTM) for their ideas, discussions, and contributions to this effort.

REFERENCES

[1] Martin De Prycker, "ATM Switching on Demand", (IEEE Network Magazine, March 1992)

[2] Andrew Day, "International Standardization of B-ISDN", (IEEE LTS, August 1991)

[3] A. Paglialunga, M. Siviero, "Signalling Protocol Evolution: from Narrowband ISDN to Target Broadband ISDN", (Globecom 1992)

[4] M. Testa, A. Paglialunga, " Evolution of the Signalling Systems architecture and Protocols", (submitted to Supercom-ICC 1994)

[5] Guy Reyniers, "Resource Control in the MAGIC functional Model", (ICC 1994)

[6] Stephen M. Walters, "A new direction for Broadband ISDN", (IEEE Communications Magazine, September 1991)

[7] A. Brill, Dr Manfred N. Huber, B. Petri, "Designing Signalling for Broadband", (Telecommunications, July 1993)

[8] J. Conlisk, "Is ATM the total answer?", (Telephony, June 21, 1993)

[9] W.R. Byrne, G. Clapp, H.J. Kafka, G.W.R. Luderer, B.L. Nelson, "Evolution of MAN to Broadband ISDN, (IEEE Communications Magazine, January 1991)

[10] CCITT Recommendation Q.1400: "Architecture Framework for the Development of Signalling and OAM Protocols Using OSI ALS Concepts"

[11] ITU-T SG11 "ISCP Baseline Document (Report R10 June 1993)"

[12] Rec. ISO/IEC 9545, Information technology - Open Systems Interconnection - Application Layer structure

[13] CCITT Recommendation X.852: Commitment, Concurrency and Recovery (CCR)

Broadband Communications, II (C-24)
S. Tohmé and A. Casaca (Editors)
Elsevier Science B.V. (North-Holland)

The Management of a Pan-European ATM Network - The X Interface

T.Barker (a), A.Macht (b), J.Johansen (c), I.Rodriguez (d), J.Rodrigues (e) and J.V.Ruymbeke (f)

(a) BT, United Kingdom

(b) DBP Telekom, Germany

(c) TeleDenmark Research, Denmark

(d) Telefónica I+D, Spain

(e) Telecom Portugal, Portugal

(f) Belgacom, Belgium

Abstract

This paper summarizes the main aspects of a specification for the International X Interface, used for the management of a Pan-European ATM VP Network, that was carried out under the Eurescom Project P.105 - European ATM Network Studies.

Keyword Codes: C.2.1; C.2.2; C.2.3
Keywords: Network Architecture an Design; Network Protocols; Network Operations

1. Introduction

The decision taken by 18 European PNOs of signing a Memorandum of Understanding for implementing a Pan-European ATM Network in beginning 94, based upon VP Crossconnects, implied in a more accurate way the need to have standardized procedures among these PNOs for managing this network.

A decision was taken in this MoU to have in a first phase local management and manual procedures across the PNO administrative borders and in a second phase a full automated procedures. This type of procedures will based on a specification of the X Interface, the interface between Network Management Systems (NMS) belonging to different PNOs, developed by the EURESCOM Project P105 - ATM Network Studies.

This paper summarizes some of the aspects of this specification, which was developed as an application of the TMN principles to an ATM Network. It has to be mentioned that till now there is no ETSI or ITU-TS standards for the X Interface for the ATM Network. So, this specification, in spite of following the TMN methodology and taking into consideration the information models present in existing standards, has to be considered as pre-standardization work.

It is expected that at least parts of the specification will be forwarded to ETSI/NA5 STC, which is dealing with Broadband Networks, in a short term.

In the first section a reference configuration for the VP based Pan-European ATM Network is described. Afterwards the transport model for the VP Crossconnect, based on which the object model is built, is highlighted. The TMN methodology used to specify the X Interface is shortly described in section 4. In section 5 the functional model, as part of the application of the methodology, is analyzed in detail. Section 6 includes a summary of the information model. The following section contains the communication model, which constitutes the last step of the application of the methodology. Finally a conclusion of the paper is drawn.

2. Reference Configuration

Fig. 2.1 shows the Reference Configuration of the Pan - European VP Based Network.

As it can be seen in the figure, the Network is composed of several PNO Subnetworks, each one controlled by an Operation System (OS). Each PNO Subnetwork can have more than one OS Controlling their VP Crossconnects (VPCC), but from the X Interface point of view, only one OS is seen.

Each PNO Subnetwork is composed of one or more VP Crossconnects (VPCC) operating as international gateways.

For the purpose of managing the VP Connections, a distinction is made between the Subnetwork where the VP Connection is originated (O-PNO Subnetwork), the Subnetwork that provides only transit to this VP Connection (T-PNO Subnetwork) and the Subnetwork where the VP Connection is terminated (D-PNO Subnetwork).

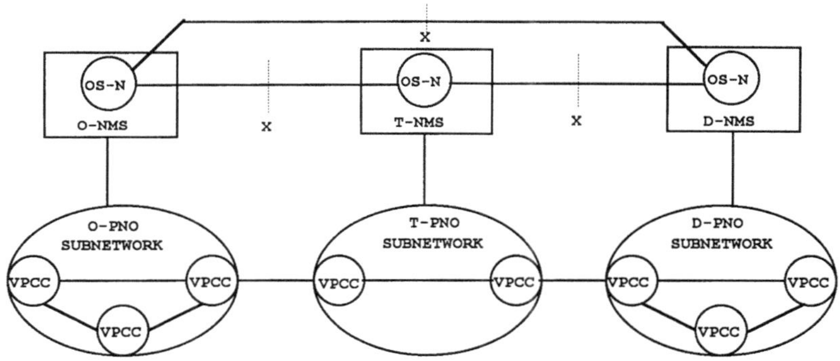

OS-N -	Operation System at Network Management Level
O - NMS -	Originating Network Management System
T - NMS -	Transit Network Management System
D - NMS	Destination Network Management system

Fig.2.1 Reference Configuration for the VP Crossconnect Network

Regarding the TMN organizational model, two options were evaluated: the Joint and the Cooperative Management.

The Joint Management would imply the existence of a Supra-PNO, with a OS that controls, at a certain level, the OSs of the other PNOs. The Cooperative Management puts all OSs of the different PNOs dialoguing at the same level.

In order to avoid the delicate situation of having a Supra-PNO, a preference was given to the Cooperative Management.

However, to overcome the complex problem associated with a complete distributed system, a ruling task was given to the NMS of the PNO where the VP Connection was originated (O-NMS). In this respect, the following assumptions were taken:

- Routing by the Originating Side-

  The choice of which subnetworks will be involved in a VP Connection will be decided by the Originating Subnetwork NMS. However this one will only have a visibility at the subnetwork level, and will not have any knowledge of the internal details and structure of each of the subnetworks involved in the VP Connection.

- Control by the Originating Side -

  The establishment/release/modification of the VP Connection will be always under the control of the Originating NMS. Any event related with the VP Connection occurred at the other PNO Subnetworks, e.g. failure indication, has to be reported to the Originating NMS, which takes the appropriate measures, e.g. reconfiguration.

The hierarchy of management is divided in four levels [1] (fig. 2.2):

Network Element - manages Network Elements (e.g. VPCCs) individually.
Network - manages the whole Network, as a concatenation of Network Elements.
Service - manages a service
Business - manages a corporate business network.

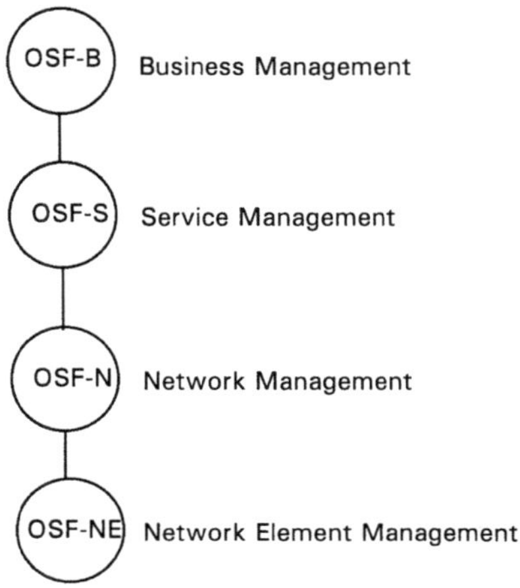

Fig.2.2 The Management Hierarchy

The specification of the X Interface was concentrated in the Network Management level, eventhoug it covers also some aspects of the Service Management level, as it will be highlighted in the following sections.

3. The Transport Model

The specification of the X Interface used the transport entities, described in the ITU-T Rec.G.803 [2] and in the ETSI DE/TM-3010 [3], for the Transport Network Architecture.

The transport entities defined in these recommendations used in the specification were the following:

- Trail - It is a transport entity responsible for the integrity of the information transferred between access points.

- Connection - It is a transport entity that transfers information, but does not insure the integrity of the transfer. A trail is made of a concatenation of several connections. Two types of connections are defined:

- Link Connection - It is a transport entity provided by the client/server association. It provides static connectivity between connection points.

- Subnetwork Connection - It is a transport entity formed by a connection across a subnetwork. It provides flexible connectivity between connection points.

Fig.3.1 shows some of the transport entities that are relevant for the X Interface specification.

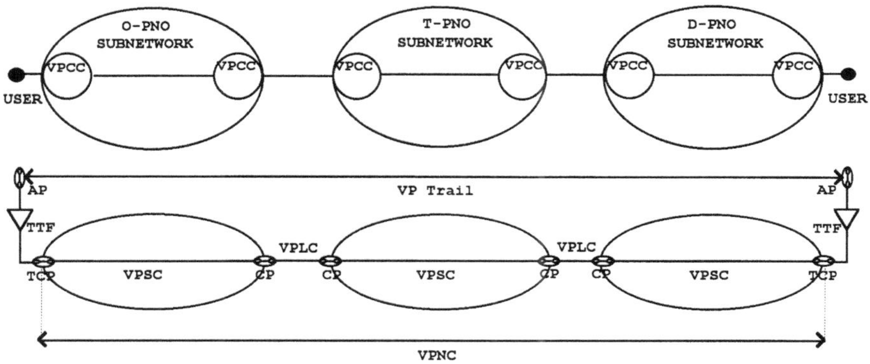

AP - Access Point
CP - Connection Point
TCP- Termination Connection Point
TTF - Trail Termination Function

VPSC - VP Subnetwork Connection
VPLC - VP Link Connection
VPNC - VP Network Connection

Fig.3.1 The Transport Model applied to the VPCC Network

According to the recommendations quoted above, a trail and in this case a VP Trail is composed of a nearend trail termination function (TTF), a network connection and a farend trail termination function.

The Trail Termination Functions (TTF) provide the insertion/extraction and processing of the overheads needed to insure the integrity of the information transferred between access points. In the case of the VP Trail the overhead flow is the F4.

According to DE/TM-3010, when the Subnetworks define administrative boundaries between PNOs, which is the case for the X Interface Specification, there is a need to monitor independently each Subnetwork Connection. For that purpose segment F4 flows are used, in the case of a VPSC.

This situation is modeled by creating a sublayer network (see fig. 3.2).

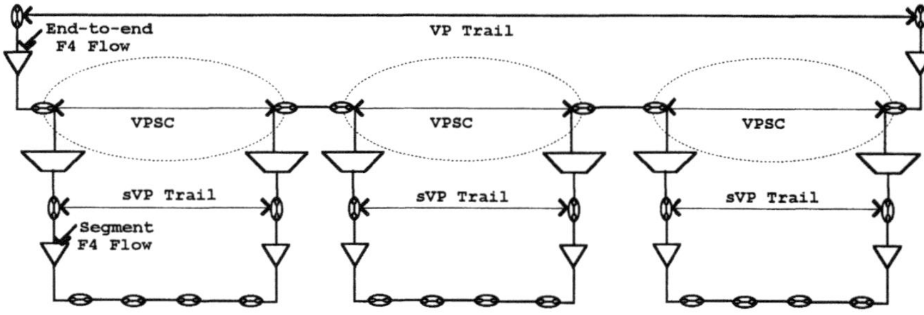

Fig.3.2 Sublayering for VPSC monitoring with Segment F4 Flow

The Trail Termination Functions at the VP sublayer network insert/extract and process the Segment F4 Flow. This flow can cross the boundary of a PNO Subnetwork, and in this case the sVP Trail extends for several subnetworks. This requires the mutual agreement among the operators involved.

To complete the Transport Model in what concerns the needs of the X Interface specification, the path layer at the interconnection of the PNO Subnetworks is shown in fig.3.3.

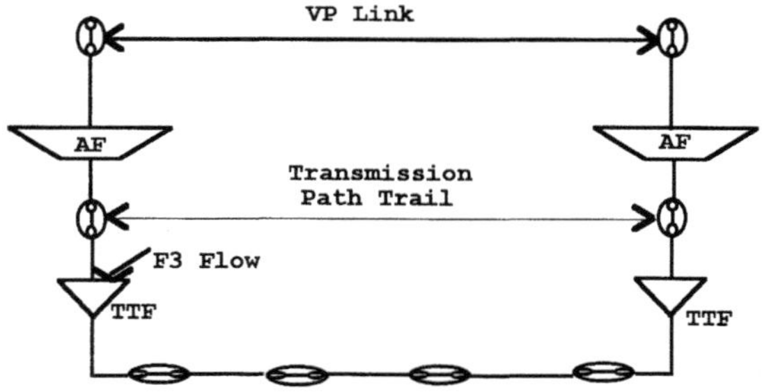

AF - Adaptation functions

Fig 3.3 Path Layer of the Transport Model

The Adaptation Function (AF) represents the association between a connection point in the client layer network, in this case the VP Layer, and an access point in the server layer network, in this case the Physical Layer. This function adapts the information from the VP layer to a form suitable for transport in the Physical Layer (e.g. Cell Rate Decoupling).

The information model described in section 6 was built upon the transport entities described in this section.

4. Applying the M.3020 Methodology

The specification of the X Interface followed the M.3020 [4] methodology.

The application of this methodology starts with the identification and description of a Management Service (MS), which is defined as an area of management activity described from the user perception of the OAM requirements.

The next step is the decomposition of the Management Service into Management Service Components (MSCs), defined as the constituent part of a MS, stating the requirements for actions to be performed on the managed network.

The MSCs are further decomposed into Management Functions, seen as the smallest part of a MS as perceived by the user of the service. These functions act upon the Managed Objects.

A further stage in the application of the methodology consists on the Object Modelling. This is made of the identification and description of managed object classes that support the functions.

The last step followed was the determination of communication requirements, selection of protocol suites for the Interface and the specification of the CMISE services.

These steps represent a simplification regarding the 13 tasks described in the M.3020 methodology.

The steps were performed in an iterative way and not in a strict sequence, which means that the results of the object modeling were feedback into the definition of the functions.

The process can be summarized as follows:

- - define the management services
- - obtain the components of the management services
- - obtain the functions required for these components
- - obtain the object model
- - map the objects to the functions (and vice versa) to consolidate the model
- - define the management communication requirements

5. Functional Model

5.1. Introduction

In order to be able to produce a full description of the X interface for the proposed reference configuration, a service had to be selected to be modeled. *VP Bearer Service Provisioning* was chosen as the initial service, as this service was seen as having the highest priority for the establishment of a Pan-European network and would involve the highest level of inter-operator management traffic across the X interface.

The M.3020 [4] methodology, briefly described above in section 4 and as applied to this project, can be shown pictorially in figure 5.1.

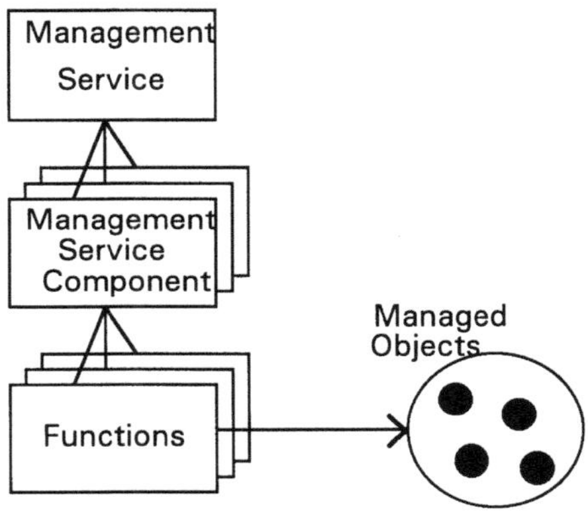

Fig.5.1 Decomposition of the Management Services

The Management Service chosen, the VP Bearer Service Provisioning, includes the establishment/release/modification of a VP Connection, as well as fault and performance management on a VP Connection already established.

5.2. Management Service Description

The 'traditional' management functional areas were chosen for the VP Bearer Service Provisioning, these being Fault, Configuration, Accounting, Performance and Security (FCAPS). This allowed the work to be aligned with existing standardization work and other management projects being undertaken in the participants company's.

The decomposition of the VP Bearer Service Provisioning into Management Service Components (MSCs) belonging to the different management areas is shown in figure 5.2.

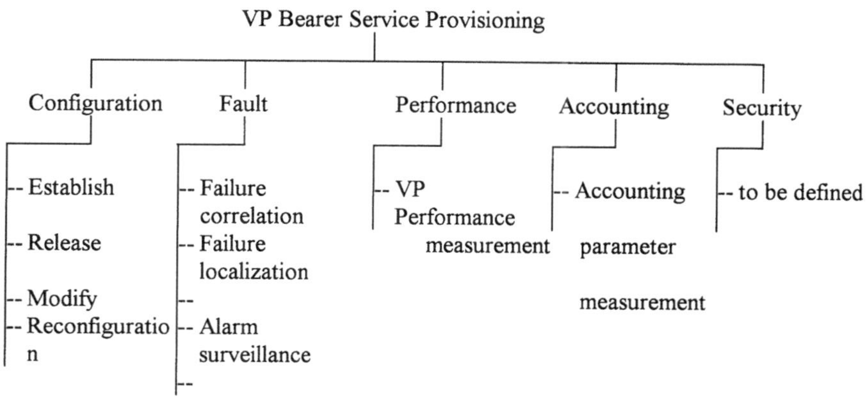

Figure 5.2 VP Bearer Service Provisioning Management Service Components

For the Configuration management functional area, four MSCs have been identified which could be briefly be described as follows:

- *Establish MSC* - It allows the reservation, activation and the initial testing of a VPC.
- *Release MSC* - It allows the cancellation of a reservation or the deactivation of a VPC.
- *Modify MSC* - It allows the modification of the characteristics of a VPC, during activation time, e.g. the cell rate.
- *Reconfiguration MSC* - It allows the reconfiguration of the whole or a part of a VPC, when a unrecoverable failure is detected.

For the Fault management functional area, four MSCs have been identified:

- *Failure Correlation MSC* - It provides information on any failure affecting the VPC and the correlation between failures.
- *Failure Localization MSC* - It allows the localization of a detected failure, which localization is unknown.
- *Alarm Surveillance MSC* - It gives information on the alarms set in the Network Elements which can affect the VPC.

For the Performance management functional area only one MSC was identified:

- *VP Performance Measuring MSC* - It allows the measuring of the performance across segments or the whole VPC, using in-built tools on the Network Elements - the OAM flows.

For the Accounting management functional area only one MSC was identified:

- *Accounting Parameter Measurement* - It allows the transfer of information of the accounting parameters measured for the VPC.

For the Security management functional area no MSCs were identified by the time the specification was completed.

Unfortunately due to lack of time the areas of Accounting and Security were not covered in the specification.

5.3. Management Service Components Description

At the X interface level each MSC has been described both in terms of input and outputs across the X interface and also by SDL diagrams showing the interaction with the other network operators.

For the Establish MSC (controlling the setup of VP connections) fig. 5.3 shows the part of the SDL diagram concerned with the initial verification of the connection with the destination network operator and the procedure for reserving the VP Subnetwork Connections. This SDL describes the process at the Originating Side. Other SDLs were produced describing the processes at the Transit and Destination PNOs.

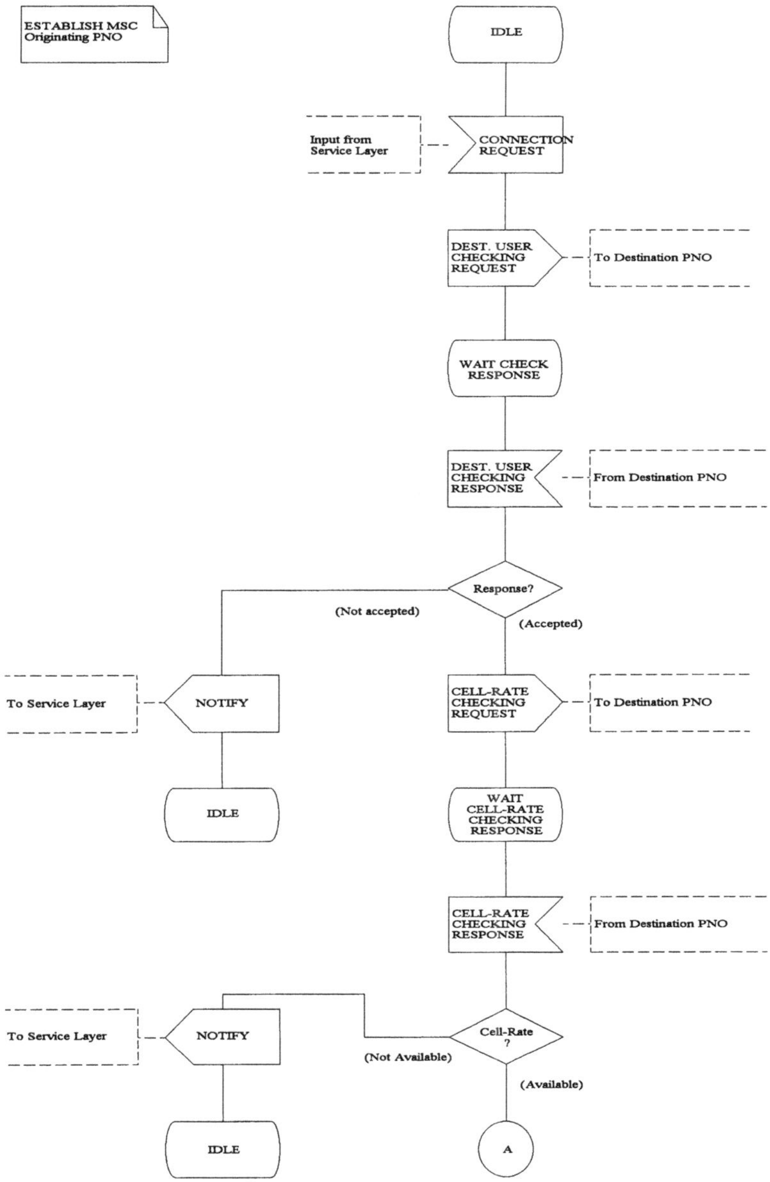

Fig.5.3 SDL describing part of the Establish MSC process

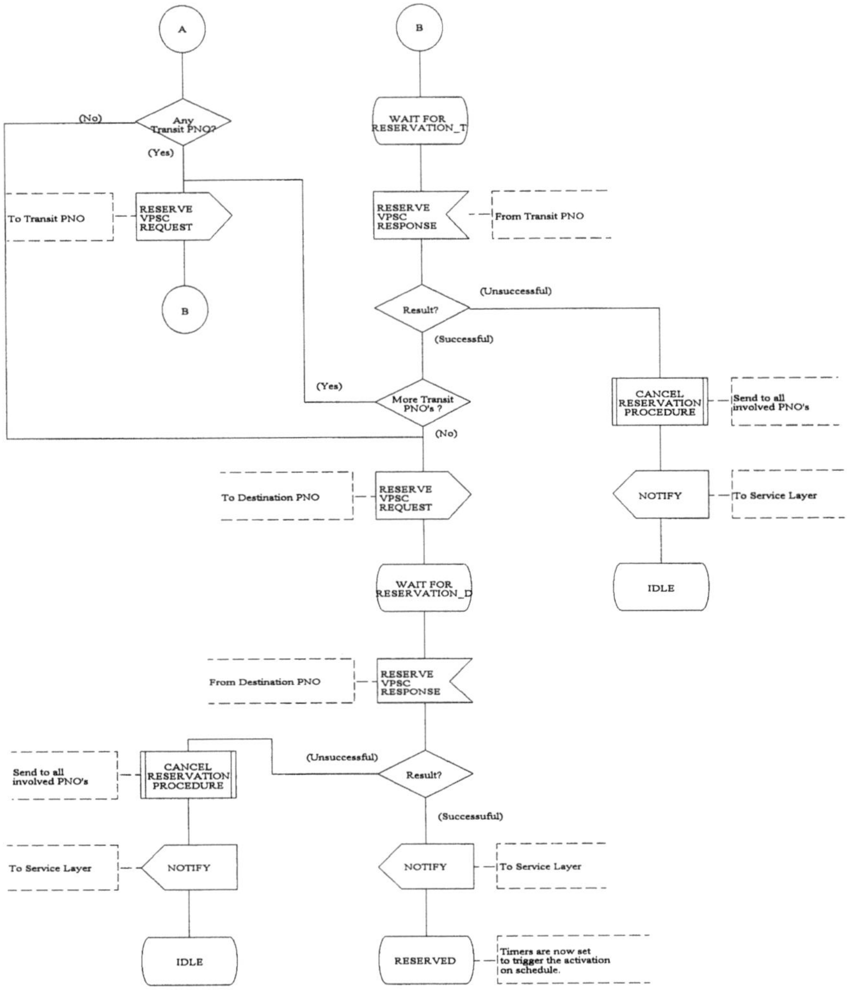

Fig.5.3 (Conclusion) SDL describing part of the Establish MSC process

5.4. Functions Description

Each Management Service Component makes use of a number of functions; a function being the lowest level of functionality of an MSC. Functions can be reused across several MSCs in order to simplify the modelling process and for this project many functions are indeed reused.

Associated with functions are request primitives and in the case of confirmed functions, response primitives. It is these primitives that are mapped to Common Management Information Service Element (CMISE) services.

The function description have been primarily concerned with the X interface. An example is given for the function *Reserve VP Subnetwork Connection* which is a part of the *Establish MSC:*

Description

This function requests the other involved PNOs to reserve a path between the specified input and output points of their subnetwork, or between the input port and the user, in case of the destination subnetwork. The other PNOs respond with the result of the reservation.

It is a confirmed type of operation. The following parameters are associated with the Request and Response primitives of the function:

Request -	VP Connection Id	Identity assigned to the connection
	Inter-PNO VP Link Connection Id (nearend)	Identification of the nearend Inter-PNO VP Link connection
	PNO Subnetwork Id (farend)	Identification of the farend PNO Subnetwork
	or Destination Address	E.164 address of destination user
	Destination VPI *(optional)*	VPI to be used by the Destination PNO only to allow user transparency of reconfiguration
	Forward traffic descriptor	Peak cell rate from source to destination
	Backward traffic descriptor	Peak cell rate from destination to source
	Forward additional traffic descriptor	Includes mean cell rate for connection
	Backward additional traffic descriptor	Includes mean cell rate for connection
	Forward QoS class	An indication of the class of QoS
	Backward QoS class	An indication of the class of QoS
	Group identifier	Group reference for this connection
	Schedule	List of activation times, dates and durations
Response	Inter-PNO VP Link Connection Id (farend)	Identification of the farend Inter-PNO VP Link Connection

Reservation result	Result of the reservation attempt. Future work could specify the details to be sent, e.g a list of intervals of time that could be reserved, in the case of negative result

Using the descriptions of the functions together with the MSC descriptions it has then been possible to perform the object modeling (described in section 6) and the communication models (described in section 7).

6. Information Model

6.1. Introduction

The physical equipment involved in a VP Connection, from an Originating User inside the responsibility area of the Originating PNO, through a Transit PNO area, to a Destination User inside the responsibility area of the Destination PNO is shown in figure 6.1.

A PNO area is connected to some of the other PNOs by bundles of one or more physical links. Inter PNO VP Connections between PNO areas can be set up inside the physical link,s identified by a VPI value.

A VP Connection from an Originating User to a Destination User consists of Inter PNO VP Connections and PNO VP Subnetwork Connections.

PNO VP Subnetwork Connections can be of three types:

1. From the Originating User to an Inter PNO VP Connection (in an Originating PNO).

2. Between two Inter PNO VP Connections (in a Transit PNO).

3. From an Inter PNO VP Connection to the Destination User (in an Destination PNO).

Two facts are important for understanding the object model at the X interface:

1. The internal structure of another PNO area is not relevant for the X interface.

2. The bundle of physical links between two PNOs is their shared responsibility and the information about a bundle has to be represented at both PNOs.

Every PNO in the Pan European Network (PEN) has knowledge about all other PNOs which are part of PEN and of the bundles interconnecting them.

In the object model a bundle of physical links is represented by an Inter PNO Topological ATM Path Trail object in both involved PNOs. An Inter PNO VP connection is represented by two VP Connection Termination Point objects one in each of the involved PNOs.

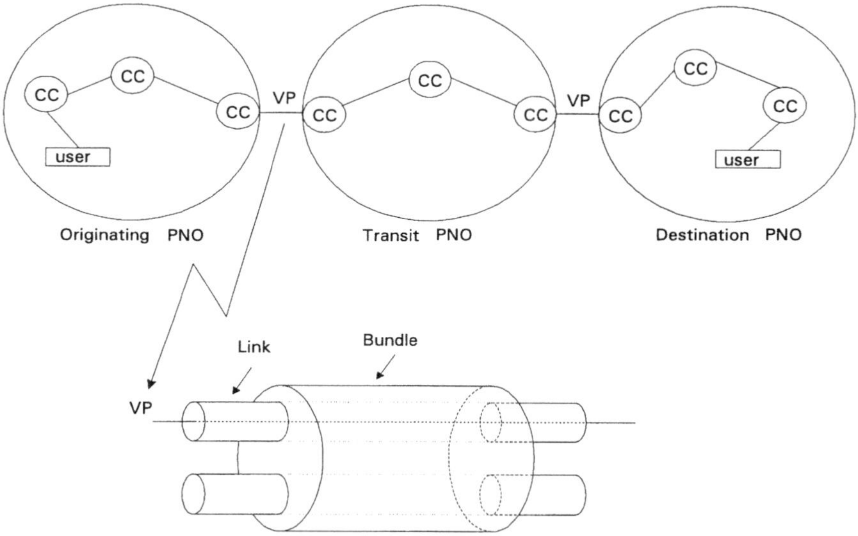

Fig. 6.1 Physical Equipment for VP connection

In the following, the view that one PNO has on the objects contained in other PNOs for different purposes are shown in figures and explained shortly to help the reader of the object descriptions.

The description follows the natural order of functions performed to set up a connection and manage it. Object names are written in bold. The big circle in all figures represents a PNO domain.

When a new PNO is connected to the Pan European VP ATM Network, all already connected PNOs are notified of the creation of one **PNO VP Subnetwork** object and a number of **Inter PNO Topological ATM Path Trail** objects.

An **Inter PNO Topological ATM Path Trail** object holds information needed for routing and represents the total number of physical links to one specific PNO. The individual physical links are represented by **PNO ATM Path Trail Termination Points** and inside a **PNO ATM Path Trail Termination Point** several **VP Connection Termination Points** can be created, which terminate one end of a **PNO VP Subnetwork Connection.**

Before the routing process starts, a compatibility check is made over the **Destination User** object in the Destination PNO, identified by the addressing information given by the Originating User. If this is successful, then an action is made over the Destination **PNO VP Subnetwork** object. This action returns information consisting of a list of **PNO ATM Path Trail Termination Points** which can accommodate a connection with the requested parameters. If there are any usable termination points in the list the routing process starts.

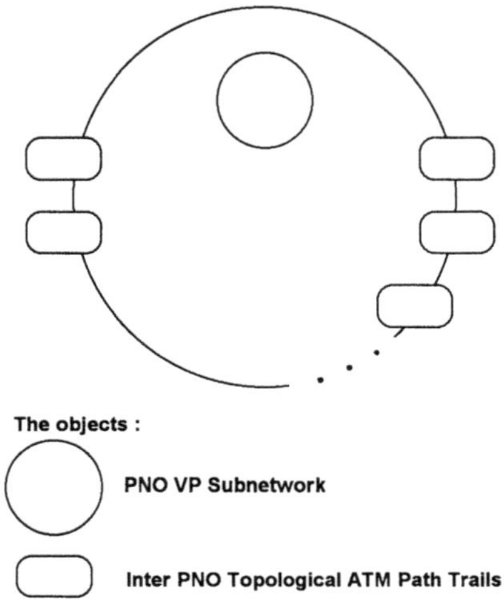

Fig. 6.2 Originating PNO view on other PNOs for routing purposes

From a routing point of view (see figure 6.2) an Originating PNO, which is requested by an Originating User to make an VP Connection to a Destination User, sees the other PNOs as entities interconnected by **Inter PNO Topological ATM Path Trails**.

An **Inter PNO Topological ATM Path Trail** has a pointer to the PNO it is connected to, and consist of one or more Inter PNO ATM Path Trails identified by their **PNO ATM Path Trail Termination Points** and their associated Maximum Available Cell Rate and QOS class.

From the information contained in the **Inter PNO Topological ATM Path Trails**, the information given by the Originating User and the information received from the destination **PNO VP Subnetwork** object, the originating PNO selects a set of **PNO ATM Path Trail Termination Points** to be used for the requested VP connection. The set of **PNO ATM Path Trail Termination Points** consist of two **PNO ATM Path Trail Termination Points** for Transit PNOs and one **PNO ATM Path Trail Termination Point** for the Destination PNO.

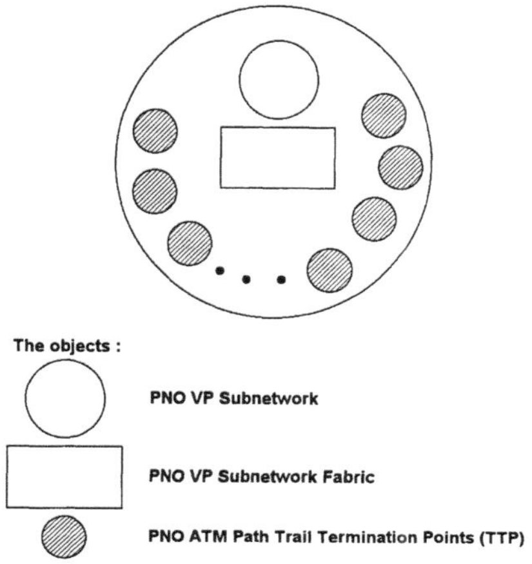

Fig. 6.3 Originating PNO view on other PNOs for the purpose of setting up a VP connection

The Originating PNO (after checking in the **PNO VP Subnetwork Fabric** objects that all subnetwork connections can be set up) orders the set up of subnetwork connections to the **PNO VP Subnetwork Fabric** object in all involved PNOs.

In the Transit PNOs a subnetwork connection is set up between two **VP Connection Termination Points** in the selected **PNO ATM Path Trail Termination Points**. In the Destination PNO between one **VP Connection Termination Point** in the selected **PNO ATM Path Trail Termination Point** and the Destination User (figure 6.3).

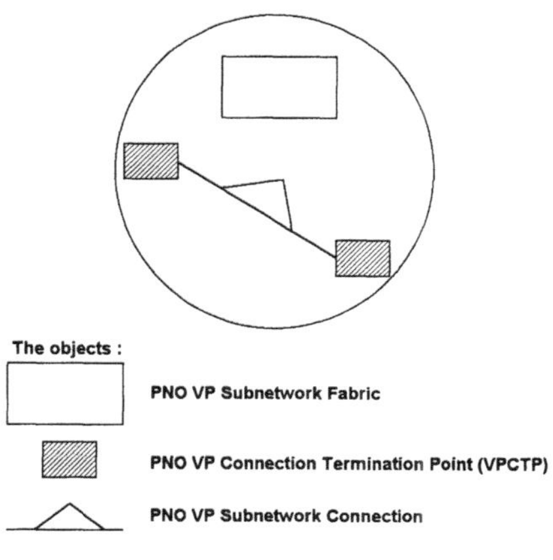

Fig. 6.4 Originating PNO view on a Transit PNO for managing a VP subnetwork connection

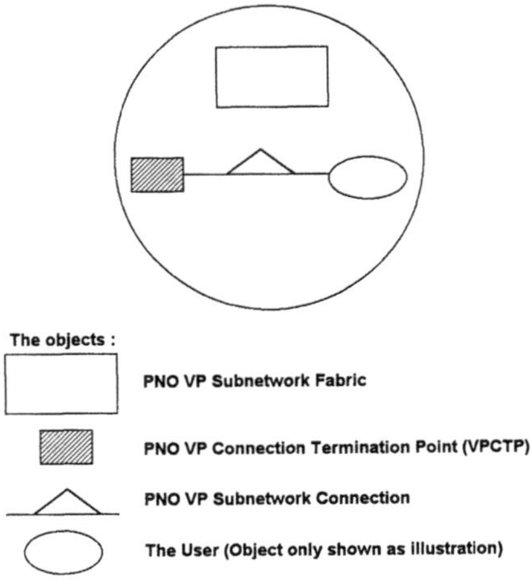

Fig. 6.5 Originating PNO view on a Destination PNO for managing a VP subnetwork connection

An Originating PNO can activate and deactivate a **PNO VP Subnetwork Connection** through the **PNO VP Subnetwork Connection** object.

Within a **VP Connection Termination Point** object, Continuity Monitor objects, Performance Monitor objects and Loopback Monitor objects can be created for performing their specific tasks.

Through the **PNO VP Subnetwork Fabric** object the parameters for an activated connection can be changed and finally a connection can be deleted (figures 6.4 and 6.5).

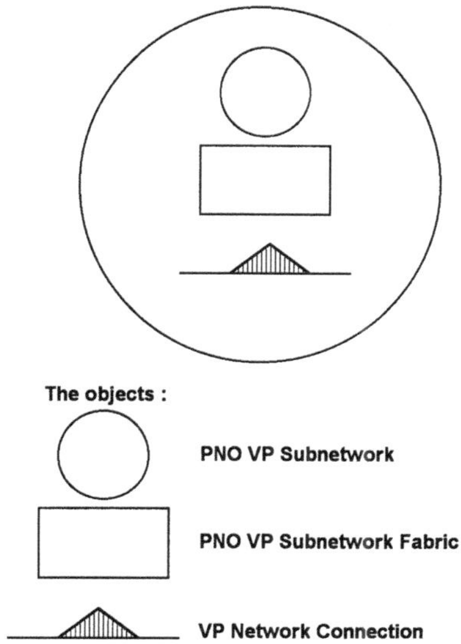

Fig. 6.6 Other PNOs view on the Originating PNO for managing a VP subnetwork connection

The **VP Network Connection** object represents a VP Connection from the Originating User to the Destination User. A **VP Network Connection** can be deleted by a Destination PNO (figure 6.6).

Notifications of faults and other events in a **PNO VP Subnetwork Connection** object are sent to the Originating PNO.

6.2. Description of the Information Model

Fig.6.7 and Fig.6.8 show the inheritance tree and the naming tree, respectively.

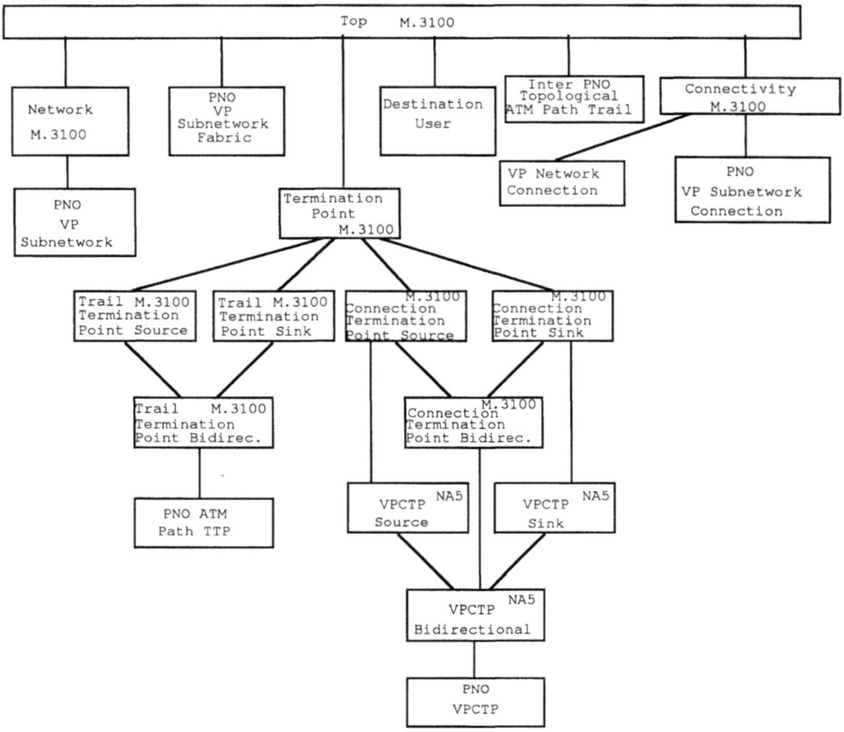

Fig.6.7 Inheritance Tree

From the inheritance tree it can be seen that many object classes are derived from those defined in Rec.M.3100 [5] (signalled with M.3100) and in the DETS DE/NA-52210 [6] (signalled with NA5).

It has been the aim of the Eurescom P105 project to create an information model where the management of an ATM connection, passing through several networks belonging to different PNOs, can be performed in an efficient way, maintaining the integrity of the involved PNOs.

Examples illustrating this are the *giveAvailableLinks* action performed on the *pnoVpSubnetwork* object and the *vpTestState* attribute of the *pnoVpSubnetworkConnection* object.

As it was referred in section 2, it was decided in the project to perform routing by the Originating Side, meaning that the PNO, where the VP Connection was originated, decides the route on its own. This is done step by step, starting from the Originating PNO requesting first a VP Subnetwork Connection to the neighbour PNO, from an available input to this PNO to an output to the next PNO network. This procedure goes on in the same way through more PNO networks until the Destination User is reached.

To assist in the successful execution of this procedure, the *giveAvailableLinks* action can be performed on the Destination PNO's *pnoVpSubnetwork* object. The result of this action is, as mentioned in the previous section, only a list of links on which the Destination PNO will accept the requested VP Connection, and it does not reveal the actual available bandwith on the Inter PNO Links to this PNO.

The *vpTestState* attribute gives to a Network Operator the freedom to decide if it will allow another Operator to use Segment F4 Flow on a particular VP Subnetwork Connection inside its own domain.

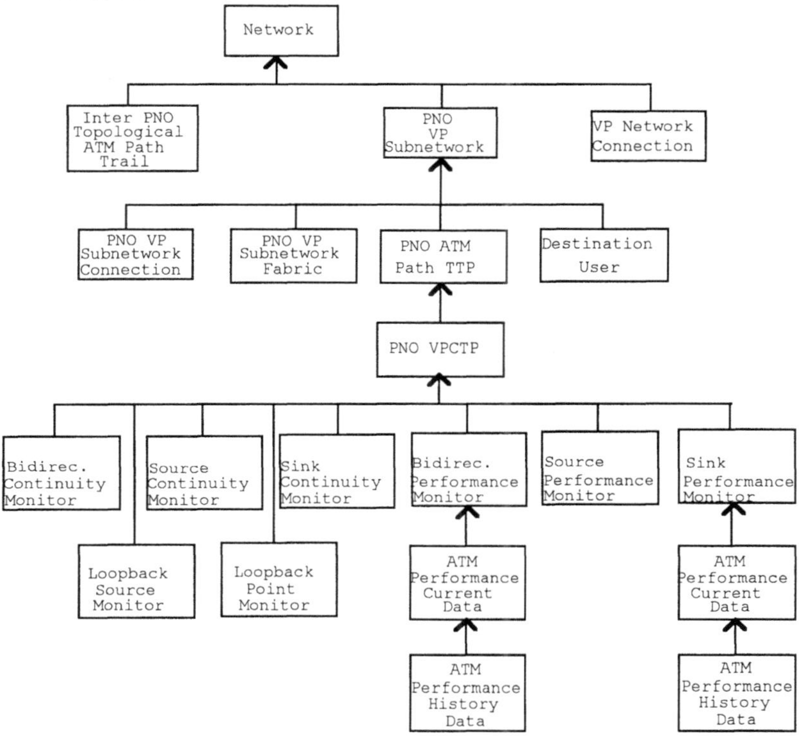

Fig.6.8 Naming Tree

As an example of an object defined, the formal description of the pnoVpSubnetworkFabric object is given:

pnoVpSubnetworkFabric MANAGED OBJECT CLASS

```
DERIVED FROM "M.3100":top;
CHARACTERIZED BY
pnoVpSubnetworkFabricPackage PACKAGE
    BEHAVIOUR pnoVpSubnetworkFabricDefinition BEHAVIOUR
        DEFINED AS
        "A pnoVpSubnetworkFabric Object represents an Operator Subnetwork fro
        the connectivity point of view. It manages the establishment (reservatior
        modification and release of Subnetwork Connections. S
        pnoVpSubnetworkConnection object instances are created when connectio
        are requested.
        There is only one instance of this object per PNO Subnetwork."
    ;;
    ATTRIBUTES
    subnetworkFabricId GET,
    "Recommendation X.721":administrativeState GET,
    "Recommendation X.721":operationalState GET;

    NOTIFICATIONS
    reportContinuityCheck,
    reportPerformanceMonitoring,
    reportLoopback;

    ACTIONS
    reservePnoVpSubnetworkConnection,
    removePnoVpSubnetworkConnection,
    changeReservation,
    cancelChange,
    activateChange;
    ;;

REGISTERED AS { ObjectClass x };
```

The behaviour of this object is expanded as follows:

When an Originating PNO wants to set up a VP Connection through another PNO, it performs the *reservePnoVpSubnetworkConnection* action indicating as parameters the identification of the nearend VP CTP (Connection Termination Point), the PNO to be reached, the requested QoS class, and the time schedule with the associated traffic descriptor.

Activation and deactivation of the VP Subnetwork Connection is done through the *administrativeState* attribute of the *pnoVpSubnetworkConnection* object.

The cancelation of the reservation of the VP Subnetwork Connection is done through the *removePnoVpSubnetworkConnection* action.

The parameters of an activated VP Subnetwork Connection can be modified. This is controlled by three actions - *changeReservation, cancelChange and activateChange.*

The procedure for performing a change starts with the Originating PNO asking all the involved PNOs (using the *changeReservation* action) if they can accommodate the requested change. If all replies are positive, the change is effectuated using the *activateChange* action. If negative responses are received, the change cannot be executed, and all involved PNOs are informed through the *cancelChange* action.

The notifications *reportContinuityCheck, reportPerformanceMonitoring, reportLoopback* are used to give to the Originating PNO the results of Continuity Check, Performance Monitoring and Loopback tests performed on its behalf.

7. Communication Model

The Communication Model developed in the project included communication scenarios between OS-Ns for supporting the management functions, the protocol suites to be applied in the X Interface and finally the specification of the CMISE services that support the management functions.

For illustrative purposes, the communication scenario for the management functions associated to the Establish MSC is described, showing the chaining of the main functions of this MSC.

The boxes represent managed object classes. The CMISE services that support the management functions are shown over the arrows.

1) Destination User Checking

This function is used for the Originating PNO to ask the Destination PNO if the Destination User is able/wants to support the VP Connection.

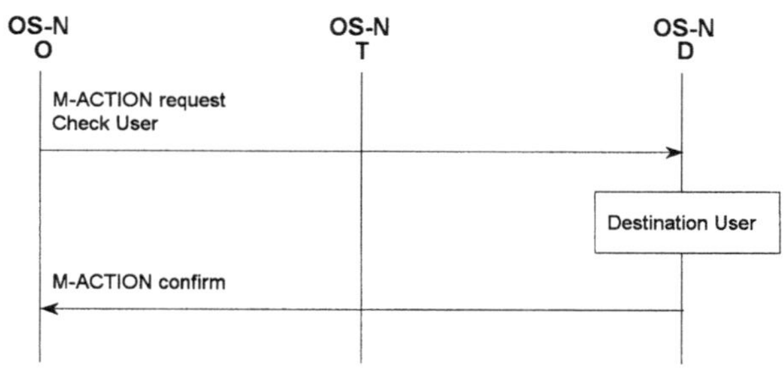

2) Check Available Cell Rate

This function is used for the Originating PNO to request the Destination PNO to give information on the inputs links that can support the VP Connection.

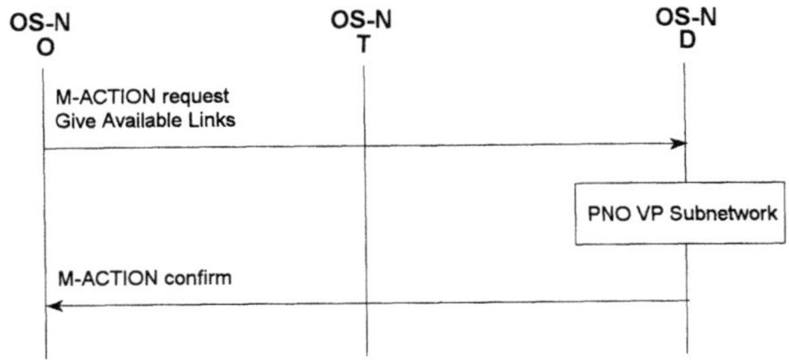

3) Reserve VP Subnetwork Connection

This function is used for the Originating PNO to request the other involved PNOs, to reserve a VP Subnetwork Connection between a specified input point and an output to other PNO or to the Destination User.

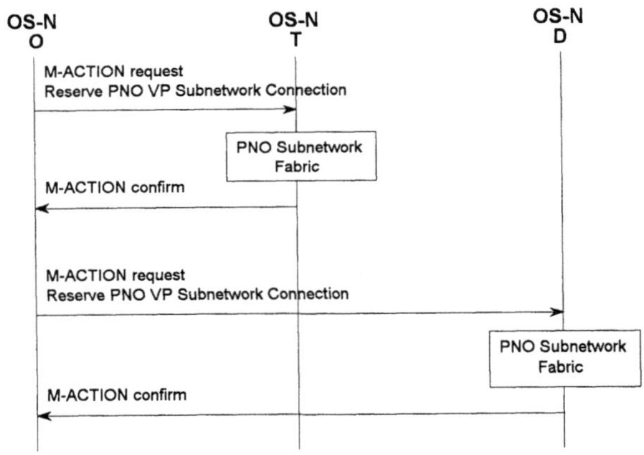

4) Activate VP Subnetwork Connection

This function is used for the Originating PNO to request the other involved PNOs, to activate a VP Subnetwork Connection that was previously reserved.

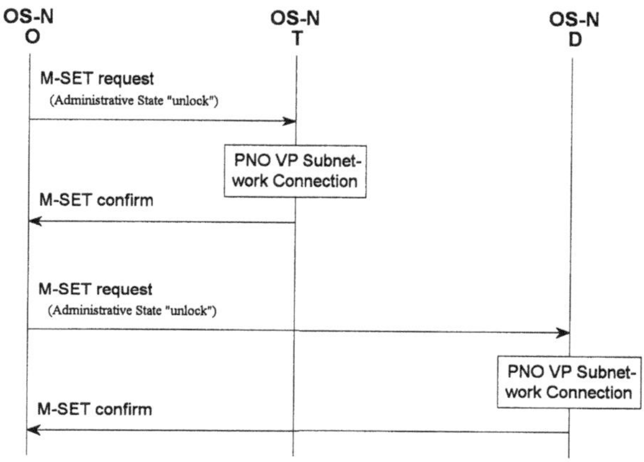

8 Conclusion

This paper gives a brief summary of the work carried out in the Eurescom project P105 in what respects the specification of the International X Interface.

This specification is covering only a Management Service, the VP Bearer Service Provisioning. The M.3020 methodology was followed in this specification and its different steps were briefly described in this paper.

For progressing the work in the specification of the X Interface for the ATM Network, further work has to be done for other Management Services, and on the functions needed between neighbour PNO for e.g reconfiguring Inter PNO Links or establishing new ones.

Acknowledgements

The authors would like to thank, in particular, the leader of the task on Management of the VP Crossconnect Network of P105, Mr.P.Chas, for his support on the work carried out in the X Interface specification, and the leader of the project P105, Mr.P.Adam, who helped also to make possible this specification.

An acknowledgement also to Eurescom, because if the project P105 did not exist this work should not be carried out at least in the same terms.

References

1. CCITT Rec.M.3010 - Principles for a Telecommunications Management Network

2. CCITT Rec.G.803 - Architectures of transport networks based on the synchronous digital hierarchy(SDH).

3. ETS DE/TM-3010 - Harmonization of Transport Network Architecture and Protocol Reference Model for the Transport of ATM Cells.

4. CCITT Rec.M.3020 - TMN Interface Specification Methodology

5. CCITT Rec. M.3100 - Generic Network Information Model.

6. ETS DE/NA-52210 - B-ISDN Management Architecture and Management Information Model for the ATM Crossconnect.

IFIP

The INTERNATIONAL FEDERATION FOR INFORMATION PROCESSING is a multinational federation of professional and technical organisations (or national groupings of such organisations) concerned with information processing. From any one country, only one such organisation – which must be representative of the national activities in the field of information processing – can be admitted as a Full Member. In addition a regional group of developing countries can be admitted as a Full Member. On 1 October 1993, 44 organisations were Full Members of the Federation, representing 65 countries.

The aims of IFIP are to promote information science and technology by:

- fostering international co-operation in the field of information processing;
- stimulating research, development and the application of information processing in science and human activity;
- furthering the dissemination and exchange of information about the subject;
- encouraging education in information processing.

IFIP is dedicated to improving worldwide communication and increased understanding among practitioners of all nations about the role information processing can play in all walks of life.

Information technology is a potent instrument in today's world, affecting people in everything from their education and work to their leisure and in their homes. It is a powerful tool in science and engineering, in commerce and industry, in education and adminstration. It is truly international in its scope and offers a significant opportunity for developing countries. IFIP helps to bring together workers at the leading edge of the technology to share their knowledge and experience, and acts as a catalyst to advance the state of the art.

IFIP came into official existence in January, 1960. It was established to meet a need identified at the first International Conference on Information Processing which was held in Paris in June, 1959, under the sponsorship of UNESCO.

Organisational Structure
The Federation is governed by a GENERAL ASSEMBLY, which meets once every year and consists of one representative from each Member organisation. The General Assembly decides on all important matters, such as general policy, the programme of activities, admissions, elections and budget.

IFIP ORGANISATION CHART

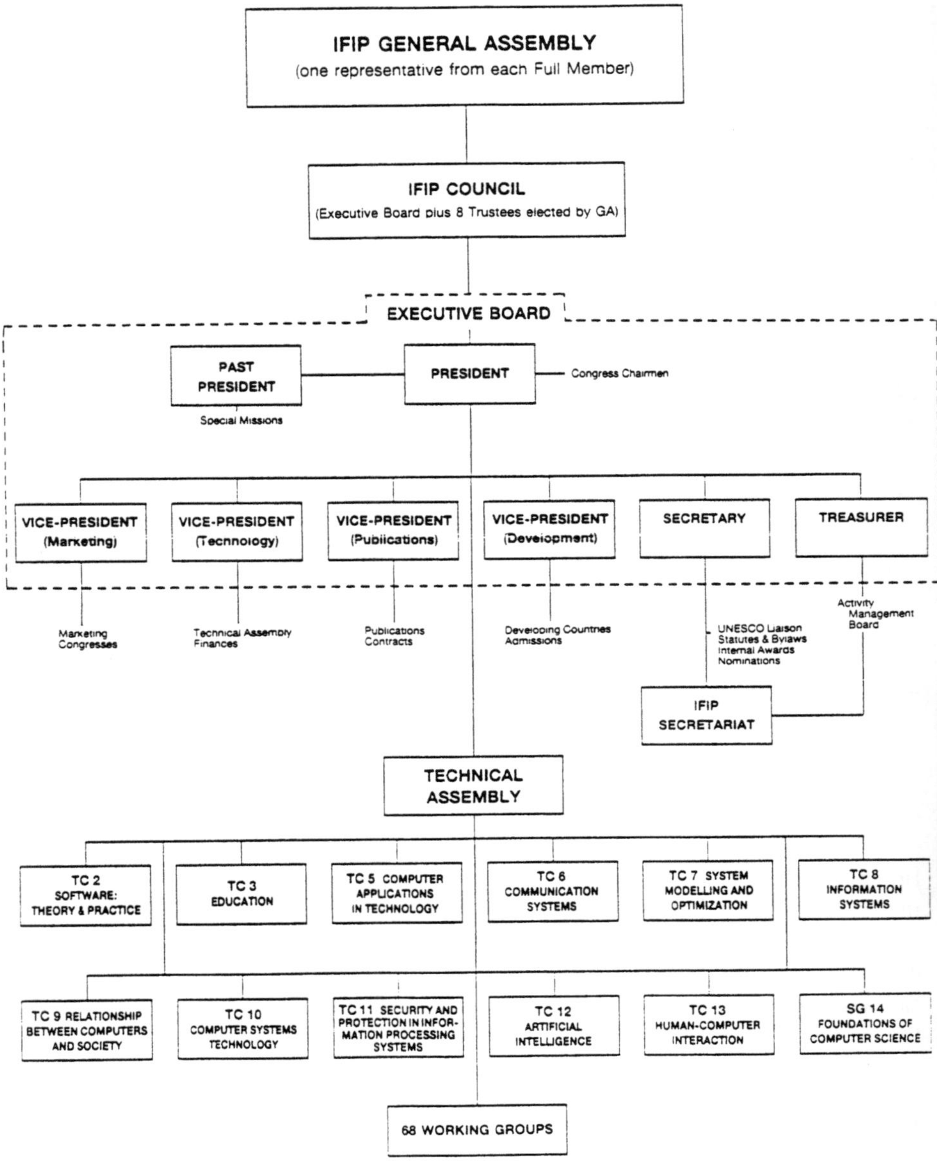

The day-to-day work of IFIP is directed by its Officers: the President, Vice-Presidents, Secretary and Treasurer, who are elected by the General Assembly and together constitute the EXECUTIVE BOARD.

The COUNCIL, consisting of the Officers and up to eight Trustees elected from the General Assembly, meets twice a year and takes decisions which become necessary between General Assembly meetings.

The headquarters of the Federation are in Geneva, Switzerland where the IFIP Secretariat administers its affairs.

For further information please contact:

IFIP Secretariat
attn. Mme. GWYNETH ROBERTS
16 Place Longemalle
CH-1204 Geneva, Switzerland
telephone: 41 (22) 310 26 49
facsimile: 41 (22) 781 23 22
Bitnet: ifip@cgeuge51

IFIP's MISSION STATEMENT

IFIP's mission is to be the leading, truly international, apolitical organisation which encourages and assists in the development, exploitation and application of Information Technology for the benefit of all people.

Principal Elements

1. To stimulate, encourage and participate in research, development and application of Information Technology (IT) and to foster international co-operation in these activities.
2. To provide a meeting place where national IT Societies can discuss and plan courses of action on issues in our field which are of international significance and thereby to forge increasingly strong links between them and with IFIP.
3. To promote international co-operation directly and through national IT Societies in a free environment between individuals, national and international governmental bodies and kindred scientific and professional organisations.
4. To pay special attention to the needs of developing countries and to assist them in appropriate ways to secure the optimum benefit from the application of IT.
5. To promote professionalism, incorporating high standards of ethics and conduct, among all IT practitioners.
6. To provide a forum for assessing the social consequences of IT applications; to campaign for the safe and beneficial development and use of IT and the protection of people from abuse through its improper application.

7. To foster and facilitate co-operation between academics, the IT industry and governmental bodies and to seek to represent the interest of users.
8. To provide a vehicle for work on the international aspects of IT development and application including the necessary preparatory work for the generation of international standards.
9. To contribute to the formulation of the education and training needed by IT practitioners, users and the public at large.

IFIP TRANSACTIONS

IFIP TRANSACTIONS is a serial consisting of 15,000 pages of valuable scientific information from leading researchers, published in 36 volumes per year. The serial includes contributed volumes, proceedings of the IFIP World Conferences, and conferences at Technical Committee and Working Group level. Mainstream areas in the IFIP TRANSACTIONS can be found in Computer Science and Technology, Computer Applications in Technology, and Communication Systems.

From 1993 onwards the IFIP TRANSACTIONS are only available as a full set.

IFIP TRANSACTIONS A:
Computer Science and Technology
1992: Volumes A1–A19
1993: Volumes A20–A40
1994: Volumes A41–A61
ISSN 0926-5473

IFIP Technical Committees that are involved in IFIP TRANSACTIONS A
Software: Theory and Practice (TC2)
Education (TC3)
System Modelling and Optimization (TC7)
Information Systems (TC8)
Relationship Between Computers and Society (TC9)
Computer Systems Technology (TC10)
Security and Protection in Information Processing Systems (TC11)
Artificial Intelligence (TC12)
Human-Computer Interaction (TC13)
Foundations of Computer Science (SG14)

IFIP TRANSACTIONS B:
Applications in Technology
1992: Volumes B1–B8
1993: Volumes B9–B14
1994: Volumes B15–B19
ISSN 0926-5481

IFIP Technical Committee that is involved in IFIP TRANSACTIONS B
Computer Applications in Technology (TC5)

IFIP TRANSACTIONS C:
Communication Systems
1992: Volumes C1–C8
1993: Volumes C9–C16
1994: Volumes C17–C26
ISSN 0926-549X

IFIP Technical Committee that is involved in IFIP TRANSACTIONS C
Communication Systems (TC6)

IFIP TRANSACTIONS FULL SET: A, B & C
1992: 35 Volumes. US $ 1892.00/Dfl. 3500.00
1993: 35 Volumes. US $ 2100.00/Dfl. 3885.00
1994: 36 Volumes. US $ 2277.00/Dfl. 4212.00
The Dutch Guilder prices (Dfl.) are definitive. The US $ prices mentioned above are for your guidance only and are subject to exchange rate fluctuations. Prices include postage and handling charges.

The volumes are also available separately in book form.

Please address all orders and correspondence to:

ELSEVIER SCIENCE B.V.
attn. PETRA VAN DER MEER
P.O. Box 103, 1000 AC Amsterdam
The Netherlands
telephone: 31 (20) 5862602
facsimile: 31 (20) 5862616
email: p.meer@elsevier.nl

www.ingramcontent.com/pod-product-compliance
Ingram Content Group UK Ltd.
Pitfield, Milton Keynes, MK11 3LW, UK
UKHW020128290926
15471UKWH00003B/347

9 780444 81834